The Earthscan Reader in Sustainable Agriculture

The Earthscan Reader in Sustainable Agriculture

Edited by

Jules Pretty

London • Sterling, VA

First published by Earthscan in the UK and USA in 2005

ISBN-13:	978-1-844072-36-1	paperback
ISBN-10:	1-84407-236-3	paperback
ISBN-13:	978-1-844072-35-4	hardback
ISBN-10:	1-84407-235-5	hardback

Typesetting by Composition and Design Services
Printed and bound in the UK by Bath Press, Bath
Cover design by Andrew Corbett

For a full list of publications please contact:

Earthscan
8–12 Camden High Street
London, NW1 0JH, UK
Tel: +44 (0)20 7387 8558
Fax: +44 (0)20 7387 8998
Email: earthinfo@earthscan.co.uk
Web: **www.earthscan.co.uk**

22883 Quicksilver Drive, Sterling, VA 20166-2012, USA

Earthscan is an imprint of James & James (Science Publishers) Ltd and publishes in association with
the International Institute for Environment and Development

A catalogue record for this book is available from the British Library

Library of Congress Cataloging-in-Publication Data
The Earthscan reader in sustainable agriculture / edited by Jules Pretty.
    p. cm.
  Includes bibliographical references.
  ISBN 1-84407-235-5 (pbk.) – ISBN 1-84 407-236-3 (hardback)
1. Sustainable agriculture. 2. Agriculture–Social aspects. 3. Agricultural ecology. I. Pretty, Jules N.
  S494.5.S86E27 2005
  631.5'8–dc22

2005011046

Contents

Part 1 Agrarian and Rural Perspectives

Part 2 Agroecological Perspectives

Part 3 Social Perspectives

Part 4 Perspectives from Industrialized Countries

Part 5 Perspectives from Developing Countries

About the Authors

Richard Bawden is currently a visiting distinguished university professor at Michigan State University; a position he assumed in 2000 following a long career at the University of Western Sydney Hawkesbury and earlier positions at the University of New England, with the Food and Agriculture Organization (FAO) in Uruguay, and in the research division of Boots Pure Drug Company in England. He was Dean of Agriculture and Rural Development at Hawkesbury from 1978–1994, was appointed professor of Systemic Development in 1988 and was foundation director of the Centre for Systemic Development there from 1996 until his retirement in 1999. He has been a visiting professor at Cornell and Rutgers universities in the US, the Open University in England and Natal University in South Africa. He is a fellow of the Royal Society of Arts.

Wendell Berry is the author of more than 40 books of essays, poetry and novels. He has worked a farm in Henry County, Kentucky since 1965. He is a former professor of English at the University of Kentucky, and a past fellow of both the Guggenheim Foundation and the Rockefeller Foundation. He has received numerous awards for his work, including an award from the National Institute and Academy of Arts and Letters in 1971, and most recently, the T. S. Eliot Award. 'My work has been motivated', Wendell Berry has written, 'by a desire to make myself responsibly at home in this world and in my native and chosen place'. Wendell Berry's books of essays include: *The Unsettling of America: Culture and Agriculture*; *The Gift of Good Land: Further Essays Cultural and Agricultural*; *Home Economics*; *Another Turn of the Crank*; *Citizenship Papers, Life is a Miracle: An Essay Against Modern Superstition*; *A Continuous Harmony: Essays Cultural and Agricultural*; *The Art of the Commonplace: The Agrarian Essays of Wendell Berry*; *A Place on Earth*; *Sex, Economy, Freedom and Community: Eight Essays*; and *What Are People For?*

Emer Borromeo is at the Philippine Rice Research Institute, and formerly was of the Plant Protection Department, International Rice Research Institute, Los Banos, the Philippines.

Martin Bourque is executive director of the Ecology Center in Berkeley, California, and at the time he wrote Chapter 25 he was programme coordinator at the Institute for Food and Development Policy Oakland, California, also known as Food First. He has an MA in Latin American Studies from the University of California at Berkeley.

Roland Bunch has worked at the grass roots in agricultural development for 37 years. He has been a pioneer in participatory technology development, farmer-to-farmer extension and the use of green manure/cover crops, which were described in his book, *Two Ears of Corn*. The book has now been published in ten languages, with multiple printings in five of them. He is presently the Coordinator for Sustainable Agriculture for World Neighbors and a member of the United Nation's Hunger Task Force, which was commissioned by Kofi Annan to plan how the world can reach the Millennium Development Goal of halving the number of hungry people in the world by the year 2015.

Robert Chambers is a research associate of the Institute of Development Studies at the University of Sussex, UK. His development work has been mainly in Sub-Saharan Africa and South Asia, in rural administration, training, research, consultancy and writing. His areas of interest have included management, professionalism, perceptions and realities of poverty and livelihoods, and scientific and participatory procedures and methodologies. He is currently working on ideas for use in development, on power and participatory approaches and methods, and on personal and institutional learning and change.

Donald Cole trained as a physician at the University of Toronto (1978). He then practised primary care, public health, occupational health and environmental health in a variety of settings in Canada and developing countries. In a community medicine residency at McMaster University he completed a Masters in Design, Measurement and Evaluation of Health Services (1991), and went on to quality as a Royal College fellow in Occupational Medicine (1990) and Community Medicine (1992). A Tri-Council Eco-Research fellowship in environmental epidemiology and the role of Interim Director of Research followed by senior scientist at the Institute for Work and Health fostered his focus on research. As a tenured associate professor of Public Health Sciences at the University of Toronto and Associate Programme Director for the Community Health and Epidemiology field of the Masters in Health Science Programme, he currently teaches, mentors, does research and contributes research evidence to public health practice both in Canada and internationally.

Gordon Conway was appointed Chief Scientific Adviser to the Department for International Development at the end of 2004. He also holds the title of professor of International Development at Imperial College, London. Prior to that he was President of The Rockefeller Foundation from 1998–2004. He was Vice-Chancellor of the University of Sussex and Chair of the Institute for Development Studies from 1992–1998, and Representative of the Ford Foundation in New Delhi from 1988–1992. He has authored *The Doubly Green Revolution: Food for all in the 21st Century* (Penguin and University Press, Cornell) and *Islamophobia: A Challenge for us All* (The Runnymede Trust).

Charles Crissman is an agricultural economist with 20 years experience in farm level research in Asia, Africa and Latin America. During the time of the research reported here, he was head of the International Potato Center (CIP) office in Ecuador and the coordinator of the pesticide research programme. He is now the Deputy Director General for Research and based at CIP Headquarters in Lima, Peru.

Michael D. Duffy is professor in the Department of Economics, Iowa State University, Ames, and is the Associate Director for the Leopold Center for Sustainable Agriculture.

He also works as an Extension Economist in farm management, and is the Professor-in-Charge of the Iowa State Beginning Farmer Center. Dr Duffy is currently responsible for the annual land value survey, cost of crop production estimates, Iowa farm costs and returns publication, and he is state leader for the Extension Farm Financial Planning Programme. His research activities include determinants of farm profitability, small farms, soil conservation, integrated pest management, switchgrass and other bioenergy sources and sustainable agriculture.

Cornelia Butler Flora is the Charles F. Curtiss Distinguished Professor of Agriculture and Sociology at Iowa State University and Director of the North Central Regional Center for Rural Development. She is past president of the Rural Sociological Society, the Community Development Society, and the Society for Agriculture, Food and Human Values. Her recent books include *Interactions Between Agroecosystems and Rural Communities*, and *Rural Communities: Legacy and Change,* 2nd edition. She serves on the boards of The Consortium for the Sustainable Development of Andean Ecoregion (CONDESAN), the Midwest Assistance Programme, the Northwest Area Foundation, Winrock International, the National Community Forestry Center and the US Department of Agriculture (USDA) National Agricultural Research, Education and Economics Advisory Board.

Jan L. Flora is professor of Sociology, Extension Community Sociologist and a member of the coordinating committee of the graduate programme in Sustainable Agriculture at Iowa State University (ISU). He is also a visiting professor at the National Agrarian University-La Molina (UNALM) in Peru. His current research analyses the relationship of community social capital to economic, community and sustainable development. His extension work focuses on involving Latino immigrants in the affairs of rural Iowa communities and on disseminating applied research and articulating policy recommendations that promote a self-sufficiency wage in Iowa. He is co-director of the ISU-UNALM exchange programme that focuses on strengthening graduate training and faculty research in sustainable agriculture in both institutions. Previous positions include: fellow at Natural Resources and Environment in Victoria, Australia; programme officer in South America for the Ford Foundation, senior fellow in Agricultural Systems at the University of Minnesota and president of the Rural Sociological Society.

Kevin Gallagher is Integrated Pest Management (IPM) specialist, Global IPM Facility, FAO Plant Protection Service, Rome.

Stephen R. Gliessman. After earning his doctorate in plant ecology at the University of California, Santa Barbara, Stephen Gliessman spent nine years in Latin America. He farmed coffee and vegetables in Costa Rica, ran a nursery in Guadalajara, Mexico, and taught and did research at a small college of tropical agriculture in Tabasco, Mexico. He was founding director of the University of California, Santa Cruz (UCSC) agroecology programme and teaches natural history, ethnobotany and agroecology in the Department of Environmental Studies at UCSC. He is the Heller Endowed Chair of Agroecology at UCSC and has been a Kellogg fellow and a Fulbright scholar. He has published extensively on traditional agriculture in Mexico, agroecology and sustainable agriculture, and his textbook, *Agroecology: Ecological Processes in Sustainable Agriculture*, now appears in four languages. He leads short courses and training seminars in agroecology in many parts of the world, and also dry farms organic wine grapes and olives with his family at their family farm in Central California.

Michael Heasman is a writer and researcher on the global food industry. He was joint editor of *New Nutrition Business* (1995–2002) and *Financial Times Food Business* (1998–2001), and is a visiting research fellow at City University, London. He is founder and editor of FoodforGood.com, which tracks and promotes ethical and socially responsible positioning.

Rachel Hine is research officer at the Centre for Environment and Society in the Department of Biological Sciences at the University of Essex. She has worked on a variety of sustainable agriculture programmes, and developed new participatory methods for community assessments.

Sir Albert Howard was the founder of the organic farming movement. He worked for 25 years as an agricultural investigator in India, first as Agricultural Adviser to states in central India, then as director of the Institute of Plant Industry at Indore, where he developed the famed Indore composting process, which put the ancient art of composting on a firm scientific basis. Early in his career he abandoned the restrictions of conventional agricultural science with its increasing over-specialization and set out to learn how to grow a healthy crop in typical conditions in the field, rather than the usual untypical conditions in laboratories and test-plots.

Dana L. Jackson is the associate director of the Land Stewardship Project (LSP), a 22-year-old, non-profit membership organization based in Minnesota. In addition to her responsibilities in administration and fund raising, Dana directs the Farm and City Food Connections Programme. She is the co-editor with her daughter Laura Jackson of *The Farm as Natural Habitat: Reconnecting Food Systems with Ecosystems*, a collection of readings about nature and farming published in 2002. She received a Chevron Conservation Award in 1989 and a Pew Scholar's Award in Conservation and the Environment in 1990. She was a co-founder of The Land Institute in Salina, Kansas and was on the founding board of directors of the Kansas Rural Center. She served two terms on the board of directors of the Minnesota Institute for Sustainable Agriculture, and is currently on the board of directors of the Wild Farm Alliance, an organization that promotes agriculture that helps to protect and restore wild nature.

Wes Jackson, president of the Land Institute (founded in 1976) was born in 1936. After attending Kansas Wesleyan (BA Biology, 1958), he studied botany (MA University of Kansas, 1960) and genetics (PhD North Carolina State University, 1967). He was a professor of biology at Kansas Wesleyan and established the environmental studies programme at California State University, Sacramento, where he became a tenured full professor. His writings include both papers and books. His most recent works are *Rooted in the Land: Essays on Community and Place* (1996), co-edited with William Vitek, *Becoming Native to this Place* (1994) and *Altars of Unhewn Stone* (1987). *Meeting the Expectations of the Land* (1984) was edited with Wendell Berry and Bruce Colman. *New Roots for Agricul-*

ture (1980) outlines the basis for agricultural research at the Land Institute. The work of the Land Institute has been featured extensively in the popular media, including *The Atlantic Monthly, Audubon,* 'The MacNeil-Lehrer News Hour' and National Public Radio's (NPR's) 'All Things Considered'. Life magazine named Wes Jackson as one of 18 individuals they predict will be among the 100 'most important Americans of the 20th century'. He is a recipient of the Pew Conservation Scholars award (1990) and a MacArthur fellowship (1992).

Peter Kenmore is coordinator, Global IPM Facility, FAO Plant Protection Service, Rome.

Jan-Willem Ketelaar is coordinator, FAO Vegetable IPM Programme for South East Asia, Bangkok.

David Kline operates a mixed farm in Fredericksburg, Ohio that is rich in wildlife. One reason this 35-cow Jersey farm attracts so many birds and other wildlife is that wildlife is never viewed as a threat. As a result, David has honed a keen appreciation and understanding of nature. He has shared his astute observations through countless articles and two books (*Great Possessions* and *Scratching the Woodchuck*). Entwined among these various stories that reveal the rewards of a slower paced life. This lifestyle supports not only families and caring communities, but also a people comfortable with themselves. Economically, the farm serves as a model that blends older traditional methods with newer technologies in a manner that is environmentally friendly and adequately profitable. The average farm in Holmes County, David's home, is 122 acres and horse powered. According to Steven Stoll, of Yale University, the mean household income from these farms beats the mean household income for the county by 26 per cent. A typical observation by a visitor to Holmes County is 'a sense of wellbeing'.

Jack Kloppenburg teaches in the Department of Rural Sociology at the University of Wisconsin, Madison. He is a food ecologist who works for sustainable, self reliant, local/ regional food production, which in turn is founded on the regional reinvestment of capital and local job creation, the strength of community institutions and direct democratic participation in the local food economy.

Tim Lang is professor of Food Policy at City University, London. He specializes in how policy affects the shape of the food supply chain, what people eat and the societal, health and environmental outcomes. He is a fellow of the Faculty of Public Health, and chair of Sustain, the UK non-governmental organization (NGO) alliance.

Aldo Leopold was a conservationist, forester, philosopher, educator, writer and outdoor enthusiast. Graduating from the Yale Forest School in 1909, he pursued a career with the newly established US Forest Service in Arizona and New Mexico. Following a transfer to Madison, Wisconsin, he published the first textbook in the field of wildlife management, and in 1935, he and his family began their own ecological restoration of a worn out farm along the Wisconsin River outside Baraboo. He planted thousands of pine trees, restored tall grass prairie, and documented the changes in the flora and fauna.

A prolific writer, Leopold conceived of a book geared for general audiences examining humanity's relationship to the natural world. Unfortunately, just one week after receiving word that his manuscript would be published, Leopold suffered a heart attack and died on April 21, 1948 while fighting a neighbour's grass fire that threatened the Leopold farm and surrounding properties. A little more than a year after his death Leopold's collection of essays, *A Sand County Almanac*, was published, and has since sold over two million copies.

Gabinò López worked his way up from being a smallholder participant in a World Neighbors programme in Guatemala, to being an agricultural promoter, a vegetable store manager, a coordinator of three different rural development programmes, an international consultant and, most recently, the coordinator of a Honduran NGO. As a consultant, Gabinò has undertaken more than 50 consultancies in 16 nations of North America, Central America, South America, Asia and Australia.

Tom Mew is head of the Plant Protection Department, International Rice Research Institute, Los Banos, the Philippines.

James Morison is reader in plant environmental physiology in the Department of Biological Sciences at the University of Essex, UK. His research on how plants are affected by, and respond to, their environment spans a range of ecosystems and organizational scales, from cellular to whole landscapes. He is particularly interested in all aspects of water and plant growth, from cellular mechanisms of drought tolerance to water use efficiency of tree stands. He has also worked on the impact of climate change on crops and natural ecosystems. He was previously lecturer in agricultural meteorology at the University of Reading UK and, while there, worked with agroclimate specialists from many countries through MSc and short courses on characterizing and optimizing the climate for crop production.

Peter Ooi is coordinator of the Regional Cotton IPM Programme, FAO Regional Office, Bangkok.

David Orr is the Paul Sears Distinguished Professor of Environmental Studies at Oberlin College, and author of *Ecological Literacy* (1992), *Earth in Mind* (1994), *The Nature of Design* (2002), *The Last Refuge* (2004) and a forthcoming book, *The Fifth Revolution*. He is perhaps best known for his pioneering work on environmental literacy in higher education and his recent work in ecological design. He raised funds for and spearheaded the effort to design and build a US$7.2 million Environmental Studies Center at Oberlin College, a building described by the *New York Times* as the most remarkable of a new generation of college buildings and selected as one of 30 milestone buildings in the 20th century by the US Department of Energy. He was named one of 25 'Environmental Champions for 2004' by *Interiors and Sources Magazine,* awarded a Bioneers Award in 2002, a *National Conservation Achievement Award* by the National Wildlife Federation in 1993, a *Lyndhurst Prize* in 1992 awarded by the Lyndhurst Foundation 'to recognize the educational, cultural, and charitable activities of particular individuals of exceptional talent, character, and moral vision', and the *Benton Box Award* from Clemson

University for his work in Environmental Education (1995). In a special citation, the Connecticut General Assembly noted Orr's 'vision, dedication, and personal passion' in promoting the principles of sustainability. The Cleveland *Plain Dealer* described him as 'one of those who will shape our lives'. David Orr is contributing editor of *Conservation Biology*, and is trustee and board member of several organizations.

Myriam Paredes has worked with farmer led development efforts in Central and South America for over ten years in the technical areas of soil and pest management. Her PhD research at Wageningen University and Research Centre, the Netherlands, is on the social interfaces between development interventions, technology and rural people. Presently she is a private consultant based in Quito, Ecuador.

Sergio Pinheiro is an agronomist and agricultural researcher with the Santa Catarina State Institution for Agricultural Research and Rural Extension (Epagri), where he has been since 1982, working mainly in rural administration and socio-economics, participatory research and development, agroecology and sustainable development projects. He is participant professor for a Masters course in agroecosystems offered by the Federal University of Santa Catarina, Brazil, where he teaches 'Agricultural Systems: Theory and Practice', and the author of many articles about rural administration, participatory research, agroecology and sustainable development. He collaborates with The Ecological Farmers Association of the Hillsides of Santa Catarina State (Agreco), is past president of the Brazilian Latin American Farming Systems Society (2001–2004) and coordinator of the V Latin American and Brazilian Farming Systems Symposium. He is a member of the organizing committee for the III Brazilian Congress of Agroecology in Florianópolis in October 2005.

Jules Pretty is head of the Department of Biological Sciences and professor of environment and society at the University of Essex. His books include *The Pesticide Detox* (2004), *Agri-Culture: Reconnecting People, Land and Nature* (2002), *Guide to a Green Planet* (edited, 2002), *The Living Land* (1998), *Regenerating Agriculture* (1995), *Fertile Ground* (1999, co-authored), *The Trainers Guide for Participatory Learning and Action* (1995, co-authored), *The Hidden Harvest* (1992, co-authored) and *Unwelcome Harvest* (1991, co-authored). He is deputy chair of the UK government's Advisory Committee on Releases to the Environment (ACRE), and has served on government advisory committees for the Department for the Environment, Food and Rural Affairs (DEFRA), the UK Department for International Development (DFID), the Cabinet Office and the Department for Trade and Industry (DTI). He is a regular speaker, contributor to media and was presenter of the 1999 BBC Radio 4 series *Ploughing Eden* and contributor and writer for the 2001 BBC TV Correspondent programme *The Magic Bean*. He received a 1997 award from the Indian Ecological Society for 'International Contributions to Sustainable and Ecological Agriculture', and was runner up for the 2002 European Sicco Mansholt Prize for agricultural science. He was appointed A. D. White Professor-at-Large by Cornell University for six years from 2001. He was appointed to the International Jury for the Slow Food Award in 2002, is chief editor of the *International Journal of Agricultural Sustainability* and is a fellow of the Institute of Biology and the Royal Society for Arts.

Niels Röling is emeritus professor of agricultural knowledge systems, Wageningen University. He is a teacher and writer on interactive learning based approaches to innovation. Initially focused on extension and smallholder development, he shifted to knowledge systems and social learning for sustainable natural resource management. His books include *Extension Science* (1988) and *Facilitating Sustainable Agriculture* (1998). He is currently involved in research (pathways of agricultural science in West Africa, and social learning for water management in the Netherlands), and in supervising 15 PhD students. Current interest: cognitive agency in fostering second order emergence. Latest diploma: 'Handling the chain saw' (when pollarding willows in the water meadows of the Rhine).

Peter Rosset is a researcher at the Center for the Study of Change in the Mexican Countryside (CECCAM) in Oaxaca, Mexico. He is a research associate at the Center for the Study of the Americas (CENSA) in Berkeley, CA, US, and at the time he wrote the chapter in this book, was co-director of the Institute for Food and Development Policy in Oakland, California, also known as Food First. He has a PhD in agricultural ecology from the University of Michigan.

Pedro Sanchez is director of tropical agriculture and senior research scholar at the Earth Institute of Columbia University in New York City. He received the World Food Prize in 2003 and a MacArthur fellowship in 2004. Sanchez is co-chair of the Hunger Task Force of the United Nations Millennium Project. He served as director general of the World Agroforestry Center (ICRAF) in Nairobi, Kenya from 1991–2001. Sanchez is also professor emeritus of soil science and forestry at North Carolina State University, was a visiting professor at the University of California, Berkeley and is adjunct professor at the Department of Ecology, Evolution and Environmental Biology at Columbia University. He is author of Properties and Management of Soils of the Tropics, and is currently writing a second edition. He is a fellow of the American Society of Agronomy and the Soil Science Society of America. He has received decorations from the governments of Colombia and Peru, and was awarded the International Soil Science Award, the International Service in Agronomy Award and the Crop Science Society of America Presidential Award. The Luo community of Western Kenya anointed Sanchez a Luo Elder with the name of Odera Kang'o.

Stephen Sherwood has been supporting knowledge based, grass-roots initiatives in sustainable agriculture and community based natural resource management in Central and South America since the mid-1980s. His PhD research at Wangingen University and Research Centre, the Netherlands, is on the socio-biological developments that led up to pesticide dependency and ecosystem crisis in the Northern Andes. He is Andes Area Representative for World Neighbors and is based in Quito, Ecuador.

Erin M. Tegtmeier directs the Experiment in Rural Cooperation, the Southeast Minnesota Regional Sustainable Development Partnership. The Regional Partnership initiative is administered by the University of Minnesota and driven by regional boards of citizen volunteers. It is an outreach effort to link University resources with community identified, regionally based projects in sustainable agriculture, natural resource use and rural economic development. She holds a Masters in sustainable agriculture from Iowa

State University and has research interests in local food systems, consumer behaviour and the external costs of agriculture. She is a geographer with past experience as a cartographer and geographic information systems analyst.

Norman Uphoff, a professor of government and international agriculture at Cornell University, was director of the Cornell International Institute for Food, Agriculture and Development between 1990 and 2005. His involvement with the System of Rice Intensification (SRI) developed in Madagascar 20 years ago has drawn him deeply into the domain of agroecology since 1997. He is managing editor for a collaborative book on *Biological Approaches to Sustainable Soil Systems*, published by CRC Press in 2005. His previous involvements have been with participatory rural development and particularly with irrigation management and community based natural resource management, as well as social capital.

Chapter Sources

Part 1 Agrarian and Rural Perspectives

1 Howard, A. (1945) 'The post-war task', in Howard, A. *Farming and Gardening for Health or Disease*, Faber, London, pp24–27

2 Leopold, A. (1949) 'Thinking like a mountain', in Leopold, A. *A Sand County Almanac and Sketches Here and There*, Oxford University Press, Oxford and Ballantine Books, New York (1974 edition), pp137–141

3 Berry, W. (1986) *The Unsettling of America*, Sierra Club Books, San Francisco, pp3–14

4 Orr, D. (1992) *Ecological Literacy*, SUNY Press, Albany, pp85–97

5 Kline, D. (1996) 'An Amish perspective', in Vitek, W. and Jackson, W. (eds) *Rooted in the Land: Essays on Community and Place*, Yale University Press, Haven and London, pp35–39

6 Jackson, W. (1994) *Becoming Native to this Place*, University Press of Kentucky, Lexington, pp53–60

7 Flora, C. B. and Flora, J. L. (1996) 'Creating social capital', in Vitek, W. and Jackson, W. (eds) *Rooted in the Land: Essays on Community and Place*, Yale University Press, Haven and London, pp217–225

Part 2 Agroecological Perspectives

8 Pretty, J. (2002) 'Reality cheques', in Pretty, J. *Agri-culture: Reconnecting People, Land and Nature*, Earthscan, London, pp52–60

9 Tegtmeier, E. M. and Duffy, M. D. (2004) 'The external costs of agricultural production in the United States', *International Journal of Agricultural Sustainability*, vol 2, pp55–175

10 Sherwood, S., Cole, D., Crissman, C. and Paredes, M. (2004) 'Transforming potato systems in the Andes', in Pretty, J. (ed) *The Pesticide Detox*, Earthscan, London, pp139–148, 153

11 Gliessman, S. R. (2004) 'Agroecology and agroecosystems', in Rickerl, D. and Francis, C. (eds) *Agroecosystem Analysis*. Agronomy Monograph Series, no 43, American Society of Agronomy Monographs, pp19–29

12 Conway, G. R. (1997) *The Doubly Green Revolution*, Penguin Books, Harmondsworth, pp205–218

Part 3 Social Perspectives

Part 4 Perspectives from Industrialized Countries

Part 5 Perspectives from Developing Countries

25 Rosset, P. and Bourque, M. (2002) 'Lessons of Cuban resistance', in Funes, F., Garcia, L., Bourque, M. Perez, N. and Rosset, P. (eds) *Sustainable Agriculture and Resistance*, Food First Books, Oakland, CA, ppxiv–xx

26 Sanchez, P. A. (2002) 'Benefits from agroforestry in Africa, with examples from Kenya and Zambia', in Uphoff, N. (ed) *Agroecological Innovations*, Earthscan, London, pp109–114

27 Pretty, J., Morison, J. I. L. and Hine, R. E. (2003) 'Reducing food poverty by increasing agricultural sustainability in developing countries', *Agriculture, Ecosystems and Environment*, vol 95, pp217–234

List of Acronyms and Abbreviations

ACF	Australian Conservation Federation
ACIAR	Australian Centre for International Agriculture Research
Agreco	The Ecological Farmers Association of the Hillsides of Santa Catarina State
APC	Agricultural Production Cooperatives
APCSA	Ecuadorian Association for the Protection of Crops and Animal Health (now called Crop Life Ecuador)
BPH	brown planthopper
BSE	Bovine Spongiform Encephalopathy
BMI	body mass index
BUCP	Basic Units of Cooperative Production
CAE	College of Advanced Education
CAFOs	concentrated animal feeding operations
CAST	Council for Agricultural Science and Technology
CDC	Centers for Disease Control and Prevention
CDR	complex, diverse and risk-prone
CECCAM	Center for the Study of Change in the Mexican Countryside
CENSA	Center for the Study of the Americas
CH_4	methane
CIIFAD	Cornell International Institute for Food, Agriculture, and Development
CIMMYT	International Centre for the Improvement of Wheat and Maize
CIP	International Potato Center (Equador)
CJD	Creutzfeld–Jakob Disease
CONDESAN	Consortium for the Sustainable Development of Andean Ecoregion
COSECHA	Associación de Consejeros una Agricultura Sostenible, Ecológica y Humana
CRP	Conservation Reserve Programme
CSA	Community-supported Agriculture
DEFRA	Department for the Environment, Food and Rural Affairs (UK)
DFID	Department for International Development (UK)
DTI	Department for Trade and Industry
EPA	Environment Protection Agency (US)
EPAGRI	Santa Catarina State Institution for Agricultural Research and Rural Extension
ERS	Economic Research Service

ESD	Ecologically Sustainable Development
FAO	Food and Agriculture Organization
FEMA	Federal Emergency Management Agency
FFS	Farmer Field School
FSIS	Food Safety and Inspection Service
GDP	gross domestic product
HACCP	Hazard Analysis and Critical Control Point
ICRAF	International Centre for Research in Agroforestry
ICRAF	World Agroforestry Center (Nairobi)
IESWTR	Interim Enhanced Surface Water Treatment Rule
IHC	Interfaith Hunger Coalition (Southern California)
IMF	International Monetary Fund
INIAP	National Institute of Agricultural Research from Ecuador
INRA	National Institute for Agricultural Research (France)
IPM	Integrated Pest Management
IPPM	Integrated Product and Pest Management
IRRI	Independent Research Institute
ISU	Iowa State University
kcal	kilocalories
LSP	Land Stewardship Project
MAS	Multi Agent Systems
NARS	national agriculture research system
NFF	National Farmers' Federation
NGO	non-governmental organization
NH_3	ammonia
NH_4	ammonium
N_2O	nitrous oxide
NO	nitric oxide
NO_x	nitrogen oxide
NPR	National Public Radio
PCB	Polychlorinated Biphenol
PPE	personal protective equipment
PRA	participatory rural appraised
Pronaf	National Programme to Empower Small Family Farmers
RAAKS	Rapid Appraisal of Agricultural Knowledge Systems
rBGH	recombinant bovine growth hormone
SDWA	Safe Drinking Water Act
SESA	Ecuadorian Plant and Animal Health Service
SRI	System of Rice Intensification
SSM	Soft Systems Methodology
SUP	safe use of pesticides
$t\ ha^{-1}$	tonnes per hectare
$t\ yr^{-1}$	tonnes per year
TOA	Trade-off Analysis
TOT	transfer of technology
UCSC	University of California, Santa Cruz

UNALM	National Agrarian University-La Molina
UNDP	United Nations Development Programme
UNEP	United Nations Environment Programme
UNESCO	United Nations Educational, Scientific and Cultural Organization
USACE	United States Army Corps of Engineers
USDA	US Department of Agriculture
WTAC	willingness to accept compensation
WTP	willingness to pay

Part 1

Agrarian and Rural Perspectives

Introduction to Part 1:
Agrarian and Rural Perspectives

Jules Pretty

Part 1 of this *Reader in Sustainable Agriculture* focuses on seven agrarian and rural perspectives on agricultural sustainability by Albert Howard, Aldo Leopold, Wendell Berry, David Orr, David Kline, Wes Jackson and Cornelia Butler Flora and Jan Flora.

Albert Howard is seen by many as one of the founders of the modern organic movement and he worked as a scientist in both India and the UK. His most influential book was *The Agricultural Testament*, where he set out many of the scientific principles for organic farming. In the excerpt included here, from his *Farming and Gardening for Health and Disease*, he makes an early critical link between the state of agriculture and the health of the public. He draws on the work of Robert McCarrison who documented the health and longevity of the people of the Hunza Valley in the western Himalayas, indicating how healthy food comes from healthy agricultural systems. He then goes on to summarize his life's work in seven key principles, stating that a 'healthy population will be no mean achievement', and suggesting that citizens should grow their own food, insist on healthy public meals (in schools especially) and use their votes wisely to shape the food system.

Aldo Leopold is widely acknowledged as one of the most influential conservation writers of the 20th century. He died at the time that his *Sand County Almanac* was published in the late 1940s. Some 60 years later, this still remains a timely and insightful book. In this short essay, *Thinking Like a Mountain*, he makes the case for a more holistic way of thinking about people and nature. One of his first jobs as a practical conservationist was to shoot wolves. But he describes how one day he shot an elderly female playing with her pups. He reached the wolf in time to 'watch a fierce green fire dying in her eyes'. His revelation was to shape his thinking: 'I reached then, and have known ever since, that there was something new to me in those eyes – something known only to her and to the mountain.' Remove the wolves, and the deer expand in numbers, which results in a destruction of vegetation and the erosion of the soil. Thinking like a mountain, and thus the whole system, remains a challenge today.

Wendell Berry is one of the best known writers on agrarian pasts and presents in the US. He is a practising farmer, poet and author of many books. In this excerpt from his 1976 book, *The Unsettling of America*, he tells the story of the change in culture and agriculture in a few short generations of frontier invasion, spread and modernization. America was not settled, but unsettled, and it resulted in the exploitation of the land and its people. It ultimately, too, undermined the environmental security and health of the settlers. The production treadmill based on competition and degradation resulted in the loss of family farmers from the land – a process that has continued at a faster pace since the writing of this chapter in the 1970s. As he says, 'the care of the earth is out most ancient and most worthy and, after all, our most pleasing responsibility'. To cherish what remains of it, and to foster renewal, is our only legitimate hope.

David Orr is professor of environmental studies at Oberlin College, and author of a number of highly respected books on the relations between people and nature. In this article, Orr explores what he calls *Ecological Literacy* – the capacity to ask 'what then?' Dialogue with nature cannot be rushed, yet we have been failing to teach the basics about the earth and how it works. Orr says, 'we are in fact teaching a large amount of stuff that is simply wrong'. A generation has passed through formal schooling 'without a clue why the colour of the water in their rivers is related to their food supply, or why storms are becoming more severe as the planet warms'. These same people will create businesses, vote, have families and consume without knowing the outcomes of their actions.

Ecological literacy is more demanding than learning from books, as it requires that we observe nature with insight, a merger of both landscape and mindscape. If this intimate knowledge disappears, then our mental landscapes will be impoverished too. Several factors are working against ecological literacy in industrialized countries, including losing the ability to think broadly (or at right angles, as Leopold put it); the belief that education is solely an indoor activity; and the decline in capacity for aesthetic appreciation. This is all about seeing wholeness and connectedness. As Orr says, 'And this is the heart of the matter. To see things in their wholeness is politically threatening.' Real ecological literacy is, therefore, radical in that it should encourage humanity to reckon with the roots of its ailments, not just with the symptoms.

David Kline is an Amish farmer and author of many books and articles drawing on his experience of farming the rolling hills of southern Ohio. In his essay, he reflects on his community's rootedness to the land. His careful use of sensitive, or sustainable, farming methods has resulted in nature being restored on his farm. He uses diverse rotation patterns, grows and raises many crops and animals, and still farms with horses. He says, 'the blending of domestic and wild works so well'. His decisions are shaped by the seasons and the weather, a rhythm lost when farming becomes a factory experience. Throughout, he strives for a balance of utility and beauty.

Wes Jackson is the founder of the Land Institute on Kansas, and has written widely about rural communities and the land. In this excerpt from his book, *Becoming Native to this Place*, he describes what is left of Matfield Green, a town of some 50 people in the rural plains of Kansas. As he says, this town is typical of countless communities throughout the midwest and Great Plains – places once of great hope, but which are now increasingly abandoned and rundown. How in such a short period of time did such decay set in, and what can be done for renewal? A large part of the land around this town is never ploughed prairie grassland – and Jackson imagines that bison could roam here one day. He tells the story of the town, the sad, slow and creeping loss of one household and service after another. Recovery can, of course, come is such communities strike it lucky and can attract in external sources of money and jobs. But will this last; and what of the communities that fail to attract such good fortune? As Jackson indicates, renewal requires thinking differently, creating a new mindscape first before we can do anything about the outside landscape.

In the final article of this part of the Reader, Cornelia Butler Flora and Jan Flora, both of Iowa State University, succinctly set out how social capital can be created in

Post-industrial rural communities. Two processes are occurring: the inside decay noted by Berry and Jackson, and the incursion from outside as the suburban and disconnected sprawl brings people with different worldviews and values. Those with affluent incomes can ignore investing in social capital, as they can substitute financial capital – yet this simply results in more goods and services being imported from outside and leaves communities struggling. This article indicates that social capital can be horizontal, hierarchical or non-existent – and different patterns define different outcomes for rural communities. Diverse networks that enhance lateral learning can lead to dynamic communities able to develop new models of agriculture.

1

The Post-war Task

Albert Howard

The problem of disease and health took on a wider scope. In March 1939 new ground was broken. The Local Medical and Panel Committees of Cheshire, summing up their experience of the working of the National Health Insurance Act for over a quarter of a century in the county, did not hesitate to link up their judgement on the unsatisfactory state of health of the human population under their care with the problem of nutrition, tracing the line of fault right back to an impoverished soil and supporting their contentions by reference to the ideas which I had for some time been advocating. Their arguments were powerfully supported by the results obtained at the Peckham Health Centre and by the work, already published, of Sir Robert McCarrison, which latter told the story from the other side of the world and from a precisely opposite angle – he was able to instance an Eastern people, the Hunzas, who were the direct embodiment of an ideal of health and whose food was derived from soil kept in a state of the highest natural fertility.

By these contemporaneous pioneering efforts the way was blazed for treating the whole problem of health in soil, plant, animal and man as one great subject, calling for a boldly revised point of view and entirely fresh investigations.

By this time sufficient evidence had accumulated for setting out the case for soil fertility in book form. This was published in June 1940 by the Oxford University Press under the title of An Agricultural Testament. This book, now in its fourth English and second American edition, set forth the whole gamut of connected problems as far as can at present be done – what wider revelations the future holds is not yet fully disclosed. In it I summed up my life's work and advanced the following views:

1   The birthright of all living things is health
2   This law is true for soil, plant, animal, and man: the health of these four is one connected chain
3   Any weakness or defect in the health of any earlier link in the chain is carried on to the next and succeeding links, until it reaches the last, namely, man
4   The widespread vegetable and animal pests and diseases, which are such a bane to modern agriculture, are evidence of a great failure of health in the second (plant) and third (animal) links of the chain

Note: Reprinted from *Farming and Gardening for Health or Disease* by Howard, A., copyright © (1945), with permission from Faber and Faber, London

5   The impaired health of human populations (the fourth link) in modern civilized countries is a consequence of this failure in the second and third links

6   This general failure in the last three links is to be attributed to failure in the first link, the soil: the undernourishment of the soil is at the root of all. The failure to maintain a healthy agriculture has largely cancelled out all the advantages we have gained from our improvements in hygiene, in housing and our medical discoveries

7   To retrace our steps is not really difficult if once we set our minds to the problem. We have to bear in mind Nature's dictates, and we must conform to her imperious demand: (a) for the return of all wastes to the land; (b) for the mixture of the animal and vegetable existence; (c) for the maintaining of an adequate reserve system of feeding the plant, that is we must not interrupt the mycorrhizal association. If we are willing so far to conform to natural law, we shall rapidly reap our reward not only in a flourishing agriculture, but in the immense asset of an abounding health in ourselves and in our children's children.

These ideas, straightforward as they appear when set forth in the form given above, conflict with a number of vested interests. It has been my self-appointed task during the last few years of my life to join hands with those who are convinced of their truth to fight the forces impeding progress. So large has been the flow of evidence accumulating that in 1941 it was decided to publish a News-Letter on Compost, embodying the most interesting of the facts and opinions reaching me or others in the campaign. The News-Letter, which appears three times a year under the aegis of the Cheshire Local Medical and Panel Committees, has grown from 8 to 64 pages and is daily gaining new readers.

The general thesis that no one generation has a right to exhaust the soil from which humanity must draw its sustenance has received further powerful support from religious bodies. The clearest short exposition of this idea is contained in one of the five fundamental principles adopted by the recent Malvern Conference of the Christian Churches held with the support of the late Archbishop of Canterbury, Dr Temple. It is as follows: 'The resources of the earth should be used as God's gifts to the whole human race and used with due consideration for the needs of the present and future generations.'

Food is the chief necessity of life. The plans for social security which are now being discussed merely guarantee to the population a share in a variable and, in present circumstances, an uncertain quantity of food, most of it of very doubtful quality. Real security against want and ill health can only be assured by an abundant supply of fresh food properly grown in soil in good heart. The first place in post-war plans of reconstruction must be given to soil fertility in every part of the world. The land of this country and the Colonial Empire, which is the direct responsibility of Parliament, must be raised to a higher level of productivity by a rational system of farming which puts a stop to the exploitation of land for the purpose of profit and takes into account the importance of humus in producing food of good quality. The electorate alone has the power of enforcing this and to do so it must first realize the full implications of the problem.

They and they alone possess the power to insist that every boy and every girl shall enter into their birthright – health, and that efficiency, wellbeing and contentment which depend thereon. One of the objects of this book is to show the man in the street how this England of ours can be born again. He can help in this task, which depends at least as much on the plain efforts of the plain man in his own farm, garden, or allotment as on all the expensive paraphernalia, apparatus and elaboration of the modern scientist: more so in all probability, inasmuch as one small example always outweighs a tonne of theory. If this sort of effort can be made and the main outline of the problems at stake are [sic] grasped, nothing can stop an immense advance in the wellbeing of this island. A healthy population will be no mean achievement, for our greatest possession is ourselves.

The man in the street will have to do three things:

1   He must create in his own farm, garden, or allotment examples without end of what a fertile soil can do
2   He must insist that the public meals in which he is directly interested, such as those served in boarding schools, in the canteens of day schools and of factories, in popular restaurants and tea shops, and at the seaside resorts at which he takes his holidays are composed of the fresh produce of fertile soil
3   He must use his vote to compel his various representatives – municipal, county, and parliamentary to see to it: (a) that the soil of this island is made fertile and maintained in this condition; (b) that the public health system of the future is based on the fresh produce of land in good heart.

This introduction started with the training of an agricultural investigator: it ends with the principles underlying the public health system of tomorrow. It has, therefore, covered much ground in describing what is nothing less than an adventure in scientific research. One lesson must be stressed. The difficulties met with and overcome in the official portion of this journey were not part of the subject investigated. They were man made and created by the research organization itself. More time and energy had to be expended in side-tracking the lets and hindrances freely strewn along the road by the various well-meaning agencies which controlled discovery than in conducting the investigations themselves. When the day of retirement came, all these obstacles vanished and the delights of complete freedom were enjoyed. Progress was instantly accelerated. Results were soon obtained throughout the length and breadth of the English-speaking world, which make crystal clear the great role which soil fertility must play in the future of mankind.

The real Arsenal of Democracy is a fertile soil, the fresh produce of which is the birthright of the nations.

Thinking Like a Mountain

Aldo Leopold

A deep chesty bawl echoes from rimrock to rimrock, rolls down the mountain, and fades into the far blackness of the night. It is an outburst of wild defiant sorrow, and of contempt for all the adversities of the world.

Every living thing (and perhaps many a dead one as well) pays heed to that call. To the deer it is a reminder of the way of all flesh, to the pine a forecast of midnight scuffles and of blood upon the snow, to the coyote a promise of gleanings to come, to the cowman a threat of red ink at the bank, to the hunter a challenge of fang against bullet. Yet behind these obvious and immediate hopes and fears there lies a deeper meaning, known only to the mountain itself. Only the mountain has lived long enough to listen objectively to the howl of a wolf.

Those unable to decipher the hidden meaning know nevertheless that it is there, for it is felt in all wolf country, and distinguishes that country from all other land. It tingles in the spine of all who hear wolves by night, or who scan their tracks by day. Even without sight or sound of wolf, it is implicit in a hundred small events: the midnight whinny of a pack horse, the rattle of rolling rocks, the bound of a fleeing deer, the way shadows lie under the spruces. Only the ineducable tyro can fail to sense the presence or absence of wolves, or the fact that mountains have a secret opinion about them.

My own conviction on this score dates from the day I saw a wolf die. We were eating lunch on a high rimrock, at the foot of which a turbulent river elbowed its way. We saw what we thought was a doe fording the torrent, her breast awash in white water. When she climbed the bank toward us and shook out her tail, we realized our error: it was a wolf. A half-dozen others, evidently grown pups, sprang from the willows and all joined in a welcoming melee of wagging tails and playful maulings. What was literally a pile of wolves writhed and tumbled in the centre of an open flat at the foot of our rimrock.

In those days we had never heard of passing up a chance to kill a wolf. In a second we were pumping lead into the pack, but with more excitement than accuracy: how to aim a steep downhill shot is always confusing. When our rifles were empty, the old wolf was down, and a pup was dragging a leg into impassable slide-rocks.

We reached the old wolf in time to watch a fierce green fire dying in her eyes. I realized then, and have known ever since, that there was something new to me in those eyes – something known only to her and to the mountain. I was young then, and full of

Note: Reprinted from 'Thinking like a Mountain', pp137–141 from *A Sand Country Almanac and Sketches Here and There* by Aldo Leopold, copyright © 1949, 1953, 1966, renewed 1977, 1981 by Oxford University Press, Inc. Used by permission of Oxford University Press, Inc

trigger-itch; I thought that because fewer wolves meant more deer, that no wolves would mean hunters' paradise. But after seeing the green fire die, I sensed that neither the wolf nor the mountain agreed with such a view.

Since then I have lived to see state after state extirpate its wolves. I have watched the face of many a newly wolfless mountain, and seen the south-facing slopes wrinkle with a maze of new deer trails. I have seen every edible bush and seedling browsed, first to anaemic desuetude, and then to death. I have seen every edible tree defoliated to the height of a saddlehorn. Such a mountain looks as if someone had given God a new pruning shears, and forbidden Him all other exercise. In the end the starved bones of the hoped-for deer herd, dead of its own too-much, bleach with the bones of the dead sage, or moulder under the high-lined junipers.

I now suspect that just as a deer herd lives in mortal fear of its wolves, so does a mountain live in mortal fear of its deer. And perhaps with better cause, for while a buck pulled down by wolves can be replaced in two or three years, a range pulled down by too many deer may fail of replacement in as many decades.

So also with cows. The cowman who cleans his range of wolves does not realize that he is taking over the wolf's job of trimming the herd, to fit the range. He has not learned to think like a mountain. Hence we have dustbowls, and rivers washing the future into the sea.

We all strive for safety, prosperity, comfort, long life and dullness. The deer strives with his supple legs, the cowman with trap and poison, the statesman with pen, the most of us with machines, votes and dollars, but it all comes to the same thing: peace in our time. A measure of success in this is all well enough, and perhaps is a requisite to objective thinking, but too much safety seems to yield only danger in the long run. Perhaps this is behind Thoreau's dictum: In wildness is the salvation of the world. Perhaps this is the hidden-meaning in the howl of the wolf, long known among mountains, but seldom perceived among men.

The Unsettling of America

Wendell Berry

One of the peculiarities of the white race's presence in America is how little intention has been applied to it. As a people, wherever we have been, we have never really intended to be. The continent is said to have been discovered by an Italian who was on his way to India. The earliest explorers were looking for gold, which was, after an early streak of luck in Mexico, always somewhere farther on. Conquests and foundings were incidental to this search – which did not, and could not, end until the continent was finally laid open in an orgy of goldseeking in the middle of the last century. Once the unknown of geography was mapped, the industrial marketplace became the new frontier, and we continued, with largely the same motives and with increasing haste and anxiety, to displace ourselves – no longer with unity of direction, like a migrant flock, but like the refugees from a broken ant hill. In our own time we have invaded foreign lands and the moon with the high-toned patriotism of the conquistadors, and with the same mixture of fantasy and avarice.

That is too simply put. It is substantially true, however, as a description of the dominant tendency in American history. The temptation, once that has been said, is to ascend altogether into rhetoric and inveigh equally against all our forebears and all present holders of office. To be just, however, it is necessary to remember that there has been another tendency: the tendency to stay put, to say, 'No farther. This is the place.' So far, this has been the weaker tendency, less glamorous, certainly less successful. It is also the older of these tendencies, having been the dominant one among the Indians.

The Indians did, of course, experience movements of population, but in general their relation to place was based upon old usage and association, upon inherited memory, tradition, veneration. The land was their homeland. The first and greatest American revolution, which has never been superseded, was the coming of people who did *not* look upon the land as a homeland. But there were always those among the newcomers who saw that they had come to a good place and who saw its domestic possibilities. Very early, for instance, there were men who wished to establish agricultural settlements rather than quest for gold, or exploit the Indian trade. Later, we know that every advance of the frontier left behind families and communities who intended to remain and prosper where they were.

But we know also that these intentions have been almost systematically overthrown. Generation after generation, those who intended to remain and prosper where they were

Note: Reprinted from *The Unsettling of America: Culture and Agriculture*, by Wendell Berry. Copyright © (1977) by Wendell Berry. Reprinted by permission of Sierra Club Books.

have been dispossessed and driven out, or subverted and exploited where they were, by those who were carrying out some version of the search for El Dorado. Time after time, in place after place, these conquerors have fragmented and demolished traditional communities, the beginnings of domestic cultures. They have always said that what they destroyed was outdated, provincial and contemptible. And with alarming frequency they have been believed and trusted by their victims, especially when their victims were other white people.

If there is any law that has been consistently operative in American history, it is that the members of any *established* people or group or community sooner or later become 'redskins' – that is, they become the designated victims of an utterly ruthless, officially sanctioned and subsidized exploitation. The colonists who drove off the Indians came to be intolerably exploited by their imperial governments. And that alien imperialism was thrown off only to be succeeded by a domestic version of the same thing; the class of independent small farmers who fought the war of independence has been exploited by, and recruited into, the industrial society until by now it is almost extinct. Today, the most numerous heirs of the farmers of Lexington and Concord are the little groups scattered all over the country whose names begin with 'Save': Save Our Land, Save the Valley, Save Our Mountains, Save Our Streams, Save Our Farmland. As so often before, these are *designated* victims – people without official sanction, often without official friends, who are struggling to preserve their places, their values and their lives as they know them and prefer to live them against the agencies of their own government which are using their own tax moneys against them.

The only escape from this destiny of victimization has been to 'succeed' – that is, to 'make it' into the class of exploiters, and then to remain so specialized and so 'mobile' as to be unconscious of the effects of one's life or livelihood. This escape is, of course, illusory, for one man's producer is another's consumer, and even the richest and most mobile will soon find it hard to escape the noxious effluents and fumes of their various public services.

Let me emphasize that I am not talking about an evil that is merely contemporary or 'modern', but one that is as old in America as the white man's presence here. It is an intention that was *organized* here almost from the start. 'The New World', Bernard DeVoto wrote in *The Course of Empire*, 'was a constantly expanding market... Its value in gold was enormous but it had still greater value in that it expanded and integrated the industrial systems of Europe.'

And he continues: 'The first belt-knife given by a European to an Indian was a portent as great as the cloud that mushroomed over Hiroshima... Instantly the man of 6000BC was bound fast to a way of life that had developed seven and a half millennia beyond his own. He began to live better and he began to die.'

The principal European trade goods were tools, cloth, weapons, ornaments, novelties, and alcohol. The sudden availability of these things produced a revolution that 'affected every aspect of Indian life. The struggle for existence...became easier. Immemorial handicrafts grew obsolescent, then obsolete. Methods of hunting were transformed. So were methods – and the purposes – of war. As war became deadlier in purpose and armament a surplus of women developed, so that marriage customs changed and polygamy became common. The increased usefulness of women in the preparation of pelts worked to the same end... Standards of wealth, prestige and honour changed. The Indians acquired commercial values and developed business cults. They became more mobile...

'In the sum it was cataclysmic. A culture was forced to change much faster than change could be adjusted to. All corruptions of culture produce breakdowns of morale, of communal integrity, and of personality, and this force was as strong as any other in the white man's subjugation, of the red man.'

I have quoted these sentences from DeVoto because, the obvious differences aside, he is so clearly describing a revolution that did not stop with the subjugation of the Indians, but went on to impose substantially the same catastrophe upon the small farms and the farm communities, upon the shops of small local tradesmen of all sorts, upon the workshops of independent craftsmen and upon the households of citizens. It is a revolution that is still going on. The economy is still substantially that of the fur trade, still based on the same general kinds of commercial items: technology, weapons, ornaments, novelties and drugs. The one great difference is that by now the revolution has deprived the mass of consumers of any independent access to the staples of life: clothing, shelter, food, even water. Air remains the only necessity that the average user can still get for himself, and the revolution has imposed a heavy tax on that by way of pollution. Commercial conquest is far more thorough and final than military defeat. The Indian became a redskin, not by loss in battle, but by accepting a dependence on traders that made *necessities* of industrial goods. This is not merely history. It is a parable.

DeVoto makes it clear that the imperial powers, having made themselves willing to impose this exploitive industrial economy upon the Indians, could not then keep it from contaminating their own best intentions: 'More than four-fifths of the wealth of New France was furs, the rest was fish, and it had no agricultural wealth. One trouble was that whereas the crown's imperial policy required it to develop the country's agriculture, the crown's economy required the colony's furs, an adverse interest.' And La Salle's dream of developing Louisiana (agriculturally and otherwise) was frustrated because 'The interest of the court in Louisiana colonization was to secure a bridgehead for an attack on the silver mines of northern Mexico...'

One cannot help but see the similarity between this foreign colonialism and the domestic colonialism that, by policy, converts productive farm, forest and grazing lands into strip mines. Now, as then, we see the abstract values of an industrial economy preying upon the native productivity of land and people. The fur trade was only the first establishment on this continent of a mentality whose triumph is its catastrophe.

My purposes in beginning with this survey of history are (1) to show how deeply rooted in our past is the mentality of exploitation; (2) to show how fundamentally revolutionary it is; and (3) to show how crucial to our history – hence, to our own minds – is the question of how we will relate to our land. This question, now that the corporate revolution has so determinedly invaded the farmland, returns us to our oldest crisis.

We can understand a great deal of our history – from Cortes' destruction of Tenochtitlán in 1521 to the bulldozer attack on the coalfields four-and-a-half centuries later – by thinking of ourselves as divided into conquerors and victims. In order to understand our own time and predicament and the work that is to be done, we would do well to shift the terms and say that we are divided between exploitation and nurture. The first set of terms is too simple for the purpose because, in any given situation, it proposes to divide people into two mutually exclusive groups; it becomes complicated only when we are dealing with situations in succession – as when a colonist who persecuted the Indians then resisted persecution by the crown. The terms exploitation and nurture, on the other hand, describe a division not only between persons but also within persons. We are all to some extent the products of an exploitive society, and it would be foolish and self-defeating to pretend that we do not bear its stamp.

Let me outline as briefly as I can what seem to me the characteristics of these opposite kinds of mind. I conceive a strip-miner to be a model exploiter, and as a model nurturer I take the old-fashioned idea or ideal of a farmer. The exploiter is a specialist, an expert; the nurturer is not. The standard of the exploiter is efficiency; the standard of the nurturer is care. The exploiter's goal is money, profit; the nurturer's goal is health – his land's health, his own, his family's, his community's, his country's. Whereas the exploiter asks of a piece of land only how much and how quickly it can be made to produce, the nurturer asks a question that is much more complex and difficult: What is its carrying capacity? (That is: How much can be taken from it without diminishing it? What can it produce *dependably* for an indefinite time?) The exploiter wishes to earn as much as possible by as little work as possible; the nurturer expects, certainly, to have a decent living from his work, but his characteristic wish is to work *as well* as possible, the competence of the exploiter is in organization that of the nurturer is in order – a human order, that is, that accommodates itself both to other order and to mystery. The exploiter typically serves an institution or organization; the nurturer serves land, household, community place. The exploiter thinks in terms of-numbers, quantities, 'hard facts'; the nurturer in terms of character, condition, quality, kind.

It seems likely that all the 'movements' of recent years have been representing various claims that nurture has to make against exploitation. The women's movement, for example, when its energies are most accurately placed, is arguing the cause of nurture; other times it is arguing the right of women to be exploiters – which men have no *right* to be. The exploiter is clearly the prototype of the 'masculine' man – the wheeler-dealer whose 'practical' goals require the sacrifice of flesh, feeling and principle. The nurturer, on the other hand, has always passed with ease across the boundaries of the so-called sexual roles. Of necessity and without apology, the preserver of seed, the planter, becomes midwife and nurse. Breeder is always metamorphosing into brooder and back again. Over and over again, spring after spring, the questing mind, idealist and visionary, must pass through the planting to become nurturer of the real. The farmer, sometimes known as husbandman, is by definition half mother; the only question is how good a mother he or she is. And the land itself is not mother or father only, but both. Depending on crop and season, it is at one time receiver of seed, bearer and nurturer of young; at another, raiser of seed-stalk, bearer and shedder of seed. And in response to these changes, the farmer crosses back and forth from one zone of spousehood to another, first as planter and then as gatherer. Farmer and land are thus involved in a sort of dance in which the partners are always at opposite sexual poles, and the lead keeps changing: the farmer, as seed-bearer, causes growth; the land, as seed-bearer, causes the harvest.

The exploitive always involves the abuse or the perversion of nurture and ultimately its destruction. Thus, we saw how far the exploitive revolution had penetrated the official character when our recent secretary of agriculture remarked that 'Food is a weapon'. This was given a fearful symmetry indeed when, in discussing the possible use of nuclear weapons, a secretary of defense spoke of 'palatable' levels of devastation. Consider the associations that have since ancient times clustered around the idea of food – associations of mutual care, generosity, neighbourliness, festivity, communal joy, religious ceremony – and you will see that these two secretaries represent a cultural catastrophe. The concerns of farming and those of war, once thought to be diametrically opposed, have become identical. Here we have art example of men who have been made vicious, not presumably by nature or circumstance, but by their *values*.

Food is *not* a weapon. To use it as such – to foster a mentality willing to use it as such – is to prepare, in the human character and community, the destruction of the sources of food. The first casualties of the exploitive revolution are character and community. When those fundamental integrities are devalued and broken, then perhaps it is inevitable that food will be looked upon as a weapon, just as it is inevitable that the earth will be looked upon as fuel and people as numbers or machines. But character and community – that is, culture in the broadest, richest sense – constitute, just as much as nature, the source of food. Neither nature nor people alone can produce human sustenance, but only the two together, culturally wedded. The poet Edwin Muir said it unforgettably:

> Men are made of what is made,
> The meat, the drink, the life, the corn,
> Laid up by them, in them reborn
> And self-begotten cycles close
> About our way; indigenous art
> And simple spells make unafraid
> The haunted labyrinths of the heart
> And with out wild succession braid
> The resurrection of the rose.

To think of food as a weapon, or of a weapon as food, may give an illusory security and wealth to a few, but it strikes directly at the life of all.

The concept of food-as-weapon is not surprisingly the doctrine of a Department of Agriculture that is being used as an instrument of foreign political and economic speculation. This militarizing of food is the greatest threat so far raised against the farmland and the farm communities of this country. If present attitudes continue, we may expect government policies that will encourage the destruction, by overuse, of farmland. This, of course, has already begun. To answer the official call for more production – evidently to be used to bait or bribe foreign countries – farmers are plowing their waterways and permanent pastures; lands that ought to remain in grass are being planted in row crops. Contour ploughing, crop rotation, and other conservation measures seem to have gone out of favour or fashion in official circles and are practiced less and less on the farm. This exclusive emphasis on production will accelerate the mechanization and chemicalization of farming, increase the price of land, increase overhead and operating costs, and thereby further diminish the farm population. Thus the tendency, if not the intention, of Mr Butz's confusion of farming and war, is to complete the deliverance of American agriculture into the hands of corporations.

The cost of this corporate totalitarianism in energy, land and social disruption will be enormous. It will lead to the exhaustion of farmland and farm culture. Husbandry will become an extractive industry; because maintenance will entirely give way to production, the fertility of the soil will become a limited, unrenewable resource like coal or oil.

This may not happen. It *need* not happen. But it is necessary to recognize that it *can* happen. That it can happen is made evident not only by the words of such men as Mr Butz, but more clearly by the large-scale industrial destruction of farmland already in progress. If it does happen, we are familiar enough with the nature of American salesmanship to know that it will be done in the name of the starving millions, in the name of liberty, justice, democracy and brotherhood, and to free the world from communism. We must, I think, be prepared to see, and to stand by, the truth: that the land should not be destroyed for *any* reason, not even for any apparently good reason. We must be prepared to say that enough food, year after year, is possible only for a limited number of people, and that this possibility can be preserved only by the steadfast, knowledgeable *care* of those people. Such 'crash programmes' as apparently have been contemplated by the Department of Agriculture in recent years will, in the long run, cause more starvation than they can remedy.

Meanwhile, the dust clouds rise again over Texas and Oklahoma. 'Snirt' is falling in Kansas. Snow drifts in Iowa and the Dakotas are black with blown soil. The fields lose their humus and porosity, become less retentive of water, depend more on pesticides, herbicides, chemical fertilizers. Bigger tractors become necessary because the compacted soils are harder to work – and their greater weight further compacts the soil. More and bigger machines, more chemical and methodological shortcuts are needed because of the shortage of man-power on the farm – and the problems of overcrowding and unemployment increase in the cities. It is estimated that it now costs (by erosion) two bushels of Iowa topsoil to grow one bushel of corn. It is variously estimated that from 5 to 12 calories of fossil fuel energy are required to produce 1 calorie of hybrid corn energy. An official of the National Farmers Union says that 'a farmer who earns US$10,000 to $12,000 a year typically leaves an estate valued at about '$320,000' – which means that when that farm is financed again, either by a purchaser or by an heir (to pay the inheritance taxes), it simply cannot support its new owner and pay for itself. And the *Progressive Farmer* predicts the disappearance of 200,000 to 400,000 farms each year during the next 20 years if the present trend continues.

The first principle of the exploitive mind is to divide and conquer. And surely there has never been a people more ominously and painfully divided than we are – both against each other and within ourselves. Once the revolution of exploitation is under way, statesmanship and craftsmanship are gradually replaced by salesmanship.[1] Its stock in trade in politics is to sell despotism and avarice as freedom and democracy. In business it sells sham and frustration as luxury and satisfaction. The 'constantly expanding market' first opened in the New World by the fur traders is still expanding – no longer so much by expansions of territory or population, but by the calculated outdating, outmoding and degradation of goods and by the hysterical self-dissatisfaction of consumers that is indigenous to an exploitive economy.

This gluttonous enterprise of ugliness, waste and fraud thrives in the disastrous breach it has helped to make between our bodies and our souls. As a people, we have lost sight of the profound communion – even the union – of the inner with the outer life. Confucius said: 'If a man have not order within him, he can not spread order about him...'. Surrounded as we are by evidence of the disorders of our souls and our world, we feel the strong truth in those words as well as the possibility of healing that is in them. We see the likelihood that our surroundings, from our clothes to our countryside, are the products of our inward life – our spirit, our vision – as much as they are products of nature and work. If this is true, then we cannot live as we do and be as we would like to be. There is nothing more absurd, to give an example that is only apparently trivial, than the millions who wish to live in luxury and idleness and yet be slender and good-looking. We have millions, too, whose livelihoods, amusements, and comforts are all destructive, who nevertheless wish to live in a healthy environment; they want to run their recreational engines in clean, fresh air. There is now, in fact, no 'benefit' that is not associated with disaster. That is because power can be disposed morally or harmlessly only by thoroughly unified characters and communities.

What caused these divisions? There are no doubt many causes, complex both in themselves and in their interaction. But pertinent to all of them, I think, is our attitude toward work. The growth of the exploiters' revolution on this continent has been accompanied by the growth of the idea that work is beneath human dignity, particularly any form of hand work. We have made it our overriding ambition to escape work, and as a consequence have debased work until it is only fit to escape from. We have debased the products of work and have been, in turn, debased by them. Out of this contempt for work arose the idea of a nigger: at first some person, and later some thing, to be used to relieve us of the burden of work. If we began by making niggers of people, we have ended by making a nigger of the world. We have taken the irreplaceable energies and materials of the world and turned them into gimcrack 'labour-saving devices'. We have made of the rivers and oceans and winds niggers to carry away our refuse, which we think we are too good to dispose of decently ourselves. And in doing this to the world that is our common heritage and bond, we have returned to making niggers of people: we have become each other's niggers.

But is work something that we have a right to escape? And can we escape it with impunity? We are probably the first entire people ever to think so. All the ancient wisdom that has come down to us counsels otherwise. It tells us that work is necessary to us, as much a part of our condition as mortality; that good work is our salvation and our joy; that shoddy or dishonest or self-serving work is our curse and our doom. We have tried to escape the sweat and sorrow promised in Genesis – only to find that, in order to do so, we must forswear love and excellence, health and joy.

Thus we can see growing out of our history a condition that is physically dangerous, morally repugnant, ugly. Contrary to the blandishments of the salesmen, it is not particularly comfortable or happy, It is not even affluent in any meaningful sense, because its abundance is dependent on sources that are being rapidly exhausted by its methods. To see these things is to come up against the question: Then what *is* desirable?

One possibility is just to tag along with the fantasists in government and industry who would have us believe that we can pursue our ideals of affluence, comfort, mobility and leisure indefinitely. This curious faith is predicated on the notion that we will soon develop unlimited new sources of energy: domestic oil fields, shale oil, gasified coal, nuclear power, solar energy, and so on. This is fantastical because the basic cause of the energy crisis is not scarcity; it is moral ignorance and weakness of character. We don't know *how* to use energy, or what to use it *for* and we cannot restrain ourselves. Our time is characterized as much by the abuse and waste of human energy as it is by the abuse and waste of fossil fuel energy. Nuclear power, if we are to believe its advocates, is presumably going to be well used by the same mentality that has egregiously devalued and misapplied man- and womanpower. If we had an unlimited supply of solar or wind power, we would use that destructively, too, for the same reasons.

Perhaps all of those sources of energy are going to be developed. Perhaps all of them can sooner or later be developed without threatening our survival. But not all of them together can guarantee our survival, and they cannot define what is desirable. We will not find those answers in Washington, DC, or in the laboratories of oil companies. In order to find them, we will have to look closer to ourselves.

I believe that the answers are to be found in our history: in its until now subordinate tendency of settlement, of domestic permanence. This was the ambition of thousands of immigrants; it is formulated eloquently in some of the letters of Thomas Jefferson; it was the dream of the freed slaves; it was written into law in the Homestead Act of 1862. There are few of us whose families have not at some time been moved to see its vision and to attempt to enact its possibility. I am talking about the idea that as many as possible should share in the ownership of the land and thus be bound to it by economic interest, by the investment of love and work, by family loyalty, by memory and tradition. How much land this should be is a question, and the answer will vary with geography. The Homestead Act said 160 acres. The freedmen of the 1860s hoped for 40. We know that, particularly in other countries, families have lived decently on far fewer acres than that.

The old idea is still full of promise. It is potent with healing and with health. It has the power to turn each person away from the big-time promising and planning of the government, to confront in himself, in the immediacy of his own circumstances and where-abouts, the question of what methods and ways are best. It proposes an economy of neces-sities rather than an economy based upon anxiety, fantasy, luxury and idle wishing. It proposes the independent, free-standing citizenry that Jefferson thought to be the surest safeguard of democratic liberty. And perhaps most important of all, it proposes an agri-culture based upon intensive work, local energies, care and long-living communities – that is, to state the matter from a consumer's point of view: a dependable, long-term food supply.

This is a possibility that is obviously imperiled – by antipathy in high places, by adverse public fashions and attitudes, by the deterioration of our present farm commu-nities and traditions, by the flawed education and the inexperience of our young people. Yet it alone can promise us the continuity of attention and devotion without which the human life of the earth is impossible.

Sixty years ago, in another time of crisis, Thomas Hardy wrote these stanzas:

Only a man harrowing clods
In a slow silent walk
With an old horse that stumbles and nods
Half asleep as they stalk.
Only thin smoke without flame
From the heaps of couch-grass;
Yet this will go onward the same
Though Dynasties pass.

Today most of our people are so conditioned that they do not wish to harrow clods either with an old horse or with a new tractor. Yet Hardy's vision has come to be more urgently true than ever. The great difference these 60 years have made is that, though we feel that this work *must* go onward, we are not so certain that it will. But the care of the earth is our most ancient and most worthy and, after all, our most pleasing responsibility. To cherish what remains of it, and to foster its renewal, is our only legitimate hope.

Note

1   The craft of persuading people to buy what they do not need, and do not want, for more than it is worth.

4

Ecological Literacy

David Orr

Literacy is the ability to read. Numeracy is the ability to count. Ecological literacy, according to Garrett Hardin, is the ability to ask 'What then?' Considerable attention is properly being given to our shortcomings in teaching the young to read, count and compute, but not nearly enough to ecological literacy. Reading, after all, is an ancient skill. And for most of the 20th century we have been busy adding, subtracting, multiplying, dividing, and now computing. But 'What then?' questions have not come easy for us despite all of our formidable advances in other areas. Napoleon did not ask the question, I gather, until he had reached the outskirts of Moscow, by which time no one could give a good answer except 'Let's go back home'. If Custer asked the question, we have no record of it. His last known words at Little Big Horn were, 'Hurrah, boys, now we have them', a stirring if dubious pronouncement. And economists, who are certainly both numerate and numerous, have not asked the question often enough. Asking 'What then?' on the west side of the Niemeh River, or at Fort Laramie, would have saved a lot of trouble. For the same reason, 'What then?' is also an appropriate question to ask before the last rain forests disappear, before the growth economy consumes itself into oblivion, and before we have warmed the planet intolerably.

The failure to develop ecological literacy is a sin of omission and of commission. Not only are we failing to teach the basics about the earth and how it works, but we are in fact teaching a large amount of stuff that is simply wrong. By failing to include ecological perspectives in any number of subjects, students are taught that ecology is unimportant for history, politics, economics, society, and so forth. And through television they learn that the Earth is theirs for the taking. The result is a generation of ecological yahoos without a clue why the colour of the water in their rivers is related to their food supply, or why storms are becoming more severe as the planet warms. The same persons as adults will create businesses, vote, have families and, above all, consume. If they come to reflect on the discrepancy between the splendour of their private lives in a hotter, more toxic and violent world, as ecological illiterates they will have roughly the same success as one trying to balance a chequebook without knowing arithmatic.

Note: Reprinted from *Ecological Literacy* by Orr, D., copyright © (1992), with permission from SUNY press, Albany

Formation of attitudes

To become ecologically literate one must certainly be able to read and, I think, even like to read. Ecological literacy also presumes an ability to use numbers, and the ability to know what is countable and what is not, which is to say the limits of numbers. But these are indoor skills. Ecological literacy also requires the more demanding capacity to observe nature with insight, a merger of landscape and mindscape. 'The interior landscape', in Barry Lopez's words, 'responds to the character and subtlety of an exterior landscape; the shape of the individual mind is affected by land as it is by genes' (Lopez, 1989). The quality of thought is related to the ability to relate to 'where on this earth one goes, what one touches, the patterns one observes in nature – the intricate history of one's life in the land, even a life in the city, where wind, the chirp of birds, the line of a falling leaf, are known'. The fact that this kind of intimate knowledge of our landscapes is rapidly disappearing can only impoverish our mental landscapes as well. People who do not know the ground on which they stand miss one of the elements of good thinking which is the capacity to distinguish between health and disease in natural systems and their relation to health and disease in human ones.

If literacy is driven by the search for knowledge, ecological literacy is driven by the sense of wonder, the sheer delight in being alive in a beautiful, mysterious, bountiful world. The darkness and disorder that we have brought to that world give ecological literacy an urgency it lacked a century ago. We can now look over the abyss and see the end of it all. Ecological literacy begins in childhood. 'To keep alive his inborn sense of wonder', a child, in Rachel Carson's words, needs the 'companionship of at least one adult who can share it, rediscovering with him the joy, excitement and mystery of the world we live in' (Carson, 1984). The sense of wonder is rooted in the emotions or what E. O. Wilson has called 'biophilia', which is simply the affinity for the living world (Wilson, 1984). The nourishment of that affinity is the beginning point for the sense of kinship with life, without which literacy of any sort will not help much. This is to say that even a thorough knowledge of the facts of life and of the threats to it will not save us in the absence of the feeling of kinship with life of the sort that cannot entirely be put into words.

There are, I think, several reasons why ecological literacy has been so difficult for western culture. First, it implies the ability to think broadly, to know something of what is hitched to what. This ability is being lost in an age of specialization. Scientists of the quality of Rachel Carson or Aldo Leopold are rarities who must buck the pressures toward narrowness and also endure a great deal of professional rejection and hostility. By inquiring into the relationship between chlorinated hydrocarbon pesticides and bird populations, Rachel Carson was asking an ecolate question. Many others failed to ask, not because they did not like birds, but because they had not, for whatever reasons, thought beyond the conventional categories. To do so would have required that they relate their food system to the decline in the number of birds in their neighbourhood. This means that they would have had some direct knowledge of farms and farming practices, as well as a comprehension of ornithology. To think in ecolate fashion presumes a breadth of experience with healthy natural systems, both of which are increasingly rare. It also presumes that the persons be willing and able to 'think at right angles' to their particular specializations, as Leopold put it.

Ecological literacy is difficult, second, because we have come to believe that education is solely an indoor activity. A good part of it, of necessity, must be, but there is a
price. William Morton Wheeler once compared the naturalist with the professional biologist in these words: '[The naturalist] is primarily an observer and fond of outdoor life,
a collector, a classifier, a describer, deeply impressed by the overwhelming intricacy of
natural phenomena and revelling in their very complexity.' The biologist, on the other
hand, 'is oriented toward and dominated by ideas, and rather terrified or oppressed by
the intricate hurly-burly of concrete, sensuous reality ... he is a denizen of the laboratory.
His besetting sin is oversimplification and the tendency to undue isolation of the organisms he studies from their natural environment' (Curtis and Greenslet, 1962). Since,
Wheeler wrote, ecology has become increasingly specialized 'and, one suspects, remote
from its subject matter. Ecology, like most learning worthy of the effort, is an applied
subject. Its goal is not just a comprehension of how the world works, but, in the light of
that knowledge, a life lived accordingly. The same is true of theology, sociology, political
science, and most other subjects that grace the conventional curriculum.

The decline in the capacity for aesthetic appreciation is a third factor working against
ecological literacy. We have become comfortable with all kinds of ugliness and seem inca-
pable of effective protest against its purveyors: urban developers, businessmen, government officials, television executives, timber and mining companies, utilities and advertisers.
Rene Dubos once stated that our greatest disservice to our children was to give them the
belief that ugliness was somehow normal. But disordered landscapes are not just an aesthetic
problem. Ugliness signifies a more fundamental disharmony between people and between
people and the land. Ugliness is, I think, the surest sign of disease, or what is now being
called 'unsustainability'. Show me the hamburger stands, neon ticky-tacky strips leading
toward every city in America, and the shopping malls, and I'll show you devastated rain
forests, a decaying countryside, a politically dependent population and toxic waste
dumps. It is all of a fabric.

And this is the heart of the matter. To see things in their wholeness is politically
threatening. To understand that our manner of living, so comfortable for some, is linked
to cancer rates in migrant labourers in California, the disappearance of tropical rain forests, 50,000 toxic dumps across the US, and the depletion of the ozone layer is to see the
need for a change in our way of life. To see things whole is to see both the wounds we
have inflicted on the natural world in the name of mastery and those we have inflicted
on ourselves and on our children for no good reason, whatever our stated intentions.
Real ecological literacy is radicalizing in that it forces us to reckon with the roots of our
ailments, not just with their symptoms. For this reason, I think it leads to a revitalization
and broadening of the concept of citizenship to include membership in a planetwide
community of humans and living things.

And how does this striving for community come into being? I doubt that there is a
single path, but there are certain common elements. First, in the lives of most if not all
people who define themselves as environmentalists, there is experience in the natural
world at an early age. Leopold came to know birds and wildlife in the marshes and fields
around his home in Burlington, Iowa before his teens. David Brower, as a young boy on
long walks over the Berkeley hills, learned to describe the flora to his nearly blind mother.
Second, and not surprisingly, there is often an older teacher or mentor as a role model:
a grandfather, a neighbour, an older brother, a parent, or teacher. Third, there are semi-

nal books that explain, heighten, and say what we have felt deeply, but not said so well. In my own life, Rene Dubos and Loren Eiseley served this function of helping to bring feelings to articulate consciousness.

Ecological literacy is becoming more difficult, I believe, not because there are fewer books about nature, but because there is less opportunity for the direct experience of it. Fewer people grow up on farms or in rural areas where access is easy and where it is easy to learn a degree of competence and self-confidence toward the natural world. Where the ratio between the human created environment to the purely natural world exceeds some point, the sense of place can only be a sense of habitat. One finds the habitat familiar and/or likeable but without any real sense of belonging in the natural world. A sense of place requires more direct contact with the natural aspects of a place, with soils, landscape, and wildlife. This sense is lost as we move down the continuum toward the totalized urban environment where nature exists in tiny, isolated fragments by permission only. Said differently, this is an argument for more urban parks, summer camps, green belts, wilderness areas, public seashores. If we must live in an increasingly urban world, let's make it one of well designed compact green cities that include trees, river parks, meandering green belts, and urban farms where people can see, touch, and experience nature in a variety of ways. In fact, no other cities will be sustainable in a greenhouse world.

Ecological literacy and formal education

The goal of ecological literacy as I have described it has striking implications for that part of education that must occur in classrooms, libraries and laboratories. To the extent that most educators have noticed the environment, they have regarded it as a set of problems which are: (1) solvable (unlike dilemmas, which are not) by (2) the analytic tools and methods of reductionist science which (3) create value-neutral, technological remedies that will not create even worse side effects. Solutions, therefore, originate at the top of society, from governments and corporations, and are passed down to a passive citizenry in the form of laws, policies and technologies. The results, it is assumed, will be socially, ethically, politically and humanly desirable, and the will to live and to sustain a humane culture can be preserved in a technocratic society. In other words, business can go on as usual. Since there is no particular need for an ecologically literate and ecologically competent public, environmental education is most often regarded as an extra in the curriculum, not as a core requirement or as an aspect pervading the entire educational process.

Clearly, some parts of the crisis can be accurately described as problems. Some of these can be solved by technology, particularly those that require increased resource efficiency. It is a mistake, however, to think that all we need is better technology, not an ecologically literate and caring public willing to help reduce the scale of problems by reducing its demands on the environment and to accept (even demand) public policies that require sacrifices. It all comes down to whether the public understands the relation between its well being and the health of the natural systems.

For this to occur, we must rethink both the substance and the process of education at all levels. What does it mean to educate people to live sustainably, going, in Aldo Leopold's words, from 'conqueror of the land community to plain member and citizen of it'? (Leopold, 1966) However it is applied in practice, the answer will rest on six foundations.

The first is the recognition that *all education is environmental education.* By what is included or excluded, emphasized or ignored, students learn that they are a part of or apart from the natural world. Through all education we inculcate the ideas of careful stewardship or carelessness. Conventional education, by and large, has been a celebration of all that is human to the exclusion of our dependence on nature. As a result, students frequently resemble what Wendell Berry has called 'itinerant professional vandals', persons devoid of any sense of place or stewardship, or inkling of why these are important (Berry, 1987).

Second, *environmental issues are complex and cannot be understood through a single discipline or department.* Despite a decade or more of discussion and experimentation, interdisciplinary education remains an unfulfilled promise. The failure occurred, I submit, because it was tried within discipline-centric institutions. A more promising approach is to reshape institutions to function as transdisciplinary laboratories that include components such as agriculture, solar technologies, forestry, land management, wildlife, waste cycling, architectural design and economics (Caldwell, 1983). Part of the task, then, of Earth-centred education is the study of interactions across the boundaries of conventional knowledge and experience.

Third, *for inhabitants, education occurs in part as a dialogue with a place and has the characteristics of good conversation.* Formal education happens mostly as a monologue of human interest, desires and accomplishments that drowns out all other sounds. It is the logical outcome of the belief that we are alone in a dead world of inanimate matter, energy flows and biogeochemical cycles. But true conversation can occur only if we acknowledge the existence and interests of the other. In conversation, we define ourselves, but in relation to another. The quality of conversation does not rest on the brilliance of one or the other person. It is more like a dance in which the artistry is mutual.

In good conversation, words represent reality faithfully. And words have power. They can enliven or deaden, elevate or degrade, but they are never neutral, because they affect our perception and ultimately our behavior. The use of words such as 'resources', 'manage', 'channelizes', 'engineer' and 'produce' makes our relation to nature a monologue rather than a conversation. The language of nature includes the sounds of animals, whales, birds, insects, wind and water – a language more ancient and basic than human speech. Its books are the etchings of life on the face of the land. To hear this language requires patient, disciplined study of the natural world. But it is a language for which we have an affinity.

Good conversation is unhurried. It has its own rhythm and pace. Dialogue with nature cannot be rushed. It will be governed by cycles of day and night, the seasons, the pace of procreation, and by the larger rhythm of evolutionary and geologic time. Human sense of time is increasingly frenetic, driven by clocks, computers, and revolutions in transportation and communication.

Good conversation has form, structure, and purpose. Conversation with nature has the purpose of establishing, in Wendell Berry's words: 'What is here? What will nature permit here? What will nature help us do here?' (Berry, 1987). The form and structure of any conversation with the natural world is that of the discipline of ecology as a restorative process and healing art.

Fourth, it follows that *the way education occurs is as important as its content*. Students taught environmental awareness in a setting that does not alter their relationship to basic life support systems learn that it is sufficient to intellectualize, emote, or posture about such things without having to live differently. Environmental education ought to change the way people live, not just how they talk. This understanding of education is drawn from the writings of John Dewey, Alfred North Whitehead, J. Glenn Gray, Paulo Friere, Ivan Illich and Eliot Wigginton. Learning in this view best occurs in response to real needs and the life situation of the learner. The radical distinctions typically drawn between teacher and student, between the school and the community, and those between areas of knowledge, are dissolved. Real learning is participatory and experiential, not just didactic. The flow can be two ways between teachers, who best function as facilitators, and students who are expected to be active agents in defining what is learned and how.

Fifth, *experience in the natural world is both an essential part of understanding the environment, and conducive to good thinking*. Experience, properly conceived, trains the intellect to observe the land carefully and to distinguish between health and its opposite. Direct experience is an antidote to indoor, abstract learning. It is also a well-spring of good thinking. Understanding nature demands a disciplined and observant intellect. But nature, in Emerson's words, is also 'the vehicle of thought' as a source of language, metaphor and symbol. Natural diversity may well be the source of much of human creativity and intelligence. If so, the simplification and homogenization of ecosystems can only result in a lowering of human intelligence.

Sixth, *education relevant to the challenge of building a sustainable society will enhance the learner's competence with natural systems*. For reasons once explained by Whitehead and Dewey, practical competence is an indispensable source of good thinking. Good thinking proceeds from the friction between reflective thought and real problems. Aside from its effects on thinking, practical competence will be essential if sustainability requires, as I think it does, that people must take an active part in rebuilding their homes, businesses, neighbourhoods, communities and towns. Shortening supply lines for food, energy, water and materials – while recycling waste locally – implies a high degree of competence not necessary in a society dependent on central vendors and experts.

The aim: Ecological literacy

If these can be taken as the foundations of Earth-centred education, what can be said of its larger purpose? In a phrase, it is that quality of mind that seeks out connections. It is the opposite of the specialization and narrowness characteristic of most education. The ecologically literate person has the knowledge necessary to comprehend interrelatedness, and an attitude of care or stewardship. Such a person would also have the practical competence required to act on the basis of knowledge and feeling. Competence can only be derived from the experience of doing and the mastery of what Alasdair MacIntyre describes as a 'practice' (MacIntyre, 1981). Knowing, caring, and practical competence constitute the basis of ecological literacy.

Ecological literacy, further, implies a broad understanding of how people and societies relate to each other and to natural systems, and how they might do so sustainably. It presumes both an awareness of the interrelatedness of life and knowledge of how the world works as a physical system. To ask, let alone answer, 'What then?' questions presumes an understanding of concepts such as carrying capacity, overshoot, Liebig's Law of the minimum, thermodynamics, trophic levels, energetics and succession. Ecological literacy presumes that we understand our place in the story of evolution. It is to know that our health, well being, and ultimately our survival depend on working with, not against, natural forces. The basis for ecological literacy, then, is the comprehension of the interrelatedness of life grounded in the study of natural history, ecology and thermodynamics. It is to understand that: ''There ain't no such thing as a free lunch'; 'You can never throw anything away'; and 'The first law of intelligent tinkering is to keep all of the pieces.' It is also to understand, with Leopold, that we live in a world of wounds senselessly inflicted on nature and on ourselves.

A second stage in ecological literacy is to know something of the speed of the crisis that is upon us. It is to know magnitudes, rates, and trends of population growth, species extinction, soil loss, deforestation, desertification, climate change, ozone depletion, resource exhaustion, air and water pollution, toxic and radioactive contamination, resource and energy use – in short, the vital signs of the planet and its ecosystems. Becoming ecologically literate is to understand the human enterprise for what it is: a sudden eruption in the enormity of evolutionary time.

Ecological literacy requires a comprehension of the dynamics of the modern world. The best starting place is to read the original rationale for the domination of nature found in the writings of Bacon, Descartes and Galileo. Here one finds the justification for the union of science with power and the case for separating ourselves from nature in order to control it more fully. To comprehend the idea of controlling nature, one must fathom the sources of the urge to power and the paradox of rational means harnessed to insane ends portrayed in Marlowe's *Doctor Faustus*, Mary Shelley's *Frankenstein*, Melville's *Moby Dick*, and Dostoevsky's 'Legend of the Grand Inquisitor'.

Ecological literacy, then, requires a thorough understanding of the ways in which people and whole societies have become destructive. The ecologically literate person will appreciate something of how social structures, religion, science, politics, technology, patriarchy, culture, agriculture and human cussedness combine as causes of our predicament.

The diagnosis of the causes of our plight is only half of the issue. But before we can address solutions there are several issues that demand clarification, 'Nature', for example, is variously portrayed as 'red in tooth and claw', or, like the film 'Bambi', full of sweet little critters. Economists see nature as natural resources to be used; the backpacker as a wellspring of transcendent values. We are no longer clear about our own nature, whether we are made in the image of God, or are merely a machine or computer, or animal. These are not trivial, academic issues. Unless we can make reasonable distinctions between what is natural and what is not, and why that difference is important, we are liable to be at the mercy of the engineers who want to remake all of nature, including our own.

Environmental literacy also requires a broad familiarity with the development of ecological consciousness. The best history of the concept of ecology is Donald Worster's *Nature's Economy* (Worster, 1985). It is unclear whether the science of ecology will be 'the last of the old sciences, or the first of the new'. As the former, ecology is the science of efficient resource management. As the first of the new sciences, ecology is the basis for a broader search for pattern and meaning. As such it cannot avoid issues of values, and the ethical questions raised most succinctly in Leopold's 'The Land Ethic'.

The study of environmental problems is an exercise in despair unless it is regarded as only a preface to the study, design and implementation of solutions. The concept of sustainability implies a radical change in the institutions and patterns that we have come to accept as normal. It begins with ecology as the basis for the redesign of technology, cities, farms, and educational institutions, and with a change in metaphors from mechanical to organic, industrial to biological. As part of the change we will need alternative measures of well being such as those proposed by Amory Lovins (least-cost end-use analysis) (Lovins, 1977), H. T. Odum (energy accounting) (Hall et al, 1986) and John Cobb (index of sustainable welfare) (Daly and Cobb, 1990). Sustainability also implies a different approach to technology. One that gives greater priority to those that are smaller in scale, less environmentally destructive, and rely on the free services of natural systems. Not infrequently, technologies with these characteristics are also highly cost-effective, especially when subsidies for competing technologies are leveled out.

If sustainability represents a minority tradition, it is nonetheless a long one dating back at least to Jefferson. Students should not be considered ecologically literate until they have read Thoreau, Kropotkin, Muir, Albert Howard, Alfred North Whitehead, Gandhi, Schweitzer, Aldo Leopold, Lewis Mumford, Rachel Carson, E. F. Schumacher and Wendell Berry. There are alternatives to the present patterns that have remained dormant or isolated, not because they did not work, were poorly thought out, or were impractical, but because they were not tried. In contrast to the directions of modern society, this tradition emphasizes democratic participation, the extension of ethical obligations to the land community, careful ecological design, simplicity, widespread competence with natural systems, the sense of place, holism, decentralization of whatever can best be decentralized, and human scaled technologies and communities. It is a tradition dedicated to the search for patterns, unity, connections between people of all ages, races, nationalities and generations, and between people and the natural world. This is a tradition grounded in the belief that life is sacred and not to be carelessly expended on the ephemeral. It is a tradition that challenges militarism, injustice, ecological destruction and authoritarianism, while supporting all of those actions that lead to real peace, fairness, sustainability and people's right to participate in those decisions that affect their lives. Ultimately, it is a tradition built on a view of ourselves as finite and fallible creatures living in a world limited by natural laws. The contrasting Promethean view, given force by the success of technology, holds that we should remove all limits, whether imposed by nature, human nature or morality. Its slogan is found emblazoned on the advertisements of the age: 'You can have it all' (Michelob Beer), or 'Your world should know no limits' (Merrill Lynch). The ecologically literate citizen will recognize these immediately for what they are: the stuff of epitaphs. Ecological literacy leads in other, and more durable, directions toward prudence, stewardship and the celebration of the Creation.

References

Berry, W. (1987) *Home Economics*, North Point Press, San Francisco
Caldwell, L. K. (1983) 'Environmental studies: Discipline or metadiscipline?' *The Environmental Professional*, pp247–259

Carson, R. (1984) *The Sense of Wonder*, Harper and Row, New York, p45

Curtis, C. P., Jr and Greenslet, F. (eds) (1962) *The Practical Cogitator*, 3rd edition, Houghton Mifflin Co, Boston, pp226–229

Daly, H. and Cobb, J. (1990) For *the Common Good*, Beacon Press, Boston, pp401–455

Hall, C. et al (1986) *Energy and Resource Quality*, Wiley and Sons, New York, pp3–151

Leopold, A. (1966) *A Sand County Almanac*, Ballantine, New York, p240

Lopez, B. (1989) *Crossing Open Ground*, Vintage, New York, p65

Lovins, A. (1977) *Soft Energy Paths*, Ballinger, Cambridge

MacIntyre, A. (1981) *After Virtue*, Notre Dame University Press, South Bend, p168–189

Wilson, E. O. (1984) *Biophilia*, Harvard University Press, Harvard

Worster, D. (1985) *Nature's Economy*, Sierra Club Books, San Francisco (1977); reissued by Cambridge University Press, Cambridge

5

An Amish Perspective

David Kline

I want to talk about our farm. It is on the 120 acres of rolling Ohio land that the county courthouse records show belongs to my wife and me and our family. It is here that I can do great harm to nature or where I can live, or at least try to live, in peaceful coexistence with the land. Here on our farm I can exploit nature or nurture it. Here is where nature giveth and nature taketh away.

But first we need to look at the past. We of northern European stock were not the first humans on the land that is now our farm. There is ample evidence otherwise.

Last fall as I was harrowing the last round in the field that was to be sown to wheat, something on the ground caught my eye. Stopping the team, I reached down and picked up what looked like a piece of chert. It was a small arrowhead, perfect except for a small chip broken from its tip. While the horses rested, I turned and sat on the harrow to admire my find.

Less than an inch in length, the triangular piece of flint was obviously crafted by a human. As I turned it over in my hand and wiped it clean of soil, I pondered on how much the previous owner of the arrowhead had depended on flint – cryptocrystalline quartz – for survival. It occurred to me that I too depend on a tiny speck of quartz, not to slay a deer, but to tell me when it is time to unhitch.

Let's suppose that in early October of 1492 a small band of Erie hunters from the Cat nation left their village along the shore of the lake now bearing their name and travelled south to gather provisions for the coming winter months. (Early historians wrote that all kinds of game, wild fruits and succulent roots abounded in this part of Ohio.) One of the hunters may have waited in ambush for a black bear or a white-tailed deer to leave the cover of the marshland below and travel to the ridge to feed on acorns from the white oaks and nuts from the beeches. As the animal passed the hunter's little coign of vantage, the Erie's bowstring snapped and the arrow streaked toward its target. But en route the arrow nicked a twig, deflecting it enough that it missed the animal and disappeared into the leaf litter of the autumn woods.

This Erie hunter, from the time called the Woodland Indian period, was not the first Native American to visit our land. I have found blades and spear points from the Archaic period, but until a few years ago I had never found any evidence from the Paleo-Indian period while working the fields. Then one spring our youngest son, Michael, and

---

Note: Reprinted from *Rooted in the Land: Essays on Community and Place* by Vitek, W. and Jackson, W. (eds), copyright © (1996), with permission from Yale University Press, Haven

I were crossing Salt Creek with the horse and hack, on our way to repair fence along the bottom pasture, when we bounced across some rocks. We stopped and looked over the dashboard at the boulders below. My astonished eyes saw, an inch behind the horse's foot, the most beautifully knapped spear point I had ever been fortunate enough to find. The blade of black flint had not a nick on it and surely had been washed away from the creek bank only recently by high water. I am told it is from the late Paleo-Indian period. So there is evidence that the land we call our farm has been inhabited by humans, at least periodically, for many centuries.

For over 300 years the small dart I held in my hand had lain undisturbed and, covered by an annual fall of leaves, had sunk deeper into the humus. Sheltered by the great oaks whose roots knitted the soil together and whose leaves added fertility, the arrowhead was not aware of the changes to come.

Nor was the Erie hunter to know that his own people, the powerful Cat nation, would be annihilated by the Iroquois, the Five Nation Confederacy, in 1656. For more than 50 years after the demise of the Eries, no Indians hunted on this land.

Then sometime in the early 1700s the Iroquois, now the Six Nations, gave this part of Ohio to the Delaware Indians, who were being pushed out of the east by European settlers. But even with the increased human activity, the flora and fauna of the land changed little. Occasionally centuries-old oaks died or were blown down by storms. The openings in the forests soon flourished with new seedlings, new trees that now number more than 250 growth rings.

Following the war of 1812 this land was visited by fewer and fewer Native Americans until by the early 1820s none at all came. The Indians who did not die from disease epidemics (which were arrogantly interpreted by some Europeans as God's way of clearing the land for His Elect) were forced to move west.

In 1824 our farm was settled and in 1825 a deed, signed by President Monroe, was given to that pioneer family. A deed, as Robert Frost wrote, that followed many deeds of war. Now the family had the right to pay taxes and to post 'No Trespassing' signs. Other settlers came. Our farm was joined on the north by the Pomerene family. West were the Walters, east the Rosses, and to the northeast were the French Catholic settlers of the soon-to-be-prosperous community of Calmoutier.

I can only speculate what happened in the next 50 to 75 years. Quite obviously a struggle to control nature took place. From the sketchy history available, most of the vast forests – white and red oaks, chestnuts, hickories, beeches and other hardwoods – were felled into windrows, left to dry for a year or two, and then burned. For many weeks of the year in the early to mid-1800s the skies over this part of Ohio were hazy from burning hardwoods, millions of board feet of the finest oaks on earth. As the trees went, so did much of the wildlife. Around 1850 the state legislature passed a law forbidding the burning of timber to open the land.

My small arrowhead escaped from being in a windrow of torched trees. The intense heat would have fractured the flint. With the opening of the land, the pioneer family began plowing between the stumps and sowing and planting crops of wheat, rye, oats and corn.

For the first 50 years after being deeded, the farm, with the high fertility of its woodlands soils (Wooster silt loam), produced enough wealth for the family to prosper. Then starting around 1875 things began to go awry. The health of the soil deteriorated.

The marriage between the steward and the land was in trouble. After being mined for too long, the fiddler's pay came due. And payment was taken in the reduced yields of the crops.

Soon after 1900 not enough income was generated from the farm to pay for the taxes. So the county sheriff ordered the 'farmed-out' land and the buildings to be sold at public auction to the highest bidder. It was the old truism: 'shirtsleeves to shirtsleeves in three generations'. One could say that nature won the battle for the farm. But in reality, both humans and nature lost. Both were poorer for the experience.

What caused the farm to fail?

From a number of possible reasons let me suggest just a few. First, the depression of the 1890s did not help matters, I am sure. A second reason was the lack of legumes growing on the farm. When my uncle moved here in 1918 the primary plants growing were Canada thistles and timothy. Legumes play a vital part in the health of a sustainable farm, since the rhizobium bacteria in legumes have the ability to convert atmospheric nitrogen into plant food. The conversion nourishes the leguminous plants as well as companion plants of other species. For instance, blue-grass and white clover in a pasture field are the perfect pair.

As bluegrass and white clover are to each other, so should be our connectedness with nature, a connection of life that is difficult to put into words. I like the way the 18th century Mennonite clover farmer David Mellinger framed it. He lived in the Palantinate, the section of southwestern Germany squeezed between the Rhine River and the French region of Alsace and Lorraine. This area is where the Mennonites and the Amish learned how important clover is, if you farm, in getting along with nature. Menninger said: 'I should have already given princes and other great lords a description of my operation and how I achieved it, but I cannot tell it so easily ... one thing leads to another. It is like a clockwork, where one wheel grabs hold of another, and then the work continues without my even being able to know or describe how I brought the machine into gear.' Maybe this wheel of life is one of the mysteries of God.

The Apostle Paul wrote that we should serve faithfully as stewards of the mysteries of God. Perhaps the inference is that the stewards are the ministers of the Word. But if we lift our eyes not away from the earth, the mysteries of God can also be His creation. We should be stewards or caretakers of creation.

Science, I think, has attempted to gather these mysteries and set them in concrete. This is something of which religion also may be guilty. Once we rely too much on human cleverness, nature suffers and we become alienated from it. Johann Goethe thought the worst thing that can happen to man is alienation from nature.

As dressers and keepers of the earth, I believe we need some unconcreted mysteries. We need the delight of the unknown and the unexplainable in nature, what Rachel Carson called the sense of wonder. How can the bobolinks find their way back to the farm after spending our winter in the austral summer of Argentina? When in late April the first returning male bobolink's bubbling, cheerful flight song drifts across the hayfield, my faith in the mysteries of God is renewed. And who can fail to feel a sense of wonder in the presence of a spectacular display of northern lights, as were seen here in early November last year?

To restore a depleted farm takes time. Especially a hill farm where extensive soil erosion took place. What took years to harm takes years to heal. It is estimated that it takes nature several hundred years to build an inch of top soil (150 tonnes per acre). With poor farming practices that amount can be lost to erosion in ten years.

I can still remember as a boy in the 1950s when the last gully disappeared beneath a thick cover of sod. Now the water leaving the farm during a hard rain is clear and clean.

Quite often the most overlooked and perhaps the most abused part of nature is the life in the soil. The healing process begins there. A gram of good soil may hold as many as four billion microbiotic organisms. These can be nurtured with the growing of legumes alongside grasses, the use of animal manures mixed with straw or fodder grown on the farm, and annual crop rotations.

The rotations we use date back to the Palatinate farmers of the 18th century. The field that is hay this year will be planted to corn next year. Following the corn will be oats, and then wheat which will be seeded to hay, and the cycle will be complete.

The health of a farm and nature is not brought about by some 'heroic feat of technology, but rather by thousands of small acts and restraints handed down by generations of experience' to quote Wendell Berry. These are acts my family and I often perform as the teams are resting while plowing and planting. Maybe it is only carrying a rock off the field, or moving a piece of sod to some low spot to check possible erosion, or moving a killdeer's or horned lark's nest out of harm's way. These small acts of stewardship, multiplied a thousand times, do add up. Confucius said, 'The best fertilizer on any farm is the footsteps of the owner'.

To earn one's livelihood from a farm while at the same time caring for and preserving nature is sometimes difficult particularly in growing seasons as abnormal as the past four years have been. What might appear to be a conflict between exploiter and nurturer could be more accurately described as farmer and land involved in a sort of dance. It is an attempt to achieve a balance between give and take, without too much stepping on each other's toes.

Wendell Berry writes: 'An exploiter wishes to earn as much as possible by as little work as possible. The nurturer expects to have a decent living from his land but his characteristic wish is to work as well as possible.'

A great deal depends on whether we look at the farm as a food factory where nature suffers at the expense of profit, or we look at the farm as a place to live, where – to quote Aldo Leopold – 'there is a harmonious balance between plants, animals, and people; between the domestic and the wild: between utility and beauty'.

This balance of utility and beauty that I strive for on the farm, my wife achieves so well in the garden, a delightful blend of the domestic and the wild. She has several gardens, but the one I want to mention is practically on our doorstop and its primary food function is as a provider of fresh salad makings – lettuces, broccoli, cauliflower, radishes, onions, a few cucumber plants, and several early tomatoes – that end up on the table only minutes after leaving the garden. After all, doesn't the quality of life begin on the dinner plate? However, only about half of the garden space – the domestic – is for our culinary benefit. The rest tends toward the wild – plantings for birds and butterflies and beauty. Annuals and perennials.

While most of the perennials are native to this part of the country (the wild rose, butterfly weed, black-eyed susan, bergamot and New England aster), a few are not: the coreopsis and purple coneflower are transplanted prairie natives. The annual flowers include a row of mixed zinnias, impatiens, petunias and salvias.

The blending of domestic and wild works so well. An American goldfinch may be feeding on the stickum-like seeds of the coneflowers, while a chipping sparrow searches for cabbage loopers on the broccoli. A ruby-throated hummingbird hovers near the delicate impatiens blossoms, sipping their nectar, while swallowtail butterflies prefer the nectar of the bergamot and the zinnias.

Maybe the main reason I love this part-domestic, part-wild garden so much is that it is located between the house and the barn. Whenever I enter or leave the house, I linger for several minutes in the garden. Perhaps I eat a radish or a tender green onion, but more often I just pause to watch the life of the garden. What is going on may determine my day's work. If it is morning and there is hay to be made, I watch the wild things. If the bees and the birds are exceptionally active and the maple leaves show their undersides in a slight breeze, I figure rain is on the way and change my plans from hay mowing to corn cultivating (or maybe fishing).

Aldo Leopold once said that a good farm must be one where the wild flora and fauna have lost ground without losing their existence. Even though much has been lost here since that 1825 deed, we need to cherish what remains. And a farm is a good place to do this cherishing. Where else can one be so much a part of nature and the mysteries of God, the unfolding of the seasons, the coming and going of the birds, the pleasures of planting and the joys of harvest, the cycle of life and death? Where else can one still touch hands with an earlier people through their flint work? Sure, there are periods of hard work. But it is labour with dignity, working together with family, neighbours and friends. Here on our 120 acres I must be a steward of the mysteries of God.

6

Becoming Native to This Place

Wes Jackson

I am writing in what is left of Matfield Green, a Kansas town of some 50 people situated in a county of a few over 3,000 in an area with 33 inches of annual precipitation. It is typical of countless towns throughout the Midwest and Great Plains. People have left, people are leaving, buildings are falling down or burning down. Fourteen of the houses here that do still have people have only one person, usually a widow or widower. I purchased seven rundown houses in town for less than US$4,000. Four friends and I purchased, for $5,000, the beautiful brick elementary school, which was built in 1938. It has 10,000 square feet, including a stage and gymnasium. The Land Institute has purchased the high school gymnasium ($4,000) and 12 acres south of town ($6,000). A friend and I have purchased 38 acres north of town from the Santa Fe Railroad for $330 an acre. On this property are a bunkhouse, some large corrals to handle cattle, and around 25 acres of never-ploughed tallgrass prairie. South of town is an abandoned natural gas booster station with its numerous buildings and facilities. Situated on 80 acres, it had been part of the first long-distance pipelines that delivered natural gas from the large Hugoton field near Amarillo to Kansas City. A neighbour in his late 70s told me that his father had helped dig the basement, using a team of horses and a scoop, around 1929 (contemporary sunlight used to leverage extraction of anciently stored energy). Owned by a major pipeline company, the booster station stands as a silent monument to the extractive economy and a foreshadow of what is to come. A small area has been contaminated with Polychlorinated Biphenols (PCBs). Think of the possible practicality and symbolism if this facility, formerly devoted to transporting the high energy demanded by an extractive economy, while at the same time dumping a major environmental pollutant, could be taken over by a community and converted to a facility that would sponsor such renewable technologies as photovoltaic panels or wind machines.

Imagine this human community as an ecosystem, as a locus or primary object of study. We know that much public policy, allegedly implemented in the public interest, is partly responsible for the demise of this community. The question then becomes, how can this human community, like a natural ecosystem community, be protected from that abstraction called 'the public'. How can both kinds of communities be built back and also protected?

The effort begins, I think, with the sort of inventory and accounting that ecologists have done for natural ecosystems, a kind of accounting seldom if ever done for human

Note: Reprinted from *Becoming Native to this Place* by Jackson, W., copyright © (1994), with permission from University Press of Kentucky, Lexington

dominated ecosystems. How do we start thinking about what is involved in setting up the books for ecological community accounting that will feature humans? I emphasize accounting because our goal is renewability, sustainability.

About 85 per cent of the county in which this small town is situated is never-ploughed native prairie. Over the millennia it has featured recycling and has run on sunlight just as a forest or a marsh does. Though the prairie is fenced and is now called pasture and grass, aside from the wilderness areas of our country it is about as close to an original relationship as we will find in any human-managed system. But though the people left in town seem to have a profound affection for the place and for one another, they are as susceptible to the world as anyone else to shopping mall living, to secular materialism in general.

Nevertheless, I have imagined this as a place that could grow bison for meat, as a place where photovoltaic panels could be assembled at the old booster station, where the school could become a gathering place that would be a partial answer to the mall, a place that might attract a few retired people, including professor types, who could bring their pensions, their libraries, and their social security cheques to help set up the books to support themselves and to ecological community accounting. Essentially all academic disciplines would be an asset in such an effort, but only if confronted with a broad spectrum of ecological necessities in the face of small-town reality. Much of what must be done will be in conflict with human desires, if not human needs.

But let us allow our imaginations to wander. What *if* bison (or even cattle) could be slaughtered in the little towns on a small scale and become the answer to the massive industrialized Iowa Beef Packers plant at Emporia, a plant to which some area residents drive 50 miles and more in order to earn modest wages and risk carpel tunnel syndrome. What if the manufacture of photovoltaic panels could happen before the eyes of the children of the workers? What if the processing of livestock could happen with those children present? What if those children were allowed to exercise their strong urge to help – to work? What, finally, if shopping malls and Little League were to become less interesting than playing 'work-up' with all ages?

One of our major tasks as educators is to expand the imagination about our possibilities. When I walk through the abandoned school with its leaky gym roof and see the solid structure of a beautiful building built in 1938, with stage curtains rotting away and paint peeling off the walls, where one blackboard after another has been carried away as though the building were a slate mine, I am forced to think that the demise of this school has resulted in part from a failure of imagination but more from the tyranny of disregard by something we call 'the economy'.

I am in that town at this moment, typing in a house that in its history of some 75 years has been abandoned more times than anyone can recall. First it was home to a family in a now-abandoned oil field a few miles southeast of here. Then it was hauled intact on a wagon into town in the 1930s, set on a native stone foundation which, when I bought it recently, was crumbling and falling away. Three ceilings supported up to three inches of dirt, much of it from back in the Dust Bowl era. The expanding side walls had travelled so far off the floor joist in the middle that the ceiling was suspended. They had to be pulled back onto the floor joist and held permanently with large screws and plates. Three major holes in the roof had allowed serious water damage to flooring and studs. An oppossum had died under the bathtub long enough ago that only its bones were displayed lying at rest in an arc over the equivalent of half a bale of hay. Two of my seven houses, one empty for nearly 20 years and the other for 15, had been walked away from with their refrigerators full. Who knows when the electric meters were pulled – probably a month or two after the houses were vacated and the bill hadn't been paid. To view the contents – eggs and ham, Miracle Whip and pickles, mustard and catsup, milk, cheese, and whatnot – was more like viewing an archaeological find than a repulsive mess.

Where I now sit at my typewriter in that oil field house that relatives and friends have helped remodel, I can see an abandoned lumberyard across the street next to the abandoned hardware store. Out another window but from the same chair is the back of the old creamery that now stores junk (on its way to becoming antiques?). With Clara Jo's retirement, the half-time post office across the street is threatened with closure. From a different window, but the same seat, I can see the bank, which closed in 1929 and paid off ten cents on the dollar. (My nephew recently bought it for $500.) I can see the bank now only because I can look across the vacant lot where the cafe stood before the natural gas booster station shut down. If my front door were open, given the angle at which I sit, I would be looking right at the Hitchin' Post, the only business left, a bar that accommodates local residents and the cowhands who pull up with their pickups and horse trailers with some of the finest saddle mounts anywhere. These horses will quiver and stomp while waiting patiently as their riders stop for lunch or after work to indulge in beer, nuts and microwave sandwiches and to shoot pool. (The Hitchin' Post lacks running water, so no dishes can be washed. The outhouse is out back.) Around the corner is an abandoned service station. There were once four! Across the street is the former barber shop.

I know that this town and the surrounding farms and ranches did not sponsor perfect people. I keep finding whiskey bottles in old outhouses and garages, stashed between inner and outer walls here and there, these people's local version of a drug culture. I hear the familiar stories of infidelities 50 years back – the overalls on the clothesline hung upside down, the flower pot one day on the left side of the step, the next day on the right – signals to lovers even at the height of the Great Depression when austerity should have tightened the family unit and maybe did. There were the shootings, the failures of justice, the story of the father of a young married woman who shot and killed his son-in-law for hitting his daughter and how charges were never filed. Third-generation bitterness is common. The human drama goes on. This place still doesn't have a lifestyle, never had a lifestyle, but rather livelihoods with ordinary human foibles. Nevertheless, the graveyard now contains the remains of both the cuckolder and the cuckold, the shooter and the shot, the drunk and the sober.

This story can be repeated thousands of times across our land, and no telling will deviate even ten per cent. I am not sure what should be done here by way of community development. I do believe, however, that community development can begin with putting roofs on buildings that are now leaking, and scabbing in two-by-fours at the bottom of studs that have rotted out around their anchoring nails. I doubt, on the other hand, that a sustainable society can start with a programme sent down from Washington or from a Rural Sociology Department in some land grant university, or for that matter as a celebration of Columbus Day.

Locals and most rural sociologists alike believe the answer lies in jobs, *any* jobs so long as they don't pollute too much. Though I am in sympathy with every urgent impulse to meet human welfare, rural America – America! – does *not* need jobs that depend on the extractive economy. We need a way to arrest consumerism. We need a different form of accounting so that both sufficiency and efficiency have *standing* in our minds.

The poets and scientists who counselled that we consult nature would have understood, I think, that we might begin by looking at that old prairie, by remembering who we are as mammals, as primates, as humanoids, as animals struggling to become human by controlling the destructive and unlovely side of our animal nature while we set out to change parts of our still unlovely human mind. The mindscape of the future must have some memory of the ecological arrangements that shaped us and of the social structures that served us well. So many surprises await us even in the next quarter century. My worry is that our context then will be so remote from the ways we survived through the ages that our organizing paradigm will become chaos theory rather than ecology.

7

Creating Social Capital

Cornelia Butler Flora and Jan L. Flora

Increasingly in post-industrial North America, households choose where they live based on preference for place. A substantial number are choosing to move from urban areas to more bucolic, peaceful locations. That choice is often based on attraction to the physical aspects – scenery, amenities, recreational possibilities – as well as the infrastructure necessary to commute physically or electronically.

Local culture, including social networks, contributes to community attraction and a sense of place. These social networks are many faceted and embedded in one another. Your dentist is also your tenor in choir and fellow bowling-league member, whose spouse serves with yours on the town council and whose daughter dates your son.

The influx of upper middle class households often causes problems in the proposed destinations. The affluence of the in-migrants drives up property values. These people almost always move from areas of higher real estate values, which (coupled with the necessity of paying capital gains on the profit made from selling a home) gives them both the capital to spend above local market value and the motivation to use it for housing. Further, as they acquire lakefront or other attractive property, they tend to privatize and limit access to environmental areas that were once public and open to access (Spain, 1993). Thus, their attraction to the physical place decreases their insertion into the social space of community.

The social distance between newcomer and old-timer often increases as newcomers note that although *the* pace of life is preferable to that of their previous urban existence, rural life has certain inconveniences. On a winter morning, the roads are not ploughed early enough for them to commute to work on time. The schools do not offer foreign languages. The public library is only open from ten till four on three weekdays. The charming farm next door spreads manure from its picturesque cows on the fields – which smells. So they go to their local elected officials to protest, as they learned to do in city life. They lodge their complaint in the urban manner – specific, to the point – and imply that the official is personally responsible for obvious inefficiency in the system.

Social distance between newcomer and old-timer can increase when economic development comes up. For the old-timers, a major problem is a fixed income and increasing expenses. They seek to attract industry to decrease their property taxes, which constitute a higher proportion of the income of old-timers than of newcomers. They look with favour on fast food establishments and Wal-Marts. These retail facilities offer low-cost,

Note: Reprinted from *Rooted in the Land: Essays on Community and Place* by Vitek, W. and Jackson, W. (eds), copyright © (1996), with permission from Yale University, Haven

good-value products – and treat their customers with uniform respect. The newcomers are appalled that the environmental capital would be threatened and the picturesque nature of Main Street would be sullied by a Hardee's. With organizing skills gained from their middle-class professional training, they do their best to stop these intrusions into their personal visions of place.

In the case of Floyd, Virginia, a Hardee's was built after a bitter battle pitting the 'come-here's versus the 'been-here's. The newcomers began with an antagonistic stance of 'we don't want it here, period' – a stance that could serve as a negotiation point in an area where networks are not so embedded. Instead of becoming the basis for compromise, here that stance fueled conflict, bringing forth deep hostilities of the old-timers to the newcomers so busy watching cholesterol that they had little time for neighbours. Because of their domination of the county commission, the old-timers won and Hardee's was built, with some deference to local architecture. In an effort to heal the breach, the grand opening was made the occasion for a local bluegrass music festival – valued by all residents in Floyd. Healing takes time and effort.

Newcomers often become involved in more diversified activities for economic development and are frequently the leaders in such efforts. We have found that very often the staff of economic development corporations and chambers of commerce in rural areas are newcomer women, intent on making the community a better place in which to live, according to their experience in urban or suburban areas. The response of the old-timers is generally, 'If things were so good in San Jose, why don't you go back there?' (The other common response to a new idea is, 'We tried that 20 years ago and it didn't work.')

In all these cases, the actions of newcomers increase conflict within the community by focusing on the environmental capital of locality and ignoring the importance of social capital. And old-timers, by not recognizing the importance of social capital, do not see the newcomers as an important resource for the community and fail to integrate them into it.

Reciprocity and mutual trust are the components of community social capital. Putnam (1993, pp35–36) describes social capital as 'features of social organization, such as networks, norms, and trust, that facilitate coordination and cooperation for mutual benefit. Social capital enhances the benefits of investment in physical and human capital.' (Environmental capital encompasses the quantity and quality of water, soil, forests, biodiversity, and scenery. Physical capital includes physical infrastructure, tools and financial capital. Human capital is individual skills, experience, training and education. Capital of all kinds, as conceived here, is an input or intermediate output that contributes to individuals' or communities' wealth, income or quality of life.)

For many newcomers to rural communities, individual efforts to 'become native' create the embedded networks that can contribute to their quality of life – and make their survival strategies easier. *Affluent* in-migrants, however, can easily ignore social capital, as they are able to substitute financial capital, paying for goods and services – including police protection and security systems – that social capital otherwise offers. The patterns of interaction they bring from suburban society, where there is even greater separation of where people work, live, shop, socialize and are educated, make it seem normal to individualize leisure (bowling as individuals, not in a league) rather than engage in the kinds of activity that build reciprocal relationships of trust and interdependence.

Social capital has a number of configurations. Each variation has different implications for community and how newcomers can become native to place. Social capital can be horizontal, hierarchical or non-existent. Horizontal social capital implies egalitarian forms of reciprocity. Each member of the community is expected to give, and gains status and pleasure from doing so. On other occasions, each is expected to receive as well. Each person in the community is seen as capable of providing something of value to any other member of the community. Contributions to collective projects, from parades to the volunteer fire department and Girl Scouts, are defined as 'gifts' to all.

Both old-timers and newcomers can work to create opportunities for all residents in a locality to contribute and where all contributions are valued. For new comers, it is important to establish that they do not think they are better then the old-timers because of higher levels of income or education or more urban experience. And for old-timers it is important to make newcomers native to place by forgiving them for not having a great-grandfather born in the community and thereby acknowledging that one can be linked to place by more than kin ties.

Hierarchical social capital also builds on norms of reciprocity and mutual trust, but those networks are vertical rather than horizontal. Traditional patron–client relationships, typical of urban gangs, are created (Portes and Sensenbrenner, 1993). Those at the bottom of the hierarchy – who obviously are beholden to the few at the top – are the majority of the population in such communities. The receivers of favours owe incredible loyalty to their 'patron' when the time comes to vote for public office. As a result, horizontal networks, particularly those outside the sphere of influence of the patron, are actively discouraged. Dependency is created and mistrust of outsiders is generated. This type of social capital is prevalent in persistent poverty communities (Duncan, 1992).

Often communities with hierarchical social capital have a history of dependence on a single industry, owing to features of the environment such as mineral deposits for mining or potential sources of power, which located textile mills near rivers. Outsiders are attracted because these communities are often located in scenic areas. Newcomers, who are not dependent on the local patron for their livelihood, represent to the old elite the potential of alternative power and influence. So the local patrons often work hard to either (1) co-opt newcomers to make them allies who understand why 'these people just can't be trusted to do anything for themselves' or (2) isolate newcomers and make life extremely uncomfortable in ways that range from 'forgetting' garbage pickup to outright vandalism. The newcomers often feel most comfortable keeping apart from the community entirely or going along with the existing power structure. There are important exceptions, where alliances have been formed with those at the bottom of the hierarchy, and through hard work – and considerable risk – new forms of more horizontal social capital have been created.

Absence of social capital is characterized by extreme isolation of residents from one another. In these communities there is little trust, and, as a result, little interaction. Such communities tend to have frequent population turnover and high levels of conflict. Some are bedroom communities. When predominantly middle-class and upper-class communities lack social capital, they are able to substitute physical capital: private guards, fenced neighbourhoods, and elaborate security systems. Poorer communities often sustain high levels of crime and delinquency Newcomers can create physical barriers that substitute for social capital. There are no existing local organizations to join or local bosses to confront. Affluent newcomers can often manage fairly well in such settings, as they can isolate themselves physically and have the experiential knowledge to reach the county and state officials if they have a problem they cannot solve individually. However, if they have children and are forced to confront the low levels of social capital and the resultant poor schools, they often resort to home schooling or boarding school. Their functional needs are met, but in a place-neutral way.

When new migrants enter a community, unless they have previous ties, they personally have low levels of community based social capital. That, in turn, reduces community social capital unless positive action is taken to build trust and reciprocal relationships. Our research suggests that the best way to do that is to concentrate on building *social infrastructure*. By understanding how social infrastructure builds social capital, both newcomers and old-timers can use social construction to strengthen the community by the diversity represented by newcomers, rather than weakening it through creation of isolated groups and individuals. Despite the multiplier effects of social capital, conventionally it has received little attention in the community development literature or in practice. One reason is that it is extremely hard to measure because of its necessarily high level of abstraction; it 'inheres in the structure of relations between actors and among actors' (Coleman, 1988, pS98).

Coleman has identified social structure that facilitates social capital on the individual level. He found that social network closure (seeing the same people in more than one setting – in the case of his study, at church functions, at school functions, and as parents of your children's friends) was critical for children and families. At the community level we have identified some basic social structures within a community – entrepreneurial social infrastructure – which can be seen as contributing to the development of community-level social capital: (1) symbolic diversity, (2) widespread resource mobilization, and (3) diversity of networks.

Symbols are the source of meaning for humans. Meaning is not intrinsic in an object but is socially determined through interaction. Different human groups have different sets of shared symbols. Indeed, the same object may have very different meanings for two different groups. The meaning given to the object in turn determines how one acts toward it. Symbolic diversity within a community means that while symbolic meanings for objects and interactions may differ, there is appreciation among community members of the different meaning sets. Symbolic diversity brings a recognition of differences, but the differences are not hierarchical. 'Different than' does not mean 'better than'.

Where there is symbolic diversity, people within the community can disagree and still respect one another. There is *acceptance of controversy*. Because differences of opinion are accepted as valid, problems are raised early and alternative solutions are discussed. Members of the community are able to separate problems ('We need better medical care') from specific solutions. For instance, if the problem is defined as the need for better medical care rather than the need for more doctors, community members will recognize the possibility of more than one solution ('We need a doctor'). People feel comfortable suggesting alternative solutions without being accused of contributing to the problem. The emphasis shifts from argument about whether there is a problem to discussion of alternative solutions to a more broadly defined problem. Thus, an individual can argue for one solution, and at a later time support an alternative solution. The person's identity is not conflated with her or his position on a particular issue. Communities that view newcomers as a resource rather than a threat accept their participation in controversial questions. Newcomers in turn accept local 'etiquette' regarding how issues are addressed.

Because controversy is accepted and issues are raised early, politics are *depersonalized*. Community members do not avoid taking a public position. Stands on issues are not viewed as moral imperatives. Because problems can be addressed early, one's stand on an issue is not equated with one's moral worth. Risk of character assassination, and destruction of one's job or ruination of one's social life, are lessened for those who take on public charges. The much-discussed burnout of volunteer public officials, often related to the volume of abuse they face from their constituents, is thus less. Newcomers and old-timers avoid labelling one another with derogatory names such as 'flatlander' or 'stick-in-the-mud'.

High levels of symbolic diversity generate a *focus on process*, rather than on ends only. Instead of the sports analogy for community development (where there is an obvious beginning and end, written rules, a neat system of scoring, clearly demarcated opposing teams, and a definitive final outcome) such communities are more likely to evoke a parenting analogy: while there may be a distinct beginning, there is no steady march to the goalposts, no ending gun when losers and winners are recognized. Process collectively recognizes celebrations and concerns. Communities that focus on process tend to have many local celebrations, but also mechanisms of showing concern for those with problems. Problems are something that happen to good people, not a sign of moral weakness.

Symbolic diversity results in a *broad definition of community and permeable boundaries*. Such communities find it easy to become part of multicommunity and regional efforts, not by giving up community identity, but by expanding it.

The ability of a community to mobilize resources is critical for social capital to develop and is a vital part of community-level social infrastructure. *Resources are defined broad*, allowing a wide range of community members to contribute. For example, elderly community members may not have large quantities of cash, but they have important knowledge of community history. The experience of newcomers in addressing community issues in other locations is recognized as an asset, not a liability.

There is also *relative equality of access to resources* within a community. For example, it is assumed that every child should have the opportunity for a good education. High school dropouts are viewed as a community-level problem, not the fulfilling of one's social destiny based on parental social status. A wide variety of resources, from swimming pools to golf courses to schools, are open to all, rather than being reserved for elite social groups. Public ownership is the most common way of ensuring broad access. If newcomers form their own organizations or social clubs, that immediately creates the appearance of unequal access.

To enhance equality of access, resource mobilization includes a willingness to *invest collectively*. Communities invest in themselves through school bonds, public recreation programmes, community festivals or other fundraising events, and volunteer fire departments and emergency squads. Newcomers who participate in these activities contribute to community social capital formation. The expectation is that all will contribute in some way, and mechanisms are in place to facilitate such participation. If newcomers are wealthy or retired homeowners *and* oppose school or industrial revenue bond issues, the willingness of others to invest collectively in the community will be strongly affected in a negative way.

Finally, there is *individual investment* of private resources. Banks in these communities have high loan-to-deposit ratios, choosing to invest locally rather than in safe but distant government securities. Local entrepreneurs can obtain both equity capital and debt capital. And local people put individual dollars into local community development corporations and enterprises, frequently assuming that there will be no payback or that the payback will be in the distant future. Newcomers build social capital when they volunteer time and contribute money to community projects.

Networks are a crucial part of social capital, and one of their critical features is *diversity*. While internally homogeneous groups are often the basis for diversity within the community, there must also be formal and informal community networks that include individuals of diverse characteristics: young and old, men and women, various racial and ethnic groups, different social classes, and (often most difficult) newcomers and old-timers.

Networks that contribute to sustainable community development *link horizontally* to other communities. We refer to this as *lateral learning* (Flora and Flora, 1993). Communities that develop this kind of networking often take a diverse group of people to a community that has done something they want to emulate. They visit together, ask lots of questions, come back determined to adapt the idea – and do it even better.

Vertical networks to regional, state or national centres link a large number of community individuals and groups to resources and markets beyond community limits. Where there is a single gatekeeper between the community and the outside, no matter how well connected that person is, the concentration of power in a single individual contributes to hierarchical, not horizontal, social capital. Newcomers are likely to have a different set of vertical linkages; including them will expand the community's vertical linkages.

Finally, community networks are *inclusive*. They are participatory, not representational. It is understood that adding more people to the table can mean a larger community pie, not a pie that has to be cut into more pieces. Adding diverse groups to the leadership networks develops community social capital, and newcomers are among those diverse groups.

Building social capital and social infrastructure provides an important avenue for becoming native to place. By identifying local strengths – points of agreement about use and enhancement of physical capital or environmental capital – social capital can be increased. Frequently, community development efforts must be initiated in areas that are not related to economic development.

Samuel and Andrea Jefferson (not their real names) moved to southwest Virginia from the Ivy League community where they had been activists. They chose the location because of its beauty and the economic and social decline the area had experienced since it was cut over. They wanted to live, as they put it, 'where they could make a difference'. There was little local industry, some farming and a lot of poverty. They brought with them some financial capital and a determination to invest it to improve the community.

They bought a farm and began learning how to farm. Despite their personal distaste for smoking and its consequences, they used the tobacco allotment that came with the farm and worked with a neighbour to learn the rich tradition and agronomic practices surrounding tobacco production. They joined a local church. Andrea became a reporter for a local newspaper, while Samuel concentrated on farming and building a bed-and-breakfast on a beautiful hill on their property. They hoped to get it started, then sell it to a local couple to run as a part of tourism development for the area. Samuel became a leader of chamber of commerce programmes aimed at economic development, and their young children entered the public schools.

Each of these activities brought frustration. Other newcomers, attracted by the low housing prices, were entering the area, which made for discipline problems in the schools and on the school bus. The Jeffersons could not find a local couple able to manage the bed-and-breakfast, much less interested in taking on the financial responsibility of owning it. And there was considerable resistance by portions of the community to the newly instituted economic development efforts. Samuel and Andrea, despite persistent attempts to build social capital were still viewed by many as meddling outsiders, probably ready to leave at any moment.

Exhausted and discouraged, Samuel turned his efforts to the children and their school. Local history was an area of interest, so he organized an oral history project involving in-depth interviews with many old-timers by the schoolchildren and an ever-increasing number of parents. A project for a community historical centre and living farm emerged – with enthusiasm from all sides. It seemed a natural part of who they were, allowing all to see each other positively. While many had viewed Samuel's efforts at economic development's self-interested, the work on local history was defined as 'community work'. Samuel admitted he had dropped out of economic development. Yet levels of community social capital were increasing. New community networks were formed. People were able to discuss alternative ways of carrying out the project, engaging in controversy, not conflict, because they had to figure out together the best way to proceed in order to meet their agreed-upon goal. Even the poorest members of the community could contribute to revealing and celebrating the local history, the elderly by telling their stories, the young people and the newcomers by listening appreciatively and recording them. Symbolic diversity and resource mobilization continually increased. And the Jefferson family, including the children, who although born there, were viewed as the progeny of outsiders, all became more native to their chosen place.

Becoming native to place implies contributing to that place as well as receiving strength and renewal from it. New migrants, particularly those with college education and professional pasts, understand the need to contribute to the physical capital of a place, through giving money to local causes and paying property taxes. Such new migrants often are vitally concerned about the need to preserve and enhance environmental capital. Sometimes, particularly if they have school-age children, they are concerned about human capital creation and work hard to improve the schools.

Because it has lacked a name, social capital is frequently neglected by those who move into an area by choice. The actions they take to enhance the other kinds of capital tend to result in the destruction of social capital, as those who *are* native to that place perceive the newcomers as pushy, elitist and constantly complaining. Another set of new migrants isolate themselves – and their various forms of capital – from the community of place, not recognizing the vital role of social capital in maintaining and enhancing the very resources that attracted them to the place initially. For both groups of newcomers, the charm of the place declines. Return migration to the cities occurs, or another move to a place that will 'feel' more comfortable.

In contrast, those newcomers who set out to create social capital *first* – who participate in local activities, who help organize actions to celebrate place, both culturally and environmentally – find themselves becoming native to place. It feels right. Their neighbours know and appreciate them – and they know it, because they know and appreciate their neighbours. Social capital is a resource that is not depleted through use. While it cannot be appropriated by an individual, it can contribute to everyone's quality of life.

References

Coleman, J. (1998) 'Social capital and the creation of human capital', *American Journal of Sociology*, vol 94 (suppl), ppS95–120

Duncan, C. (1992) *Rural Poverty in America*, Auburn House, Westport

Flora, C. B. and Flora, J. (1993) 'Entrepreneurial social infrastructure: A necessary ingredient', *The Annals of the American Academy of Politics and Social Sciences*, vol 529, pp48–58

Portes, A. and Sensenbrenner, J. (1993) 'Embeddedness and immigration: Notes on the social determinants of economic action', *American Journal of Sociology*, vol 98, pp1320–1350

Putnam, R. A. (1993) 'The prosperous community: Social capital and public life', *American Prospect*, vol 13, pp35–42

Spain, D. (1993) 'Been-heres versus come-heres: Negotiating conflicting community identities', *Journal of the American Planning Association*, vol 59, pp156–171

Part 2

Agroecological Perspectives

Introduction to Part 2:
Agroecological Perspectives

Jules Pretty

Part 2 of this *Reader in Sustainable Agriculture* draws together five agroecological perspectives by Jules Pretty, Erin Tegtmeier and Michael Duffy, Steve Sherwood and colleagues, Stephen Gliessman and Gordon Conway.

In the first article, an excerpt from the 2002 book, *Agri-Culture*, Jules Pretty indicates that the real costs of food are much higher than the price paid in the shop. Environmental externalities and the diversion of tax revenue to subsidize agriculture contribute to the real cost. Agriculture, like any economic sector, has both negative and positive side-effects, and it is the movement towards a more multifunctional view of agriculture that could result in a better understanding of what contributes to agricultural sustainability. This paper summarizes the first study of the full costs of a modern agricultural sector. In the UK, these amounted to some UK£1.5 billion per year during the 1990s. These external costs are alarming – and should call into question what we mean by efficiency. Increased sustainability in agricultural systems can only happen if these external costs are substantially reduced.

In the second paper, recently published in the *International Journal of Agricultural Sustainability*, Erin Tegtmeier and Michael Duffy analyse the full cost of modern agricultural production in the US. These are of the order of US$5.6 to 16.9 billion per year (in 2002 $), arising from damage to water resources, soils, air, wildlife and biodiversity, and harm to human health. Additional annual costs of $3.7 billion arise from agency costs associated with programmes to address these problems or encourage a transition towards more sustainable systems. Following various partial studies published in the 1990s, this was the first study of the costs of the whole of the agricultural sector in the US. As the authors indicate, 'many in the US pride themselves on our cheap food. But this study demonstrates that consumers pay for food well beyond the grocery store.'

The third article by Steve Sherwood and co-authors is a chapter drawn from the 2004 book, *The Pesticide Detox*. It focuses on pesticide use and its effects in the highland region of Carchi in the northern Andes. Farmers use a wide range of pesticides, both hazardous and benign, and although local and international businesses indicate that highly toxic products can be used safely, the evidence from the ground is different. This study found that poisonings in Carchi are amongst the highest recorded in the world – an annual rate of 171 per 100,000 population for morbidity, and 21 per 100,000 for mortalities. Pesticides were found on family clothing, on food, and in children's bedding. To illustrate these pathways to local people, the researchers added fluorescent dyes to pesticides, and then used UV lights to show their presence. The challenge, now, is to develop new ways of learning about pests and diseases in such rural communities, as well as develop agricultural practices that reduce dependency on those pesticides that are harmful to humans and the environment.

The fourth article by Stephen Gliessman is an overview of agroecological approaches to the management of agricultural systems. As he indicates, 'discussions about sustainable agriculture must go beyond what happens within the fences of any individual farm'. It is the wider environmental, economic and social interactions that are critical. A practising farmer as well as a distinguished academic, Gliessman draws on a wide range of experience to set out an agroecological perspective to the flows of energy and nutrients in agroecosystems, and identifies the population regulating mechanisms and potential for developing dynamic equilibria. The paper includes a table (11.1) that summarizes the guiding principles for a process of design of and conversion to sustainable agricultural systems. Comparisons are made between traditional, conventional (or modern) and sustainable systems. This question of redesign is critical if we are to emerge different patterns of agricultural and environmental management that are able to produce both food and important environmental services.

In the final paper of this section of the Reader, drawn from his book *The Doubly Green Revolution*, Gordon Conway summarizes the ecological principles for controlling pests, and describes the history of integrated pest management approaches. Since the middle of the 20th century, the main approach to pest, pathogen and weed problems has been to spray with insecticides, fungicides and herbicides. But these are frequently costly and inefficient, let alone potentially harmful to human health and wildlife. Conway describes the successful control of cocoa pests in north Borneo following detailed examination of local ecology, which found that the elimination of some spraying allowed parasites to re-establish and control the target pests. The paper goes on to describe the growth in understanding of rice field agroecology, and how pest outbreaks seemed to be linked to pesticide use. Alternative approaches to pest management were radical in their emphasis on understanding and managing biological diversity in and around rice fields. Ultimately, this needs the full engagement and participation of farmers in the process of systems analysis and transformation.

8

Reality Cheques

Jules Pretty

The real costs of food

When we buy or bake our daily bread, do we ever wonder about how much it really costs? We like it when our food is cheap, and complain when prices rise. Indeed, riots over food prices date back at least to Roman times. Governments have long since intervened to keep food cheap in the shops, and tell us that policies designed to do exactly this are succeeding. In most industrialized countries, the proportion of the average household budget spent on food has been declining in recent decades. Food is getting cheaper relative to other goods, and many believe that this must benefit everyone as we all need to eat food. But we have come to believe a damaging myth. Food is not cheap. It only appears cheap in the shop because we are not encouraged to think of the hidden costs of damage caused to the environment and human health by certain systems of agricultural production. Thus we actually pay three times for our food. Once at the till in the shop, a second time through taxes that are used to subsidise farmers or support agricultural development, and a third time to clean up the environmental and health side-effects. Food looks cheap because we count these costs elsewhere in society. As economists put it, the real costs are not internalized in prices.[1]

This is not to say that prices in the shop should rise, as this would penalize the poor over the wealthy. Using taxes to raise money to support agricultural development is also potentially progressive, as the rich pay proportionally more in taxes, and the poor, who spend proportionally more of their budget on food, benefit if prices stay low. But this idea of fairness falters when set against the massive distortions brought about by modern agricultural systems that additionally impose large environmental and health costs throughout economies. Other people and institutions pay these costs, and this is both unfair and inefficient. If we were able to add up the real costs of producing food, we would find that modern industrialized systems of production perform poorly in comparison with sustainable systems. This is because we permit cost-shifting – the costs of ill-health, lost biodiversity and water pollution are transferred away from farmers, and so not paid by

Note: Reprinted from *Agri-culture: Reconnecting People, Land and Nature* by Pretty, J. copyright © (2002), Earthscan, London

those producing the food nor included in the price of the products sold. Until recently, though, we have lacked the methods to put a price on these side-effects.

When we conceive of agriculture as more than simply a food factory, indeed as a multifunctional activity with many side-effects, then this idea that farmers do only one thing must change. Of course, it was not always like this. It is modern agriculture that has brought a narrow view of farming, and it has led us to crisis. The rural environment in industrialized countries suffers, the food we eat is as likely to do as much harm as good, and we still think food is cheap. The following words were written more than 50 years ago, just before the advent of modern industrialized farming:

> Why is there so much controversy about Britain's agricultural policy, and why are farmers so disturbed about the future?... After the last war, the people of these islands were anxious to establish food production on a secure basis, yet, in spite of public good will, the farming industry has been through a period of insecurity and chaotic conditions.

These are the opening words to a national enquiry that could have been written about a contemporary crisis. Yet they are by Lord Astor, written in 1945 to introduce the Astor and Rowntree review of agriculture. This enquiry was critical of the replacement of mixed methods with standardized farming. They said, 'to farm properly you have got to maintain soil fertility; to maintain soil fertility you need a mixed farming system'. They believed that farming would only succeed if it maintained the health of the whole system, beginning in particular with the maintenance of soil fertility: 'obviously it is not only sound business practice but plain common sense to take steps to maintain the health and fertility of soil'.[2]

But in the enquiry, some witnesses disagreed, and called for a 'specialised and mechanised farming', though interestingly, the farming establishment at the time largely supported the idea of mixed farming. But in the end, the desire for public subsidies to encourage increases in food production took precedence, and these were more easily applied to simplified systems rather than mixed ones. The 1947 Agriculture Act was the outcome, a giant leap forward for modern, simplified agriculture, and a large step away from farming that valued nature's assets for farming. Sir George Stapledon, British scientist knighted for his research on grasslands, was another perceptive scientist well ahead of his time. He too was against monocultures and in favour of diversity, arguing in 1941 that 'senseless systems of monoculture designed to produce food and other crops at the cheapest possible cost have rendered waste literally millions of acres of once fertile or potentially fertile country'.[3] In his final years, just a decade after the 1947 Act, he said:

> today technology has begun to run riot and amazingly enough perhaps nowhere more so than on the most productive farms... Man is putting all his money on narrow specialisation and on the newly dawned age of technology has backed a wild horse which given its head is bound to get out of control.

Wise words from eminent politicians and scientists. But lost on the altar of progress. Until now, perhaps, as new ideas on agriculture have begun to emerge and gather credence.

Agriculture's unique multifunctionality

We should all now be asking: what is farming for? Clearly, in the first instance, to produce food, and we have become very good at it. A great success, but only if our measures of efficiency are narrow. Agriculture is unique as an economic sector. It does more than just produce food, fibre, oil and timber. It has a profound impact on many aspects of local, national and global economies and ecosystems. These impacts can be either positive or negative. The negative ones are worrying. Pesticides and nutrients leaching from farms have to be removed from drinking water, and these costs are paid by water consumers, not by the polluters. The polluters, therefore, benefit by not paying to clean up the mess they have created, and have no incentive to change behaviour. What also makes agriculture unique is that it affects the very assets on which it relies for success. Agricultural systems at all levels rely for their success on the value of services flowing from the total stock of assets that they control, and five types of asset, natural, social, human, physical and financial capital, are now recognized as being important.[4]

Natural capital produces nature's goods and services, and comprises food, both farmed and harvested or caught from the wild, wood and fibre; water supply and regulation; treatment, assimilation and decomposition of wastes; nutrient cycling and fixation; soil formation; biological control of pests; climate regulation; wildlife habitats; storm protection and flood control; carbon sequestration; pollination; and recreation and leisure. *Social capital* yields a flow of mutually beneficial collective action, contributing to the cohesiveness of people in their societies. The social assets comprising social capital include norms, values and attitudes that predispose people to cooperate; relations of trust, reciprocity and obligations; and common rules and sanctions mutually agreed or handed down. These are connected and structured in networks and groups.

Human capital is the total capability residing in individuals, based on their stock of knowledge skills, health and nutrition. It is enhanced by access to services that provide these, such as schools, medical services and adult training. People's productivity is increased by their capacity to interact with productive technologies and with other people. Leadership and organizational skills are particularly important in making other resources more valuable. *Physical capital* is the store of human made material resources, and comprises buildings, such as housing and factories, market infrastructure, irrigation works, roads and bridges, tools and tractors, communications, and energy and transportation systems, that make labour more productive. *Financial capital* is more of an accounting concept, as it serves as a facilitating role rather than as a source of productivity in and of itself. It represents accumulated claims on goods and services, built up through financial systems that gather savings and issue credit, such as pensions, remittances, welfare payments, grants and subsidies.

As agricultural systems shape the very assets on which they rely for inputs, a vital feedback loop occurs from outcomes to inputs. Donald Worster's three principles for good farming capture this idea. It is farming that makes people healthier, farming that promotes a more just society, and farming that preserves the earth and its networks of life. He says 'the need for a new agriculture does not absolve us from the moral duty and common-sense advice to farm in an ecologically rational way. Good farming protects the land, even when it uses it.'[5] Thus sustainable agricultural systems tend to have a positive effect on natural, social and human capital, whilst unsustainable ones feed back to deplete these assets, leaving less for future generations. For example, an agricultural system that erodes soil whilst producing food externalizes costs that others must bear. But one that sequesters carbon in soils through organic matter accumulation helps to mediate climate change. Similarly, a diverse agricultural system that enhances on-farm wildlife for pest control contributes to wider stocks of biodiversity, whilst simplified modernized systems that eliminate wildlife do not. Agricultural systems that offer labour absorption opportunities, through resource improvements or value added activities, can boost economies and help to reverse rural-to-urban migration patterns.

Agriculture is, therefore, fundamentally multifunctional. It jointly produces many unique non-food functions that cannot be produced by other economic sectors so efficiently. Clearly, a key policy challenge, for both industrialized and developing countries, is to find ways to maintain and enhance food production. But the key question is: can this be done whilst seeking both to improve the positive side-effects and to eliminate the negative ones? It will not be easy, as past agricultural development has tended to ignore both the multifunctionality of agriculture and the pervasive external costs.[6]

This leads us to a simple and clear definition for sustainable agriculture. It is farming that makes the best use of nature's goods and services whilst not damaging the environment.[7] It does this by integrating natural processes such as nutrient cycling, nitrogen fixation, soil regeneration and natural enemies of pests into food production processes. It also minimizes the use of non-renewable inputs that damage the environment or harm the health of farmers and consumers. It makes better use of the knowledge and skills of farmers, so improving their self-reliance, and it makes productive use of people's capacities to work together to solve common management problems. Through this, sustainable agriculture also contributes to a range of public goods, such as clean water, wildlife, carbon sequestration in soils, flood protection and landscape quality.

Putting monetary values on externalities

Most economic activities affect the environment, either through the use of natural resources as an input or by using the 'clean' environment as a sink for pollution. The costs of using the environment in this way are called externalities. As they are side-effects of the economic activity, they are external to markets, and so their costs are not part of the prices paid by producers or consumers. When such externalities are not included in prices, they distort the market by encouraging activities that are costly to society even if the private benefits are substantial. The types of externalities encountered in the agricultural sector have several features. Their costs are often neglected, and often occur with a time lag. They often damage groups whose interests are not represented, and the identity of the producer of the externality is not always known.[8]

In practice, there are few agreed data on the economic cost of agricultural externalities. This is partly because the costs are highly dispersed and affect many sectors of economies. It is also necessary to know about the value of nature's goods and services, and what happens when these largely unmarketed goods are lost. As the current system of economic accounting grossly underestimates the current and future value of natural capital, this makes the task even more difficult.[9] It is relatively easy, for example, to count the treatment costs following pollution, but much more difficult to value, for example, skylarks singing on a summer's day, and the costs incurred when they are lost.

Several studies have recently put a cost on the negative externalities of agriculture in China, Germany, the Netherlands, the Philippines, the UK and the US.[10] When it is possible to make the calculations, our understanding of what is the best or most efficient form of agriculture can change rapidly. In the Philippines, researchers from the International Rice Research Institute found that modern rice cultivation was costly to human health. They investigated the health status of rice farmers exposed to pesticides, and estimated the monetary costs of significantly increased incidence of eye, skin, lung and neurological disorders. By incorporating these into the economics of pest control, they found that modern, high pesticide systems suffer twice, as with nine pesticide sprays per season they returned less per hectare than the integrated pest management strategies, and cost the most in terms of ill-health. Any expected positive production benefits of applying pesticides were overwhelmed by the health costs. Rice production using natural control methods has multifunctionality in contributing positively both to human health as well as sustaining food production.[11]

At the University of Essex, we recently developed a new framework to study the negative externalities of UK agriculture. This uses seven cost categories to assess negative environmental and health costs – damage to water, air, soil and biodiversity, and damage to human health by pesticides, micro-organisms and disease agents. The analysis of damage and monitoring costs counted only external costs, as private costs borne by farmers themselves, such as from increased pest or weed resistance from pesticide overuse, are not included. We conservatively estimated that the external costs of UK agriculture, almost all of which is modernized and industrialized, to be at least £1.5 to 2 billion each year. Another study by Olivia Hartridge and David Pearce has also put the annual costs of modern agriculture in excess of £1 billion.[12] These are costs imposed on the rest of society, and effectively a hidden subsidy to the polluters.[13] The annual costs arise from damage to the atmosphere (£316 million), to water (£231 million), to biodiversity and landscapes (£126 million), to soils (£96 million) and to human health (£777 million). Using a similar framework of analysis, the external costs in the US amount to nearly £13 billion per year.[14]

How do all these costs arise? Pesticides, nitrogen and phosphorus nutrients, soil, farm wastes and micro-organisms escape from farms to pollute ground and surface water. Costs are incurred by water delivery companies, and then passed onto their customers, to remove these contaminants, to pay for restoring water courses following pollution incidents and eutrophication, and to remove soil from water. Using UK water companies' returns for both capital and operating expenditure, we estimated annual external costs to be £125 million for removal of pesticides below legal standards, £16 million for nitrate, £69 million for soil and £23 million for *Cryptosporidium*.[15] These costs would be much greater if the policy goal were complete removal of all contamination.

Agriculture also contributes to atmospheric pollution through the emissions of four gases: methane from livestock, nitrous oxide from fertilizers, ammonia from livestock wastes and some fertilizers, and carbon dioxide from energy and fossil fuel consumption and loss of soil carbon. These in turn contribute to atmospheric warming (methane, nitrous oxide and carbon dioxide), ozone loss in the stratosphere (nitrous oxide), acidification of soils and water (ammonia) and eutrophication (ammonia). The annual cost for these gases is some £310 million.[16] A healthy soil is vital for agriculture, but modern farming has accelerated erosion, primarily through the cultivation of winter cereals, the conversion of pasture to arable, the removal of field boundaries and hedgerows, and over-stocking of livestock on grasslands. Off-site costs arise when soil carried off farms by water or wind blocks ditches and roads, damages property, induces traffic accidents, increases the risk of floods, and pollutes water through sediments and associated nitrate, phos-phate and pesticides. These amount to £14 million per year. Carbon in organic matter in soils is also rapidly lost when pastures are ploughed or when agricultural land is inten-sively cultivated, and adds another £82 million to the annual external costs.

Modern farming has had a severe impact on wildlife in the UK. More than nine-tenths of wildflower-rich meadows have been lost since the 1940s, together with a half of heathland, lowland fens, and valley and basin mires, and a third to a half of ancient lowland woods and hedgerows. Species diversity is also declining in the farmed habitat itself. Increase use of drainage and fertilizers has led to grass monocultures replacing flower-rich meadows, overgrazing of uplands has reduced species diversity, and herbicides have cut diversity in arable fields. Hedgerows were removed at a rate of 8000km a year between the mid-1980s and 1990s. Farmland birds have particularly suffered, with the popula-tions of nine species falling by more than a half in the 25 years to 1995.[17] The costs of restoring species and habitats under Biodiversity Action Plans were used as a proxy for the costs of wildlife and habitat losses, and together with the costs of replacing hedge-rows, stonewalls and bee colonies, bring the annual costs to £126 million.

Pesticides can affect workers engaged in their manufacture, transport and disposal, operators who apply them in the field, and the general public. But there is still great uncer-tainty because of differing risks per product, poor understanding of chronic effects, such as in cancer causation, weak monitoring systems, and misdiagnoses by doctors.[18] For these reasons, it is very difficult to say exactly how many people are affected by pesticides each year. According to voluntary reporting to government, 100–200 incidents occur each year in the UK.[19] However, a recent government survey of 2000 pesticide users found that 5 per cent reported at least one symptom in the past year and about which they had consulted a doctor, and a further 10 per cent had been affected, mostly by headaches, but had not consulted a doctor, incurring annual costs of about £1 million. Chronic health hazards associated with pesticides are even more difficult to assess. Pesticides are ingested via food and water, and these represent some risk to the public. With current scientific knowledge, it is impossible to state categorically whether or not certain pesticides play a role in cancer causation. Other serious health problems arising from agriculture are food-borne illnesses, antibiotic resistance and Bovine Spongiform Encephalopathy-Creutzfeldt–Jakob Disease (BSE-CJD).[20]

These external costs of UK agriculture are alarming. They should call into question what we mean by efficiency. Farming receives £3 billion of public subsidies each year, yet causes another £1.5 billion of costs elsewhere in the economy. If we had no alterna-tives, then we would have to accept these costs. But in every case, there are choices. Pesticides do not have to get into water. Indeed, they do not need to be used at all in many farm systems. The pesticide market in the UK is £500 million, yet we pay £120 million just to clean them out of drinking water. We do not need farming that damages biodiversity and landscapes; we do not need intensive livestock production that encour-ages infections and overuse of antibiotics. Not all costs, though, are subject to immediate elimination with sustainable methods of production. Cows will still belch methane, until animal feed scientists find a way of amending ruminant biochemistry to prevent its emission. But it is clear that many of these massive distortions could be removed with some clear thinking, firm policy action, and brave action by farmers.[21]

Notes

1   On the cheapness of food, Donald Worster recognized this about a decade ago: 'the farm experts merely assume, on the basis of marketplace behaviour, that the public wants cheapness above all else. Cheapness, of course, is supposed to require abundance, and abundance is supposed to come from greater economies of scale, more concentrated economic organisation, and more industrialised methods. The entire basis for that assumption collapses if the marketplace is a poor or imperfect reflector of what people want.' Worster, 1993, *The Wealth of Nature*, p87.

2   See Astor and Rowntree, 1945, pp33, 47.

3   For more on George Stapledon, see Conford, 1988, *The Organic Tradition,* pp192–193, pp196–197.

4   Despite my regular use of these five terms as capitals, I agree with the misgivings that many have. Capital implies an asset, and assets should be looked after, protected and built up. But as a term, capital is problematic for two reasons. It implies measurability and transferability. Because the value of something can be assigned a single monetary value, then it appears to matter not if it is lost, as we could simply allocate the required money to buy another, or transfer it from elsewhere. But we know this must be nonsense. Nature and its cultural and social meanings is not so easily replaceable. It is not a commodity, reducible only to monetary values. Nonetheless, as terms, natural capital and social capital have their uses in helping to reshape thinking around basic questions such as what is agriculture for, and what system works best? For further discussions, Bourdieu, 1986; Coleman, 1988, 1990; Putnam, 1995; Benton, 1998; Carney, 1998; Flora, 1998; Grootaert, 1998; Pretty, 1998; Scoones, 1998; Uphoff, 1998; Costanza et al, 1999; Pretty and Ward, 2001.

5   Worster, 1993, p92. See also Michael Neuman of Texas A & M University, who resolves the concept of sustainability into four very simple and compelling ideas: the rates of consumption, rates of production, rates of accumulation and depletion, and rates of assimilation.

6   See Conway and Pretty, 1991; Altieri, 1995; Pingali and Roger, 1995; Conway, 1997; Pretty, 1998; FAO, 1999.

7   For more on sustainable agriculture definitions and principles, see Altieri, 1995, 1999; Pretty, 1995a, 1998; Thrupp, 1996; Conway, 1997; Drinkwater et al, 1998; Tilman, 1998; Hinchliffe et al, 1999; Zhu et al, 2000; Wolfe, 2000.

8   An externality is any action that affects the welfare of or opportunities available to an individual or group without direct payment or compensation, and may be positive or negative. See Baumol and Oates, 1988; Pearce and Turner, 1990; EEA, 1998; Brouwer, 1999; Pretty et al, 2000. Economists distinguish between 'technological' or physical externalities, and 'pecuniary', or price effect, externalities. Pecuniary exter-

nalities arise, for example, when individuals or firms purchase or sell large enough quantities of a good or service to affect price levels. The change in price levels affects people who are not directly involved in the original transactions, but who now face higher or lower prices as a result of those original transactions. These pecuniary externalities help some groups and hurt others, but they do not necessarily constitute a 'failure' of the market economy. An example of a pecuniary externality is the rising cost of housing for local people in rural villages that results from higher-income workers from metropolitan areas moving away from urban cores and bidding up the price of housing in those villages. Pecuniary externalities are a legitimate public concern, and may merit a public policy response. Technological externalities, however, do constitute a form of 'market failure'. Dumping pesticides sewage into a lake, without payment by the polluter to those who are adversely affected, is a classic example of a technological externality. The market 'fails' in this instance, because more pollution occurs than would be the case if the market or other institutions caused the polluter to bear the full costs of its actions. It is technological externalities that are commonly simply termed 'externalities' in most environmental literature (see Davis and Kamien, 1972; Common, 1995; Knutson et al, 1998).

9 For more on the value of nature's goods and services, see Abramovitz, 1997; Costanza et al, 1997, 1999; Daily, 1997; and the whole issue of *Ecological Economics*, 1999, volume 25, issue 1.

10 See Pimentel et al, 1992, 1995; Rola and Pingali, 1993; Evans, 1995; Pingali and Roger, 1995; Steiner et al, 1995; Fleischer and Waibel, 1998; Waibel and Fleischer, 1998; Bailey et al, 1999; Norse et al, 2000. The data from these studies are not easily comparable in their original form as different frameworks and methods of assessment have been used. Methodological concerns have also been raised about some studies. Some have noted that several effects could not be assessed in monetary terms, whilst others have appeared to be more arbitrary (eg the $2 billion cost of bird deaths in the US is arrived at by multiplying 67 million losses by $30 a bird: see Pimentel et al, 1992). The Davison et al (1996) study on Netherlands agriculture was even more arbitrary, as it added an estimate the costs farmers would incur to reach stated policy objectives, and these were based on predicted yield reductions of 10–25 per cent arising from neither cheap nor preferable technologies, which led to a large overestimate of environmental damage (see Bowles and Webster, 1995; Crosson, 1995; van der Bijl and Bleumink, 1997; Pearce and Tinch, 1998).

11 On the effects of pesticides in rice, see Rola and Pingali, 1993; Pingali and Roger, 1995.

12 Hartridge and Pearce, 2001.

13 See Pretty et al, 2000, 2001. These are likely to be conservative estimates of the real costs. Some costs are known to be substantial underestimates, such as acute and chronic pesticide poisoning of humans, monitoring costs, eutrophication of reservoirs and restoration of all hedgerow losses. Some currently cannot be calculated, such as dredging to maintain navigable water, flood defences, marine eutrophication and poisoning of domestic pets. The costs of returning the environment or human health to pristine conditions were not calculated, and treatment and prevention costs may be underestimates of how much people might be willing to pay to see positive externalities created. The data also do not account for time lags between the cause of a

problem and its expression as a cost, as some processes long since stopped may still be causing costs; some current practices may not yet have caused costs, and this study did not include the externalities arising from transporting food from farms to manufacturers, processors, retailers and finally to consumers.

14 See Pretty et al, 2001.

15 The government's Office of the Director General of Water Services sets industry price levels each five years, which determine both the maximum levels of water bills and specifies investments in water quality treatment. During the 1990s, water industry undertook pesticide and nitrate removal schemes, resulting in the construction of 120 plants for pesticide removal and 30 for nitrate removal (Ofwat, 1998). Ofwat estimates that water companies will spend a further UK£600 million between 2000 and 2005 on capital expenditure alone due to continuing deterioration of 'raw water' quality due to all factors. Ofwat predicts capital expenditure for pesticides to fall to £88 million per year at the end of the 1990s/early 2000s; and for nitrate to fall to £8.3 million/year.

Although Ofwat has sought to standardize reporting, individual companies report water treatment costs in different ways. Most do distinguish treatment for pesticides, nitrate, *Cryptosporidium* and several metals (iron, manganese and lead). The remaining treatment costs for phosphorus, soil removal, arsenic and other metals, appear under a category labelled 'other'. Of the 28 water companies in England and Wales, three report no expenditure on treatment whatsoever; and a further three do not disaggregate treatment costs, with all appearing under 'other'. Twenty companies report expenditure on removal of pesticides, 11 on nitrates and 10 on *Cryptosporidium*. It is impossible to tell from the records whether a stated zero expenditure is actually zero, or whether this has been placed in the 'other' category. Using Ofwat and water companies' returns, we estimate that 50 per cent of expenditure under the 'other' category refers to removal of agriculturally related materials.

16 We originally calculated the annual external costs of these gases to be £280 million for methane, £738 million for nitrous oxide, £47 million for carbon dioxide and £48 million for ammonia. But a more appropriate measure would have been to use an accepted policy target for these costs, such as the 25 per cent cut required to meet agreements made in the Kyoto Protocol. This would put the total annual costs at £314 million.

17 Campbell et al, 1997; Pain and Pienkowski, 1997; DETR, 1998a, 1998b; Mason, 1998; Pretty, 1998; Siriwardena et al, 1998; Krebs et al, 1999.

18 Repetto and Baliga, 1996; Pearce and Tinch, 1998; HSE, 1998a, 1998b; Pretty, 1998.

19 Fatalities from pesticides at work in Europe and the US are rare – one a decade in the UK, and eight a decade in California. In the UK, a variety of institutions collect mortality and morbidity data, but in California, where there is the most comprehensive reporting system in the world, official records show that 1200–2000 farmers, farmworkers and the general public are poisoned each year (see CDFA, *passim*; Pretty, 1998). There appears to be greater risk from pesticides in the home and garden where children are most likely to suffer. In Britain, 600–1000 people need hospital treatment each year from home poisoning.

20 On food poisoning in the UK, see Wall et al, 1996; Evans et al, 1998; PHL, 1999. For study of foodborne illnesses in Sweden, see Lindqvist et al, 2001. When BSE

was first identified in late 1986, research confirmed that it was a member of a group of transmissible diseases occurring in animals and humans. It appeared simultaneously in several places in the UK, and has since occurred in native born cattle in other countries. By mid-2001, more than 180,000 cases had been confirmed in the UK, the epidemic having reached a peak in 1992. The link between BSE and variant CJD in humans was confirmed in 1996, and 100 deaths from CJD have occurred to 2001. The annual external costs of BSE were £600 million at the end of the 1990s. See NAO, 1998; WHO, 2001. By mid 2001, there had been 181,000 cases of BSE reported in the UK, 648 in Ireland, 564 in Portugal, 381 in Switzerland, 323 in France, 81 in Germany, 46 in Spain and 34 in Belgium. For more on the important lessons of BSE, see Lobstein et al, 2001; Millstone and van Zwanenberg, 2001.

21 For an excellent review of food crises and the need for new thinking in food systems, see Lang et al, 2001. Also see Waltner-Toews and Lang, 2000.

References

Abramovitz, J. (1997) 'Valuing nature's services' in Brown L., Flavin C. and French, H. (eds) *State of the World*, Worldwatch Institute, Washington, DC

Altieri, M. A. (1995) *Agroecology: The Science of Sustainable Agriculture*, Westview Press, Boulder, Colorado

Altieri, M. A. (1999) 'Enhancing the productivity of Latin American traditional peasant farming systems through an agro-ecological approach'. Paper for Conference on Sustainable Agriculture: New Paradigms and Old Practices? Bellagio Conference Centre, Italy, 26–30 April 1999

Astor, V. and Rowntree, B. S. (1945) *Mixed Farming and Muddled Thinking: An Analysis of Current Agricultural Policy*, Macdonald and Co, London

Bailey, A. P., Rehman, T., Park, J., Keatunge, J. D. H. and Trainter, R. B. (1999) 'Towards a method for the economic evaluation of environmental indicators for UK integrated arable farming systems', *Agriculture, Ecosystems and Environment*, vol 72, pp145–158

Baumol, W. J. and Oates, W. E. (1988) *The Theory of Environmental Policy*, Cambridge University Press, Cambridge

Benton, T. (1998) 'Sustainable development and the accumulation of capital: Reconciling the irreconcilable?' in Dobson, A. (ed) *Fairness and Futurity*, Oxford University Press, Oxford

Bourdieu, P. (1986) 'The forms of capital' in Richardson, J. (ed) *Handbook of Theory and Research for the Sociology of Education*, Greenwood Press, Westport, Connecticut

Bowles, R. and Webster, J. (1995) 'Some problems associated with the analysis of the costs and benefits of pesticides', *Crop Protection*, vol 14, pp593–600

Brouwer, R. (1999) 'Market integration of agricultural externalities: a rapid assessment across EU countries'. Report for European Environment Agency, Copenhagen

Campbell, L. H., Avery, M. L., Donald, P., Evans, A. D., Green, R. E. and Wilson, J. D. (1997) 'A review of the indirect effects of pesticides on birds'. Report No 227, Joint Nature Conservation Committee, Peterborough

Carney, D. (1998) *Sustainable Rural Livelihoods*, Department for International Development, London

Coleman, J. (1988) 'Social capital and the creation of human capital', *American Journal of Sociology*, vol 94, (suppl) S95–S120

Coleman, J. (1990) *Foundations of Social Theory*, Harvard University Press, Harvard, Massachusetts

Common, M. (1995) *Sustainability and Policy*, Cambridge University Press, Cambridge

Conford, P. (ed) (1988) *The Organic Tradition: An Anthology of Writing on Organic Farming*, Green Books, Bideford, Devon

Conway, G. R. (1997) *The Doubly Green Revolution*, Penguin, London

Conway, G. R. and Pretty, J. N. (1991) *Unwelcome Harvest: Agriculture and Pollution*, Earthscan, London

Costanza, R., d'Arge, R., de Groot, R., Farber, S., Grasso, M., Hannon, B., Limburg, K., Naeem, S., O'Neil, R. V., Paruelo, J., Raskin, R. G., Sutton, P. and van den Belt, M. (1997 and 1999) 'The value of the world's ecosystem services and natural capital', *Nature*, vol 387, pp253–260; also in *Ecological Economics*, vol 25, pp3–15

Crosson, P. (1995) 'Soil erosion estimates and costs', *Science*, vol 269, pp461–464

Daily, G. (ed) (1997) *Nature's Services: Societal Dependence on Natural Ecosystems*, Island Press, Washington, DC

Davis, O. and Kamien, M. (1972) 'Externalities, information, and alternative collective action' in Dorfman, R and Dorfman, N (eds) *Economics of the Environment: Selected Readings*, W. W. Norton and Co, New York, pp69–87

Davison, M. D., van Soest, J. P., de Wit, G. and De Boo, W. (1996) *Financiële waardering van de milieuschade door de Nederlandse landbouw – een benadering op basis van de preventiekosten*. [Financial valuation of environmental hazard from Dutch agriculture – an approximation based on prevention costs]. Centre for Energy Conservation and Clean Technology (CE), Delft, the Netherlands

DETR (1998a) *The Environment in Your Pocket*, www.environment.detr.gov.uk/des20/pocket/env24.htm

DETR (1998b) *Digest of Environmental Statistics No 20*, UK Emissions of Greenhouse Gases, www.environment.detr.gov.uk/des20/chapter1/

Drinkwater, L. E., Wagoner, P. and Sarrantonio, M. (1998) 'Legume-based cropping systems have reduced carbon and nitrogen losses', *Nature*, vol 396, pp262–265

EEA (1998) 'Europe's environment: The second assessment', Report and Statistical Compendium EEA, Copenhagen

Evans, H. S., Madden, P., Douglas, C., Adak, G. K., O'Brien, S. J., Djuretic, T., Wall, P. G. and Stanwell-Smith, R. (1998) 'General outbreaks of infectious disease in England and Wales 1995–1996', *Communicable Disease and Public Health*, vol 1, pp165–171

Evans, R. (1995) *Soil Erosion and Land Use: Towards a Sustainable Policy*, Cambridge Environmental Initiative, University of Cambridge, Cambridge

FAO (1999) *The Future of Our Land*. FAO, Rome

Fleischer, G. and Waibel, H. (1998) 'Externalities by pesticide use in Germany'. Paper presented to Expert Meeting, 'The Externalities of Agriculture: What Do We Know?' EEA, Copenhagen, May 1998

Flora, J. L. (1998) 'Social capital and communities of place', *Rural Sociology*, vol 63, pp481–506

Grootaert, C. (1998) 'Social capital: The missing link', *World Bank Social Capital Initiative Working Paper No 5*, Washington, DC

Hartridge, O. and Pearce, D. (2001) *Is UK Agriculture Sustainable? Environmentally Adjusted Economic Accounts*, CSERGE, University College, London

Health and Safety Executive (HSE) (1998a) Pesticides Incidents Report 1997/8. HSE, Sudbury

HSE (1998b) 'Pesticide users and their health: Results of HSE's 1996/7 feasibility study', www.open.gov.uk/hse/hsehome.htm

Hinchcliffe, F., Thompson, J., Pretty, J., Guijt, I. and Shah, P. (eds) (1999) *Fertile Ground: The Impacts of Participatory Watershed Development*, IT Publications, London

Knutson, R., Penn, J. and Flinchbaugh, B. (1998) *Agricultural and Food Policy*, 4th edition. Prentice Hall, Upper Saddle River, New Jersey

Krebs, J. R., Wilson, J. D., Bradbury, R. B. and Siriwardena, G. M. (1999) 'The second silent spring?' *Nature*, vol 400, pp611–612

Lang, T., Barling, D. and Caraher, M. (2001) 'Food, social policy and the environment: Towards a new model', *Social Policy and Administration*, vol 35, pp538–558

Lindqvist, R., Andersson, Y., Lindback, J.,Wegscheider, M., Eriksson, Y., Tidestrom, L., Lagerqvist-Widh, A., Hedlund, K.-O., Lofdahl, S., Svensson, L. and Norinder, A. (2001) 'A one-year study of foodborne illnesses in the municipality of Uppsala, Sweden', Emerging Infectious Diseases (Centre for Disease Control), vol 7, June 2001 (suppl), pp1–10

Lobstein, T., Millstone, E., Lang, T. and van Zwanenberg, P. (2001) 'The lessons of Phillips. Questions the UK government should be asking in response to Lord Phillips' Inquiry into BSE'. Centre for Food Policy, Thames Valley University, London

Mason, C. F. (1998) *Biology of Freshwater Pollution*, 3rd edition, Addison, Wesley Longman, Harlow

Millstone, E. and van Zwanenberg, P. (2001) 'Politics of expert advice: From the early history of the BSE saga', *Science and Public Policy*, vol 28, pp99–112

National Audit Office (NAO) (1998) *BSE: The Cost of a Crisis*, NAO, London

Norse, D., Li Ji and Zhang Zheng (2000) *Environmental Costs of Rice Production in China: Lessons from Hunan and Hubei*, Aileen Press, Bethesda

Ofwat (1992–1998) 'Annual returns from water companies – water compliances and expenditure reports'. Office of Water Services, Birmingham

Ostrom, E. (1990) *Governing the Commons: The Evolution of Institutions for Collective Action*, Cambridge University Press, New York

Pain, D. J. and Pienkowski, M. W. (eds) (1997) *Farming and Birds in Europe*, Academic Press Ltd, London

Pearce, D. and Tinch, R. (1998) 'The true price of pesticides' in Vorley, W. and Keeney, D. (eds) *Bugs in the System*, Earthscan, London

Pearce, D. W. and Turner, R. H. (1990) *Economics of Natural Resources and the Environment*, Harvester Wheatsheaf, New York

PHL (1999) 'Public Health Laboratory Service – facts and figures', www.phls.co.uk/facts/

Pimentel, D., Acguay, H., Biltonen, M., Rice, P., Silva, M., Nelson, J., Lipner, V., Giordano, S., Harowitz, A. and D'Amore, M. (1992) 'Environmental and economic cost of pesticide use', *Bioscience*, vol 42, pp750–760

Pimentel, D., Harvey, C., Resosudarmo, P., Sinclair, K., Kunz, D., McNair, M., Crist, S., Shpritz, L., Fitton, L., Saffouri, R. and Blair, R. (1995) 'Environmental and economic costs of soil erosion and conservation benefits', *Science*, vol 267, pp1117–1123

Pingali, P. L. and Roger, P. A. (1995) *Impact of Pesticides on Farmers' Health and the Rice Environment*, Kluwer Academic Press, the Netherlands

Pretty, J. N. (1995a) *Regenerating Agriculture: Policies and Practice for Sustainability and Self-Reliance*, Earthscan, London; National Academy Press, Washington, DC; ActionAid, Bangalore

Pretty, J. N. (1995b) 'Participatory learning for sustainable agriculture', *World Development*, vol 23, pp1247–1263

Pretty, J. N. (1998) *The Living Land: Agriculture, Food and Community Regeneration in Rural Europe*, Earthscan, London

Pretty, J. N. and Ward, H. (2001) 'Social capital and the environment', *World Development*, vol 29, pp209–227

Pretty, J. N., Brett, C., Gee, D., Hine, R., Mason, C. F., Morison, J. I. L., Raven, H., Rayment, M. and van der, Bijl, G. (2000) 'An assessment of the total external costs of UK agriculture', *Agricultural Systems*, vol 65(2), pp113–136

Pretty, J. N., Brett, C., Gee, D., Hine, R., Mason, C., Morison, J., Rayment, M., van der Bijl, G. and Dobbs, T. (2001) 'Policy challenges and priorities for internalising the externalities of modern agriculture', *Journal of Environmental Planning and Management*, vol 44, pp263–283

Putnam, R. (1995) 'Bowling alone: America's declining social capital', *Journal of Democracy*, vol 6, pp65–78

Putnam, R. D. with Leonardi, R. and Nanetti, R. Y. (1993) *Making Democracy Work: Civic Traditions in Modern Italy*, Princeton University Press, Princeton, New Jersey

Repetto, R. and Baliga, S. S. (1996) *Pesticides and the Immune System: The Public Health Risks*, WRI, Washington, DC

Rola, A. and Pingali, P. (1993) *Pesticides, Rice Productivity and Farmers – An Economic Assessment*, IRRI, Manila and WRI, Washington, DC

Scoones, I. (1998) 'Sustainable rural livelihoods: A framework for analysis', IDS Discussion Paper, 72, University of Sussex, Falmer

Siriwardena, G. M., Ballie, S. R., Buckland, G. T., Fewster, R. M., Marchant, J. H. and Wilson, J. D. (1998) 'Trends in the abundance of farmland birds: A quantitative comparison of smoothed Common Birds Census indices', *Journal of Applied Ecology*, vol 35, pp24–43

Steiner, R., McLaughlin, L., Faeth, P. and Janke, R. (1995) 'Incorporating externality costs in productivity measures: A case study using US agriculture' in Barbett, V, Payne, R and Steiner, R (eds) *Agricultural Sustainability: Environmental and Statistical Considerations*, John Wiley, New York, pp209–230

Thrupp, L. A. (1996) *Partnerships for Sustainable Agriculture*, World Resources Institute, Washington, DC

Tilman, D. (1998) 'The greening of the green revolution', *Nature*, vol 396, pp211–212

Uphoff, N. (1998) 'Understanding social capital: Learning from the analysis and experience of participation' in Dasgupta, P and Serageldin, I (eds) *Social Capital: A Multiperspective Approach*, World Bank, Washington, DC

van der Bijl, G. and Bleumink, J. A. (1997) *Naar een Milieubalans van de Agrarische Sector* [Towards an Environmental Balance of the Agricultural Sector], Centre for Agriculture and Environment (CLM), Utrecht

Waibel, H. and Fleischer, G. (1998) *Kosten und Nutzen des chemischen Pflanzenschutzes in der Deutsen Landwirtschaft aus Gesamtwirtschaftlicher Sicht*, Vauk-Verlag, Kiel

Wall, P. G., de Louvais, J., Gilbert, R. J. and Rowe, B. (1996) 'Food poisoning: Notifications, laboratory reports and outbreaks – where do the statistics come from and what do they mean?' *Communicable Disease Report*, vol 6 (7), R94–100

Waltner-Toews, D. and Lang, T. (2000) 'A new conceptual base for food and agricultural policy', Global Change and Human Health, vol 1, pp2–16

WHO (2001) *Food and Health in Europe. A Basis for Action*, Regional Office for Europe, WHO, Copenhagen

Wolfe, M. (2000) 'Crop strength through diversity', *Nature*, vol 406, pp681–682

Worster, D. (1993) *The Wealth of Nature: Environmental History and the Ecological Imagination*, Oxford University Press, New York

Zhu, Y., Chen, H., Fen, J., Wang, Y., Li, Y., Zhen, J., Fan, J., Yang, S., Hu, L., Leaung, H., Meng, T. W., Teng, A. S., Wang, Z. and Mundt, C. C. (2000) 'Genetic diversity and disease control in rice', *Nature*, vol 406, pp718–722

9

External Costs of Agricultural Production in the United States

Erin M. Tegtmeier and Michael D. Duffy

Introduction

All agricultural practices impact the environment. Industrial agriculture is increasingly being recognized for its negative consequences on the environment, public health and rural communities. Soil loss and erosion reduce crop yields and impair natural and man-made water systems (Clark et al, 1985; Crosson, 1986; Holmes, 1988; Atwood, 1994; Pimentel et al, 1995; Evans, 1996). Runoff of agricultural chemicals from farm fields contaminates groundwater and disrupts aquatic ecosystems (Conway and Pretty, 1991; Pimentel et al, 1992; Waibel and Fleischer, 1998; USDA, 2000d; Pretty et al, 2003). Monocropping and feedlot livestock production threaten diversity and may increase food-borne pathogens and antibiotic resistance in humans, as well as pest resistance to chemical controls (National Research Council, 1989; Altieri, 1995; Iowa State University and The University of Iowa Study Group, 2002). The health of rural communities is affected negatively by declining community involvement and increased division of social classes (Bollman and Bryden, 1997; Flora et al, 2002).

The costs of impacts are external to agricultural systems and markets for products. They are borne by society at large. Assessing the monetary costs of such impacts aids in fully identifying their consequences. Cost estimates can inform and guide policy makers, researchers, consumers and agricultural producers, and may encourage a closer look at the impacts of industrial agriculture.

According to western neoclassical economics, well defined property rights ensure that an owner benefits exclusively from use of property and wholly incurs the costs of use. However, in many circumstances, costs are borne by those who are not decision makers. Impacts of agriculture involve costs to individuals and communities who are not making decisions about production methods. These consequences indicate when property rights are not well defined and they represent market failures, which lead to economic ineffi-ciencies. In an unregulated situation, a polluter will weigh the private costs and benefits

Note: Reprinted from *International Journal of Agricultural Sustainability*, vol 2, Tegtmeier, E. M. and Duffy, M. D., 'External Costs of Agricultural Production in the United States', pp55–175, copyright © (2004), with permission from Multilingual Matters

of an action, producing too much pollution with too little cleanup or producing too much product at too low a price (Miranowski and Carlson, 1993; Samuelson and Nordhaus, 1995).

Because these effects occur outside the marketplace, they are called externalities. 'Negative' externalities occur when costs are imposed; 'positive' externalities occur when others gain benefits without charge. To identify forces resulting in externalities and actions that may mitigate their effects, economists distinguish types of externalities. They can be broadly classified by the nature of their consumption (public versus private) and by their effects on resource allocation (pecuniary versus technological).

An externality is 'consumed' by those affected by it. Many externalities have the characteristic of a public good (or bad) where consumption by one individual does not reduce the good's availability to others nor the utility of consumption received by others (Baumol and Oates, 1988). For example, polluted air or scenic views are experienced in this way. They are public and undepletable and are not exchanged in the marketplace where each consumer can be charged for use. A private externality, however, is depletable. If an individual dumps trash onto another's property, this affects only the victim (Baumol and Oates, 1988). Externalities that affect public goods are of greater policy interest because there are fewer 'defensive activities' available to victims.

Externalities also are differentiated by whether the competitive marketplace can adjust to their effects. In the context of agriculture, soil erosion is a technological externality, whereas the decline of rural communities as a consequence of the character and structure of large, industrial farms is considered pecuniary. Research has described declines in purchases from local businesses, increases in crime and civil court cases and decreased property values (Flora et al, 2002). These effects, although undesirable, are not results of market failure in the neoclassical sense. They are, rather, results of the market responding to changes in supply and demand.

Economists and policy makers rely on valuation, or the process of assigning economic value, to apply the concept of externalities. A monetary metric provides a base for comparisons to aid in policy decisions. Externalities, however, often are highly complex and difficult to delineate. Even though assumptions are necessary, economists continue to refine techniques and view valuation as a way of revealing problems with the status quo.

A key assumption underlying valuation is that economic value of an object or service is derived through a function that contributes to human wellbeing and can be measured by 'establishing the link between that function and some service flow valued by people' (Freeman, 1998, p305). Measurement is based on the concepts of willingness to pay (WTP) for the improvement of an object or service or willingness to accept compensation (WTAC) for its deterioration (Hanley et al, 1997; Farber et al, 2002). Valuation approaches generally fall into two categories: direct survey methods and indirect methods (Zilberman and Marra, 1993; Hanley et al, 1997). Survey techniques seek to measure individual preferences for improvement in a situation or loss of wellbeing associated with a condition. Indirect valuation methods observe behavior in related markets and use such data as proxies.

In all valuation efforts, sufficient and reliable data is a concern. People who are surveyed often do not have well defined preferences to which they can assign value or they simply may not be familiar with the services provided by an environmental resource (Hanley et al, 1997). Also, value for many resources is composed of both use values and non-use values that may be particularly difficult to delineate (Hanley et al, 1997). Non-use values include existence value (the value of knowing a thing merely exists, regardless of intent to use) and option value (the value of preserving a resource for possible future use).

We continue to learn about the intricacies of ecosystems on a societal level, but critical data that would strengthen current indirect valuation projects often are not available. Also, environmental externalities, especially those associated with agriculture, frequently have broad spatial and temporal effects, adding to the complexity of valuation efforts.

Study framework

This study assembles available valuation data to arrive at an aggregate, national figure for particular external costs of agricultural production in the US. We focus on technological externalities with public goods characteristics. A literature review revealed data on such externalities in three broad damage categories:

1   natural resources (comprising water, soil and air subcategories)
2   wildlife and ecosystem biodiversity
3   human health (comprising pathogen and pesticide subcategories).

A study on the total external costs of agriculture in the UK (Pretty et al, 2000) guided our work. Pretty et al compiled data from various datasets and studies to estimate costs, categorized by damages to natural capital and human health. They calculated costs of UK£208 per hectare of arable land and permanent pasture. This figure is higher than the cost per cropland hectare for the US reported here. The difference, in part, may be due to the inclusion of costs of the bovine spongiform encephalopathy (BSE or 'mad cow') crisis and the difference in agricultural land area. Also, the UK study included costs to public agencies for monitoring and administering environmental and public health programmes associated with agriculture.

We collected programme costs in the form of agency budgets, but decided not to incorporate them into our total cost figure. This is not meant to diminish the research and conclusions of Pretty et al but, considering the available data for direct costs, we feel that using programme costs as proxies could be viewed as double counting. And, as Pretty et al acknowledge, such activities may be necessary for any type of agricultural production. However, programme costs would likely decrease if agriculture were more environmentally benign.

Other studies on agricultural cost accounting in the UK include Adger and Whitby (1991, 1993) and Hartridge and Pearce (2001). Estimates can be found for other European countries as well: Denmark (Schou, 1996), France (Piot-Lepetit et al, 1997; Bonnieux et al, 1998; le Goffe, 2000) and Italy (Tiezzi, 1999). A discussion on integrating agricultural externalities for a number of countries in the European Union can be found in Brouwer (1999).

For the US, work has been done by Faeth and Repetto (1991), Hrubovcak et al (2000), Smith (1992) and Steiner et al (1995). The study by Steiner et al is the most comparable with our research in that it compiles available data on national estimates of agricultural externalities. Our analysis relies on some of the same sources, indicating how the lack of current, available data limits investigation. Steiner et al (1995, p210) also acknowledge that external costs ideally should be calculated on a 'location-specific basis – which currently is impossible because of a lack of information'. We subsequently have found a dearth of local or regional data to qualify the national figures.

Steiner et al focused on externalities caused by pesticides, fertilizers and soil erosion, and included regulatory programme costs. As reported in 1987 to 1990 dollars, these costs total US$1.3 to $3.6 billion, $12 to $33 million and $5.8 to $20.3 billion, respectively. In effect, we update their study and add information on the treatment of surface water for microbial pathogens, human health costs caused by foodborne pathogens and greenhouse gas emissions. We also attempt to identify, within the scope of the damage categories, a total cost figure attributable to agriculture and a cost figure per cropland hectare.

Methods

Previous studies that assign values to specific impacts of agriculture in the US form the basis of our analysis. Cost estimates are revised and updated to reflect changes in conditions and the Consumer Price Index. Final figures are in 2002 dollars.

Two points in the methodology call for further clarification. We used the Consumer Price Index as opposed to one of the other indices available because we felt that the impact of externalities would be more directly felt by consumers than producers. A second point concerns the changes in technology or production practices that may have occurred since the original estimates were made. In our calculations of damages due to soil erosion, we deflate some of the estimates by a multiplier to address the subsequent decrease in soil erosion. However, this methodology does not fully account for the changes. There really is not a clean way to make such adjustments. This issue points to the need for more updated estimates.

Cost estimates are classified according to production type (crop or livestock) and area-based external cost figures for crop production are also calculated. Agricultural land use areas reported by the United States Department of Agriculture (USDA, 2000b) are used. Of 184.1 million hectares of cropland in the US, approximately 15.3 million are idled each year. The remaining 168.8 million hectares is used for area-based calculations. The external cost of crop production within each damage category is divided by 168.8 million hectares to arrive at cost per hectare figures. Area-based figures are not calculated for those external costs associated with livestock production, considering that production practices and the land areas they affect vary greatly and depend on the animal being raised.

Table 9.1 presents our resulting national tally. Table 9.2 summarizes programme budgets of agencies associated with agricultural activities. Following the tables, each damage category is further described with calculation details.

Results

Table 9.1 *Selected annual external costs of US agricultural production (2002, million $)*

Damage categories	Costs	C/L[a]
1 Damage to water resources		
1a Treatment of surface water for microbial pathogens	118.6	L
1b Facility infrastructure needs for nitrate treatment	188.9	C
1c Facility infrastructure needs for pesticide treatment	111.9	C
Category 1 Subtotal	**419.4**	
2 Damage to soil resources		
2a Cost to water industry	277–831.1	C
2b Cost to replace lost capacity of reservoirs	241.8–6,044.5	C
2c Water conveyance costs	268–790	C
2d Flood damages	190–548.8	C
2e Damages to recreational activities	540.1–3183.7	C
2f Cost to navigation: shipping damages, dredging	304–338.6	C
2g Instream impacts: commercial fisheries, preservation values	224.2–1218.3	C
2h Off-stream impacts: industrial users, steam power plants	197.6–439.7	C
Category 2 Subtotal	**2,242.7–13,394.7**	
3 Damage to air resources		
3a Cost of greenhouse gas emissions from cropland	283.8	C
3b Cost of greenhouse gas emissions from livestock production	166.7	L
Category 3 Subtotal	**450.5**	
4 Damage to wildlife and ecosystem biodiversity		
4a Honey bee and pollination losses from pesticide use	409.8	C
4b Loss of beneficial predators by pesticide applications	666.8	C
4c Fish kills due to pesticides	21.9–51.1	C
4d Fish kills due to manure spills	11.9	L
4e Bird kills due to pesticides	34.5	C
Category 4 Subtotal	**1,144.9–1,174.1**	
5 Damage to human health – pathogens		
5a Cost of illnesses caused by common foodborne pathogens	375.7	L
5b Cost to industry to comply with HACCP[b] rule	40.7–65.8	L
Category 5 Subtotal	**416.4–441.5**	

Table 9.1 *Selected annual external costs of US agricultural production (2002, million $)*

Damage categories	Costs	C/L[a]
6 Damage to human health – pesticides		
6a Pesticide poisonings and related illnesses	1,009.0	C
Category 6 Subtotal	**1,009.0**	
	TOTALS: 5,682.9 to 16,889.2	
	(£3,256.3 to £9,677.5 million)	

a C/L refers to production type that is main cause of impact: crop or livestock.
b HACCP, Hazard Analysis and Critical Control Point.

Table 9.2 *Associated costs: Agency budgets (million $)[a]*

Damage Categories	Costs	C/L[b]
1 Damage to water resources		
1d USEPA FY2003 budget requests for Non-point Source Programme and state grants	153.2	C&L
4 Damage to wildlife and ecosystem biodiversity		
4f USEPA FY2003 budget for Reduce Public and Ecosystem Risk from Pesticides goal	21.9	C
4g USDA FY2003 budget for Natural Resources Conservation Service	1,260.0	C&L
4h USDA FY2003 budget for Farm Service Agency Conservation Programs	1,968.0	C&L
Category 4 Subtotal	**3,249.9**	
5 Damage to human health – pathogens		
5c USDA Food Safety and Inspection Service FY2003 budget	27.2	L
5d FDA Food Safety Initiative FY2002 estimated budget	8.4	L
5e USDA ARS FY1999 budget for food safety, pathogen preharvest research	21.2	C&L
5f USDA APHIS FY2003 budget for Plant & Animal Health Monitoring	143.0	C&L
5g USDA AMS FY2003 budget for Microbiological Data Programme	1.5	C
Category 5 Subtotal	**201.3**	
6 Damage to human health – pesticides		
6b EPA Safe Food Programem FY2003 budget request	86.7	C
6c USEPA FY2003 budget for Reduce Public and Ecosystem Risk from Pesticides goal	27.7	C

Table 9.2 *Associated costs: Agency budgets (million $)[a]*

Damage Categories	Costs	C/L[b]
6d USDA AMS FY2003 budget for Pesticide Data Programme	15.0	C
Category 6 Subtotal	**129.4**	
	TOTAL: 3,733.8	
	(£2,139.5 million)	

a Contact authors for calculation and source information on programme costs.
b C/L refers to production type that is main cause of impact: crop, livestock or both.

1 Damage to water resources

Impacts on water resources are gauged by the costs of treatment necessary to control major pollutants associated with agricultural production (microbial pathogens, nitrate and pesticides).

1a Treatment for microbial pathogens

Micro-organisms in livestock waste can cause several diseases and human health problems. *Cryptosporidium* and *Giardia* are waterborne, disease causing parasites (USDA, 2000e). They are found in beef herds and *Cryptosporidium* may be prevalent among dairy operations (USDA, 1994, 2000d; Juranek, 1995). *Cryptosporidium* oocysts have been found in 67–97 per cent of surface water sampled in the US according to the Centers for Disease Control and Prevention (CDC, 1996).

The Interim Enhanced Surface Water Treatment Rule is one of the EPA's latest rulings on microbial protection addressing *Cryptosporidium* and continuing requirements for *Giardia* and viruses. According to the Environment Protection Agency's (EPA's) Office of Water, the total annualized national cost for implementing this rule is $307 million (USEPA, 1998a). There are three potential sources of both *Giardia* and *Cryptosporidium*: wildlife, domestic livestock and humans (Pell, 1997). From this, we assume that livestock causes one-third, or approximately 35 per cent, of the damages associated with these pathogens. Applying 35 per cent to $307 million, $107.5 million of the national cost to meet the ruling may be due to livestock production. Updated from 1998 to 2002 dollars, the cost is $118.6 million.

1b Treatment for nitrate

Nitrate, a compound of nitrogen, can leach into groundwater sources or be carried by soil particles into surface waters via runoff. Agricultural sources of nitrate include fertilizers, livestock waste and mineralization of crop residues. Agricultural regions have been shown to be highly vulnerable to nitrate contamination of surface and groundwater (USDA, 2000d). Nitrate impairs aquatic ecosystems and is a human health concern. It can be converted to nitrite in the gastrointestinal tract and may prevent the proper transport of oxygen in the bloodstream, causing methemoglobinaemia, or 'blue-baby syndrome' in infants (USDA, 2000d).

Human activities have doubled the amount of nitrogen in our ecosystems since the 1970s through atmospheric deposition of nitrogen compounds (USEPA, 2002b). Fossil fuel combustion is the primary source of nitrogen oxides (NO_x). Transportation related sources (engines in vehicles) account for 53 per cent of these emissions, totalling 10 to 11 million tonnes of NO_x, and large, stationary utility and industrial boilers account for 45 per cent (USEPA, 2002b). Emissions of ammonia (NH_3) from livestock and fertilized croplands contribute to atmospheric deposition of ammonium (NH_4) (Vitousek et al, 1997, as cited in Lawrence et al, 1999). Because ammonium is highly water-soluble, it tends to be deposited closer to emission sources than nitrogen oxides.

The EPA estimated, in 1995 dollars, a total investment of $200 million was needed immediately for water treatment facilities to meet federal nitrate standards. Also, an estimated $3.3 billion is needed over 20 years to replace and maintain water system infrastructure to meet surface water, coliform and nitrate standards (USEPA, 1997a). Considering the additional cost for infrastructure maintenance, we use $200 million as an annual cost. Pretty et al estimated that 80 per cent of nitrate pollution is due to agriculture. We apply this same percentage to $200 million. In 2002 dollars, the facilities cost is $188.9 million per year.

For comparison, Crutchfield et al (1997) employed WTP survey methods to estimate the value placed on reducing nitrates in drinking water for households in four regions in the US. Estimates were $314 to $351 million per year.

Water treatment costs for nitrate are associated mostly with background levels of inorganic nitrogen from fertilizers. Catastrophic manure spills occur intermittently and are not considered here. Many farmers, but not all who should, appropriately credit nitrogen applied to cropland via manure.

1c Treatment for pesticides

Pesticides from agriculture enter surface and groundwater systems through runoff and leachate and pose risks to aquatic and human health. Approximately 447 million kilograms of active ingredients from pesticides are currently used in crop production in the US (Gianessi and Marcelli, 2000) and a number of studies have detected pesticides in water supplies (USDA, 2000d).

The EPA estimated a total need of $400 million, in 1995 dollars, for treatment facilities to meet Safe Drinking Water Act (SDWA) regulations for pesticides and other chemicals (USEPA, 1997a). Approximately 30 per cent of the chemicals listed are pesticides (USEPA, 1998b). Also, agriculture's share of national, conventional pesticide usage is 79 per cent (USEPA, 1999a). So, the $400 million figure is revised using multipliers of 30 per cent and 79 per cent. Updated to 2002 dollars, the annual cost is $111.9 million. This figure does not account for many unregulated pesticides.

Category 1 summary

Total damage to water resources due to agricultural production, according to available research, is calculated to be $419.4 million per year. Crop or livestock production is associated with these costs as follows:

- Livestock – treatment for microbial pathogens ($118.6 million)
- Crop – infrastructure needs for treatment of nitrate and pesticides ($300.8 million).

Using the above cost totals and 168.8 million hectares of cropland, water resources are impacted by cropland at a level of $1.78 per hectare annually.

This is not a complete review of all impacts on water by agricultural production. Of note, the multifaceted impacts of agricultural chemicals and sedimentation on aquatic ecosystems are not included here. The next subsection on soil resources addresses effects of sedimentation on water treatment, storage and conveyance systems. Valuation also is included for fish kills due to pesticides in Subsection 4. However, these do not fully address structural disturbances to habitats and the food chain of aquatic environments.

2 Damage to soil resources

Agriculture practices result in soil erosion through tillage, cultivation and land left bare after harvest. After such disturbances, wind and water carry soil particles off the land. In 1997, average annual soil erosion due to water from cropland and land in the Conservation Reserve Programme (CRP) was 969 million tonnes, with approximately 958 million tonnes coming off of cropland. Erosion due to wind in that same year was 762 million tonnes (USDA, 2000c). Conservation efforts since 1982 have reduced soil erosion by 38 per cent on cropland and CRP land combined (USDA, 2001b), with the composition of the combined land use changing as cropland has been enrolled in the CRP. Still, agriculture remains the single largest contributor to soil erosion. To date, external costs of waterborne erosion have been studied and quantified more than those of windborne erosion. Thus, the costs that follow reflect damages due to waterborne erosion only. Because soil erosion greatly affects the condition and use of surface waters, the following costs support the need for integrated land and water policies.

Erosion reduces soil fertility, organic matter and water holding capacity and negatively affects productivity. Environmental externalities may result with increases of fertilizer and pesticide use to counteract these effects. On-farm costs of lost productivity due to soil erosion are not included here, assuming the majority of these costs are borne by the producer. Although this is not entirely true, it is beyond the scope of this study to identify on-site effects that have off-site impacts. Some estimates of annual on-farm costs due to soil loss include $500 to $600 million (Crosson, 1986), $500 million to $1.2 billion (Colacicco et al, 1989) and $27 billion (Pimentel et al, 1995).

2a Cost to water industry

Sediment causes turbidity in water supplies and transports toxic materials, including fertilizer and pesticide residues that are bound to clay and silt particles. According to Holmes (1988), sediment contributes 88 per cent of total nitrogen and 86 per cent of total phosphorus to the nation's waterways.

Annual costs of supplying water were based on Holmes' method, using a range of treatment costs multiplied by national surface water withdrawals. Updated to 2002 dollars, Holmes' treatment costs are $26.38 to $78.22 per million litres. Similarly, the EPA's Office of Water (2001c) claims that the cost to treat and deliver drinking water is approximately $527.8 per million litres, 15 per cent of which goes to treatment. According to these figures, treatment costs $79.17 per million litres.

In 1995, water withdrawn for public supply was estimated at 152.174 billion litres per day, of which 63 per cent (approximately 95.87 billion litres per day) was from surface water sources (USGS, 1998).

Holmes (1988) estimated that cropland contributes 30 per cent of total suspended solids. Therefore, costs attributed to agriculture are calculated using 30 per cent of the estimate of 95.87 billion litres per day at a cost of $26.38 to $79.17 per million litres. Our numbers, $277 to $831.1 million, are likely to be conservative because treatment of groundwater sources and erosion from pastureland are not considered. However, there may be some overlap between these costs and those to meet nitrate water standards as discussed previously.

2b Lost capacity of reservoirs
Reservoir capacity lost to sedimentation poses a complex problem. Many existing reservoirs are irreplaceable because of unique site characteristics. Dredging is almost prohibitively expensive at a minimum cost of $2.50 per cubic metre. Additionally, there are few disposal sites for dredged material. Alternative energy sources may partially alleviate the need for reservoirs for energy production, but, in terms of water storage, the problem remains (Morris and Fan, 1998).

Although building new reservoirs may not be the realistic solution, this impact is calculated in terms of construction costs to provide some valuation of the problem. Crowder's model (1987) for assessing the cost of reservoir sedimentation is updated.

Total national water storage capacity is 627.6 billion cubic metres (Graf, 1993; Morris and Fan, 1998). Crowder (1987) reported that 0.22 per cent of the nation's water storage capacity is lost annually. Atwood (1994, as cited in USDA, 1995) examined survey records of reservoirs and lakes and found an average storage loss of 5 per cent from sediment depletion.

Construction costs for new capacity from 1963 to 1981 were $243.40 to $567.70 per thousand cubic metres (Crowder, 1987). Updating the median from 1981 to 2002 dollars yields $802.60 per thousand cubic metres.

Total costs were calculated using 0.2 to 5 per cent loss of total national capacity (627.6 billion cubic metres) at the $802.60 per thousand cubic metres replacement value. According to Crowder's analysis, 24 per cent of sediment is from cropland. Reflecting this percentage, final total costs are $241.8 to $6,044.5 million.

2c Cost to water conveyance systems
Roadside ditches and irrigation canals become clogged and require sediment removal and maintenance to prevent local flooding. A cost range of $268 to $790 million is calculated by updating Ribaudo's (1989) figures for these categories and allotting 50 per cent for the contribution of sediment from cropland (Clark et al, 1985).

Subcategories 2d through 2h
These estimates are based primarily on the work of Clark et al (1985) who calculated total erosion effects and applied a multiplier for the percentage due to cropland appropriate to each category. However, erosion from cropland has decreased by 38 per cent since this work (USDA, 2001b). To reflect this improvement, the cropland erosion for each category is multiplied by 62 per cent and updated to 2002 dollars.

2d Flood damages
Sediment contributes heavily to floods and flood damages by increasing water volumes and heights and settling on property once floodwaters have abated. Figuring the percentage

of flood damages that are due to sediment, as well as the percentage of sediment that is due to agricultural practices, is highly speculative, as indicated by the range of estimates.

The estimate by Clark et al of flood damages due to cropland erosion, but not including loss of life, is revised by the method discussed above to yield a range of $184.5 to $548.8 million. Ribaudo (1989) reported a cost range of $653 to $1,546 million in 1986 dollars for annual damages due to soil erosion. Using 32 per cent due to cropland per Clark et al (1985) and updating to 2002 dollars, the revised range is $343 to $812 million, but this does not account for decreased erosion rates since the late 1980s.

The Federal Emergency Management Agency (FEMA) reports dollars and lives lost for billion-dollar weather disasters from 1980 to 1997 (FEMA, 2002). Average annual damages are estimated at $6.4 billion in 2002 dollars and 30 lives lost. Numerous studies have arrived at different estimates for the value of a life. An EPA document (1999b) reviews 26 studies and calculates a mean value for avoiding one statistical death to be $5.9 million. The annual cost of floods increases to $6.6 billion when using this valuation for each of the 30 lives lost. Applying percentages of flood damages due to sedimentation (9 to 22 per cent) and sedimentation due to cropland (32 per cent), $190 to $465 million of this $6.6 billion could be attributable to agriculture.

This last estimate calculated from FEMA data falls within the revised range of Clark et al. High and low range estimates are eliminated as potential outliers. Also, the high end of the valuation based on Ribaudo (1989) may be dropped, considering the revision does not account for the subsequent decrease in cropland erosion. So, the range of $190 to $548.8 million is used in the national tally.

2e Cost to recreational activities

As sediment builds up in lakes and rivers, surface water recreation, including fishing, decreases. Freeman (1982) determined the costs of water pollution that affect recreation. Clark et al used these cost figures and applied a proportion due to sediment as calculated by Vaughan and Russell (1982). Not included were the costs of accidental deaths and injuries caused by increased turbidity. The range revised to 2002 dollars is $540.1 to $3,183.7 million.

2f Cost to navigation

Sediment from erosion collects in navigational channels causing groundings and delays, reliance on smaller vessels and lighter loads, and damage to engines due to sand, pollution and algae.

To assess value in this category, Clark et al (1985) included only commercial shipping damages from inland groundings ($20 to $100 million) and costs for dredging by the United States Army Corps of Engineers (USACE), which we update. Accidents and fuel or cargo spills also cause injuries and deaths and damage to public health and the environment; however, these have not been assessed here. According to the Navigation Data Center (USACE, 2003), the FY2002 cost for dredging navigational channels by the Army Corps and its contractors was $922.9 million.

Commercial shipping damages, according to Clark et al, are revised and added to an estimate of national dredging costs. Taking 32 per cent of the result to account for sedimentation from cropland (Clark et al, 1985), the final costs to navigation due to agricultural activities are approximately $304 to $338.6 million.

2g Other in-stream costs: commercial fisheries and preservation values

Clark et al uses Freeman's (1982) estimates of benefits to commercial fisheries and preservation values that could be gained by controlling water pollution from all sources. Preservation values are non-user values, and, in this case, cleaner water provides non-users with aesthetic and ecological benefits and options for future use. As revised, these annual figures are $224.2 to $1,218.3 million.

Sediment, with its associated contaminants and algal blooms, negatively impacts waterfront property values. A study of lakeside properties in Ohio (Bejranonda et al, 1999) figured benefits to annual rental rates ranging from $23.22 to $115.90 per ac-ft ($1.88 to $9.40 per hundred cubic metres) were accrued by reducing the rate of sediment inflow. However, impacts of sediment on property values are not included in the tally because these values cannot be applied nationally and no other sources were found.

2h Other off-stream costs: Municipal and industrial users

Municipal and industrial users, including steam power plants, experience increased operational costs associated with dissolved minerals and salts remaining in water received from water treatment suppliers. To avoid scale and algae build up in water and boiling systems, water needs to be demineralized and treated. Again using revised calculations of Clark et al, these costs are estimated at $197.6 to $439.7 million.

Category 2 summary

According to this research, total damage to soil resources due to agricultural production is calculated to be $2,242.7 to $13,394.7 million per year. Although waterborne erosion is considerable on western rangelands, our sources focused on cropland erosion, which is associated with all of these costs.

Using the above cost totals and 168.8 million hectares of cropland, soil resources are impacted by crop production at a level of $13.29 to $79.35 per hectare annually. The external cost of the eroded soil itself can be calculated by dividing the total damages due to cropland by 958 million tonnes of erosion from cropland each year. These costs range from $2.34 to $13.98 per tonne of eroded soil.

The damage totals for impacts on soil resources are among the highest for categories covered in this study. Perhaps, this is because a great deal of research exists on soil erosion from agriculture, which has been a long-term concern. Also, the direct effects of soil erosion may be simpler to track and analyse than damages to other categories.

3 Damage to air resources

Agriculture damages air resources through:

- particulate matter released by soil erosion;
- volatilization of NH_3 from urea and manure fertilizers;
- emissions of nitric oxide (NO) and nitrous oxide (N_2O) from fertilizer applications, field burning and soil denitrification;
- hazardous pollutants from manure storage at concentrated animal feeding operations (CAFOs) (Thorne, 2002);

- emissions of methane (CH_4) from enteric fermentation and eructation (belching) of ruminant livestock and manure storage (Cavigelli et al, 1998; USEPA, 2003b).

Some of these releases are greenhouse gases, which interact with the environment and affect human and ecological health. They cause climate change through atmospheric warming, aggravate pulmonary and respiratory functioning, degrade building materials and contribute to the acidification and eutrophication of water resources.

Greenhouse gas emissions from agricultural sources in 2001 totalled 474.9 million tonnes carbon dioxide equivalents, which represents approximately 7 per cent of total greenhouse gas emissions in the US, including 70 per cent of all nitrous oxide emissions from anthropogenic activities and 25 per cent of total CH_4 emissions (USEPA, 2003b). The net impact of agriculture is lessened by the uptake of carbon by agricultural soils, and policy efforts are underway to promote practices that will increase this carbon sequestration. Agricultural soils provided a sink for 15.2 million tonnes carbon dioxide equivalents in 2001 (USEPA, 2003b).

Two sources of valuation for greenhouse gases provide a range of estimates. A study by Titus (1992) considers impacts of climate change to the US, including effects on agricultural production, increases in energy consumption, sea level rise, heat-related deaths and change in forest biomass. The study calculates that a doubling of CO_2 (and equivalents) could cost $37 to $351 billion per year (1992 dollars). Also, the marginal cost of climate change from burning one gallon of gasoline is calculated at $0.16 to $0.36, at a 3 per cent discount rate. This translates to $20 to $50 per tonne carbon dioxide equivalents (2002 dollars).

The Chicago Climate Exchange enables member corporations, municipalities and other institutions to trade greenhouse gas credits in an effort to 'determine the most cost-effective means of reducing overall emissions' (Chicago Climate Exchange, 2004). Members who have reduced emissions receive credits, which can be sold to other members. The final market price for 2003 carbon dioxide equivalents closed at $0.98 per tonne. This is much lower than the range calculated in the Titus study. This is not surprising because the trading price is what companies are willing to pay for emission reductions and does not necessarily reflect health and environmental externalities. Also, participation in the Exchange is strictly voluntary.

However, in the interest of being conservative, we use $0.98 per tonne carbon dioxide equivalents. As discussed, net emissions from agriculture in 2001 were 459.7 million tonnes carbon dioxide equivalents, according to the United States Emissions Inventory (USEPA, 2003b). Total damage from agriculture is then calculated at $450.5 million.

EPA emission data suggest that 63 per cent of this cost is from crop production ($283.8 million) and 37 per cent is from livestock sources ($166.7 million), as follows:

- Crop – soil management, burning crop residues and rice cultivation
- Livestock – enteric fermentation and manure management.

Using the above cost totals and 168.8 million hectares of cropland, air resources are impacted by cropland at a level of $1.68 per hectare annually.

4 Damage to wildlife and ecosystem biodiversity

These costs involve impacts to bird, fish and insect populations, which, in turn, influence ecosystem biodiversity. With approximately 447 million kilograms of active ingredients used in agricultural production (Gianessi and Marcelli, 2000), pesticides affect ecosystem balance.

Our primary valuation source is a study on the environmental impacts of pesticides by Pimentel et al (1992). We acknowledge that since this research was done formulations and application methods of some pesticides have changed to reduce toxicity. For example, the use of granular carbofuran has been severely restricted since 1994 (Pesticide Management Education Program, 1991). The EPA estimated in the 1980s that granular carbofuran killed one to two million birds each year. In spite of this, the restrictions continue to be challenged as evidenced by the recent emergency use request of rice growers in Louisiana. The EPA initially approved use of granular carbofuran on 4050 hectares, but this was reduced to 1010 hectares after public comments were received (American Bird Conservancy, 2002; National Coalition Against the Misuse of Pesticides, 2002).

Aside from the effects of pesticide use, we do include one calculation to value fish killed by manure spills. But, other known environmental stressors associated with agriculture are not represented here. These include inorganic fertilizer runoff and its impact on aquatic ecosystems and the suppression of biodiversity by monocultural practices. Again, impacts on natural ecosystems are difficult to track and analyse and valuation studies are few. Our coverage of this category is far from comprehensive.

4a Honeybee and pollination losses

Pollinators, especially honeybees, are fundamental to ecosystem and agricultural stability. Various studies have attempted to value the agricultural services of pollinators. Southwick and Southwick (1992) estimated $1.6 to $5.7 billion in total annual benefit to agricultural consumers in the US from honeybee pollination. Morse and Calderone (2000) claim the annual value of honeybee pollination to be $14.6 billion, in terms of increased yields and product quality.

For our purposes, the more conservative economic impact of pesticide use on honeybees as calculated by Pimentel et al (1992) is used. Their estimate of $319.6 million is figured in terms of colony losses, reduced honey production and crop pollination and the cost of bee rentals. Assuming original reporting in 1992 dollars, the annual figure is $409.8 million in 2002 dollars.

4b Loss of beneficial predators

Most pesticide applications not only affect the primary crop pest, but also natural enemies of the pest. As the population of beneficial insects drops, outbreaks of secondary pests occur, which in turn lead farmers to apply more pesticide. The cost of these additional applications and crop losses associated with secondary pests is $666.8 million, updating the figure per Pimentel et al (1992).

Although these costs could be considered on-site, they are included because the invertebrate loss due to broad-spectrum pesticides affects not only crop production, but also the ecosystem as a whole. In addition, pesticides may harm micro-organisms. The number and activity of micro-organisms in the soil are measures of soil and ecosystem health, as they break down organic matter and cycle nutrients.

4c Fish kills due to pesticides

Pesticides contaminate aquatic environments, poisoning fish and damaging their food sources and habitat. It is difficult to calculate losses in severe fish kill events and low level poisonings often are not detected. Pimentel et al (1992) use EPA data to estimate 6 to 14 million fish deaths per year due to pesticides and values of freshwater fish from the American Fisheries Society (1982), reflecting commercial hatchery production costs of various fish species. We calculate the average of these values, omitting sturgeon and paddlefish over 38cm long, at $1.67 per fish in 1980 dollars, or $3.65 in 2002 dollars. These numbers yield a damage range of $21.9 to $51.1 million.

4d Fish kills due to manure spills

Manure spills, leaks and dumping by animal feeding operations into surface waters also cause damage to aquatic environments and can be partially valued by the number of fish killed in documented events. A report by the Clean Water Network (2000) records information on feedlot spills and associated fish kills in ten states from 1995 through 1998. Most of the data were collected from state agency databases and reports. More than 13 million fish were killed in over 200 documented manure pollution events. This does not reflect the effects of smaller spills and cumulative impacts and, of course, is not a national count. However, because a high number of animal feeding operations are located in the states included in this report, these numbers are used as a rough proxy for a national estimate. Thirteen million is divided by four years and multiplied by the value of $3.65 per fish given earlier. The estimated annual cost is conservatively set at $11.9 million.

4e Bird kills due to pesticides

Birds exposed to pesticides may be poisoned directly or may ingest pesticide residues with prey and seeds. Pesticides affect the life cycle and reproductive ability of birds and their habitats. Toxicity is difficult to quantify, however, considering avian risk assessments customarily test only one to three bird species; the total number of bird species globally is estimated at 10,000, and over 800 species occur in the US and Canada (Mineau et al, 2001).

Pimentel et al (1992) figure approximately 672 million birds are directly exposed to pesticides on cropland and that 10 per cent of these birds die. The study provides values for a bird's life ranging from $0.40 to $216 to $800. These figures reflect, respectively, cost per bird for bird watching, hunting costs per bird felled and the cost of rearing and releasing a bird to the wild. The higher figures may be considered inappropriate because they are associated with species not as directly affected by agricultural pesticides. By updating the lowest, most conservative valuation to $0.51 per bird death, the cost of bird kills due to pesticides is $34.5 million. This total does not address life cycle and reproductive damages due to poisonings.

Category 4 Summary

Total annual damage to wildlife and ecosystem biodiversity due to agricultural production, according to this research, is calculated to be $1,144.9 to $1,174.1 million. Pesticide use for crop production is associated with all of the costs, except for fish kills due to manure spills from livestock operations. These external costs can be split as follows: $1,133 to $1,162.2 million in damages due to crop production and $11.9 million due to livestock production. Considering the impacts in terms of pesticide use, each kilogram of active ingredient, of 447 million kilograms applied, generates approximately $2.55 in external costs.

Using the above cost totals and 168.8 million hectares of cropland, crop production's injuries to biodiversity cost $6.80 per hectare annually.

The external costs calculated here are substantial and suggest the need for a comprehensive examination of pesticide products and application methods. To curb manure spills, regulations for manure handling at animal feeding operations should continue to be reviewed and enforced and the promotion of other options for livestock finishing should be considered.

5 Damage to human health: Pathogens

According to the Centers for Disease Control and Prevention (CDC), more than 250 food transmitted diseases cause an estimated 76 million illnesses, 325,000 hospitalizations and 5200 deaths annually in the US (CDC, 2002). A Council for Agricultural Science and Technology (CAST) task force estimated microbial foodborne disease cases to number between 6.5 and 33 million annually, with deaths possibly as high as 9000 (CAST, 1994).

Estimates for this category include costs of illnesses associated with foodborne pathogens and costs to the food industry to comply with pathogen reduction regulations. Data are not readily available for other societal costs, such as those incurred by the public health sector or from antibiotic resistance in humans. A recent CAFO air quality study in Iowa describes antibiotic resistance as 'a health threat of great concern' (Iowa State University and The University of Iowa Study Group, 2002, p1–11).

Costs of illnesses associated with waterborne pathogens are not included because states should have implemented the Interim Enhanced Surface Water Treatment Rule (IESWTR) by January 1, 2002. The avoidance benefit of the IESWTR for *Giardia* spp, and *Cryptosporidium parvum* infections due to agriculture is estimated to be between $628 million and $1 billion annually (USEPA 1997c, 1998a).

5a Cost of foodborne illnesses

Most microbial contamination stems from the processing and packaging of animal products. According to a USDA web page (2000a), 'Simple changes in food processing and handling practices can eliminate at least 90 percent of foodborne illnesses.' This suggests that 10 per cent of foodborne pathogen contamination arises from production and meal preparation. Zero contamination is not realistic and other entry points for contamination may not be identified, so we estimate that 3 per cent of the health costs in this category are attributable to agricultural production unless otherwise noted.

Pathogens causing illness may be bacterial, parasitic, fungal or viral. Cost studies by the USDA's Economic Research Service (ERS) have focused on common bacterial agents found in meat, eggs and dairy products. Other food sources include some vegetables, fruits, juices and seafood.

The ERS estimates the annual costs for five bacterial pathogens at $6.9 billion in 2000 dollars (USDA, 2001c). These pathogens are *Campylobacter* spp, *Salmonella, E. coli* O157:H7, *E. coli* non-O157 STEC and *Listeria monocytogenes.* In addition to these, Buzby et al (1997) provide damage estimates for the bacteria *Clostridium perfringens* and *Staphylococcus aureus* and the parasite *Toxoplasma gondii* totalling $4.5 billion (1995 dollars). Updating these figures and attributing 3 per cent of the totals to agricultural production, the estimate for the costs of illnesses and deaths from these common pathogens is $375.7 million annually.

This is conservative given that unidentified agents cause the majority of illnesses, and estimates have been calculated only for the common, known pathogens. The CDC (Mead et al, 1999) estimates that 82 per cent of foodborne illnesses and 65 per cent of deaths are caused by unknown pathogens. Also, many illnesses go unreported or are not diagnosed as food-related.

Furthermore, these costs include only the impacts on households, in terms of lost productivity and income, medical costs and premature death. Household costs not valued include pain and disability, travel cost for medical care, loss of work time for caregivers and chronic health complications.

5b Cost to industry to comply with HACCP rule

In 1997, USDA's Food Safety and Inspection Service (FSIS) issued the first stage of the Pathogen Reduction/Hazard Analysis and Critical Control Point (HACCP) systems rule to meet targets for microbial pathogen reduction. FSIS cites industry costs for meat and poultry plants to comply with HACCP regulations that range from $1.3 to $2.1 billion in year 2000 dollars (USDA, 2001a). These estimates are based on four scenarios of different pathogen control percentages and interest rates. The estimate for costs due to agricultural production is $40.7 to $65.8 million, which is 3 per cent of the range of industry costs and updated to 2002 dollars. Costs of complying with HACCP may be considered health costs internalized by the food processing industry, but this 3 per cent is viewed as a cost caused by agricultural production practices, which is externalized beyond the farm gate to processors and consumers.

Category 5 summary

According to this research, damage to human health from foodborne pathogens due to livestock production is calculated at $416.4 to $441.5 million per year. Although contamination often originates during processing and preparation, livestock health and production methods contribute to a large number of illnesses and should be evaluated to fully address food safety issues. Growing evidence that antibiotic use in livestock increases the resistance of foodborne pathogens reinforces the need to further explore the role of production in this health threat (Iowa State University and The University of Iowa Study Group, 2002).

6 Damage to human health: Pesticides

Pesticides endanger human health through direct exposure, release into the environment and residues on food. Exposure to pesticides, depending on toxicity and quantity, can cause poisoning, eye damage, respiratory ailments, disruption of the endocrine system

(USEPA, 2002c), birth defects, nerve damage, cancer and other effects that may develop over time (USEPA, 2001c). Of particular concern are pesticides that act as endocrine disruptors:

> The endocrine system consists of a set of glands and the hormones they produce that help guide the development, growth, reproduction, and behavior of animals including human beings… EPA is concerned about the growing body of evidence that some man-made chemicals may be interfering with normal endocrine system functioning in humans and other animals (USEPA, 1997d).

Detectable levels of pesticides have been found on approximately 35 per cent of purchased food in the US (Pimentel et al, 1992). Farm workers who handle and apply pesticides face distinct risks. More than 58,000 unintentional poisonings by agricultural pesticides were reported to the American Association of Poison Control Centers in 2002 (Watson et al, 2003).

6a Pesticide poisonings

Very little research has been done to identify and quantify health impacts of pesticides on a national scale for the US. Studies in the Philippines and Ecuador document health effects and calculate reduction in farmer productivity caused by pesticide use (Rola and Pingali, 1993; Antle and Pingali, 1994; Crissman et al, 1994; Antle et al, 1998; Cole et al, 2000). These results, however, are not transferable to agriculture in the US, considering differences in farmer training and production methods. Here, we rely on Pimentel et al (1992), who calculate the costs of pesticide poisonings and deaths based on hospitalizations, outpatient treatment, loss of work and fatalities due to accidental poisonings and treatment costs for pesticide-induced cancers. Their estimate of $787 million ($1,009 million in 2002 dollars) is based, in part, on speculation regarding the incidence of illness and death. However, it could be regarded as conservative considering the number of poisonings reported to control centres. Also, the estimate does not include unreported or misdiagnosed illnesses or costs of chronic ailments, other than cancer, associated with pesticide exposure. In addition, detection techniques are not available for the majority of pesticides used in the US and their health effects have not been determined (Pimentel et al, 1992).

Part of this valuation may be considered double counting with the water treatment costs in subsection 1c. However, water treatment processes do not prevent all waterborne exposure and associated illnesses.

Category 6 summary

The cost to human health from pesticides used in crop production is $1,009 million annually. Using this valuation and 168.8 million hectares of cropland, human health is affected by pesticide applications on cropland at a level of $5.98 per hectare annually. In terms of pesticide use, the impact to human health translates to $2.26 per kilogram active ingredient. This is a substantial external cost. The damages reported here and in subsections 1 and 4 call for increased scrutiny of the human and environmental effects of chemical use in agricultural production.

In 2002, farmers spent $8.2 billion on pesticides in the US (USDA, 2004). But, this retail cost reflects under 80 per cent of the actual cost of pesticide use, when considering

Table 9.3 *Annual external costs of crop production per hectare*

Damage category	Cost
Water resources	$1.78
Soil resources	$13.29 – $79.35
Air resources	$1.68
Biodiversity	$6.71 – $6.89
Human health – pesticides	$5.98
TOTALS	**$29.44 – $95.68 (£16.87 to £54.82)**

the $2,253.9 to $2,283.1 million in damages to water resources, wildlife and ecosystem biodiversity and human health calculated here.

Summary

Agricultural production in the US negatively impacts water, soil, air, wildlife and human health at an estimated cost of $5.7 to $16.9 billion (£3.3 to £9.7 billion) per year. This is the aggregate cost range from the studies reviewed. The breakdown of these costs by production type, as indicated in Table 9.1, is $4,969.3 to $16,150.50 million per year of impacts due to crop production and $713.6 to $738.7 million due to livestock production. With the estimate of 168.8 million hectares of cropland in the United States, total external cost per cropland hectare is calculated at $29.44 to $95.68 (£16.87 to £54.82), as shown in Table 9.3 by damage category.

These figures offer a broad, preliminary view of how the externalities of agriculture encumber society. And yet, these numbers are conservative, considering we are limited by the complexities of assigning monetary values to environmental and health impacts and the lack of related data.

Comparing our findings to a more comprehensive list of agricultural externalities illustrates the incomplete nature of our national tally. For this we turn to social and natural resource accounting efforts, which attempt to incorporate human and environmental capital assets and flows into traditional income and product measures. These assets are not priced in the current market economy and require valuation to be included in social accounts. We refer the reader to other sources for further information on systems of accounts:

- *System of National Accounts* (Commission of the European Communities et al, 1993)
- *The Handbook of National Accounting: Integrated Environmental and Economic Accounting* (United Nations et al, 2003)
- *A System of Economic Accounts for Food and Agriculture* (Food and Agriculture Organization, 1996)
- *Environmental Indicators for Agriculture* (Organisation for Economic Co-operation and Development, 2001).

Clearly, further research is needed on external costs of agriculture, including detailed studies in each impact category, by geographical region and by production type. Comparative valuation studies also would be instructive, that is, examinations of grazing versus feedlot production of livestock or monocropping versus diverse cropping systems. In comparing production methods, trade-offs should be taken into account. For instance, lower pesticide use often requires increased tillage and possibly causes more soil erosion. Also of interest would be an examination of positive, or beneficial, externalities provided by agriculture, *ie*, carbon sequestration, wildlife habitat and aesthetics. Pricing these services may open the door to policy decisions that compensate producers for such 'products'.

Conclusion

Many in the US pride themselves on our 'cheap' food. But, this study demonstrates that consumers pay for food well beyond the grocery store checkout. We pay for food in our utility bills and taxes and in our declining environmental and personal health. These costs total, conservatively, $5.7 to $16.9 billion (£3.3 to £9.7 billion) each year. We also support at least $3.7 billion (£2.1 billion) annually in efforts to regulate the present system and mitigate damages. Additional public costs of agricultural production in the US include direct subsidies and other support mechanisms for farmers. These are not included in our final tally but must be considered in the true cost of food.

What can be done? By using 'ecological' or 'sustainable' methods, some agricultural producers claim to be internalizing many of these external costs. However, the market and policy structure in which most producers operate offers narrow return margins and discourages changes in production methods. Considering this, the partial estimate of damage costs presented here promotes responsible, creative policy actions to acknowledge and internalize the externalities of production practices that are generally accepted and widespread.

Furthermore, the estimates presented in this paper are conservative for reasons beyond the need for more valuation data. Many industrial agricultural practices present us with environmental risks that have unknown potential consequences. Potentialities are difficult to define because effects are diffuse in time and location. Some of these risks have been acknowledged scientifically but not necessarily politically, that is, ecosystem behaviour in a monocropped environment, antibiotic resistance in humans, loss of pollinators.

Political intention is required to reassess and reform agricultural policy. Programmes that highlight sustainable methods rather than destructive, risky practices would be a start at internalizing the true costs of the present system.

References

Adger, W. N. and Whitby, M. C. (1991) 'National accounts and environmental degradation: Accounting for the impact of agriculture and forestry on environmental quality'. *European Economic Review*, vol 35, pp629–641

Adger, W. N. and Whitby, M. C. (1993) 'Natural-resource accounting in the land-use sector: Theory and practice', *European Review of Agricultural Economics*, vol 20, pp77–97

Altieri, M. (1995) *Agroecology: The Science of Sustainable Agriculture*, Westview Press, Boulder, Colorado

Antle, J. M., Cole, D. C. and Crissman, C. C. (1998) 'Further evidence on pesticides, productivity and farmer health: Potato production in Ecuador'. *Agricultural Economics*, vol 18, pp199–207

Antle, J. M. and Pingali, P. (1994) 'Pesticides, productivity, and farmer health: A Philippine case study', *American Journal of Agricultural Economics*, vol 76, pp418–430

American Bird Conservancy (2002) 'News release: Conservation groups prevent use of eagle-killing pesticide' 29 July. Available at www.abcbirds.org/media/releases/carbofuran_victory_release.htm

American Fisheries Society (1982) *Monetary Values of Freshwater Fish and Fish-kill Counting Guidelines*, special publication No 13, American Fisheries Society, Bethesda, Maryland

Atwood, J. (1994) *RCA Reservoir Sediment Data Reports 1-5*, Soil Conservation Service, Washington, DC

Baumol, W. J. and Oates, W. E. (1988) *The Theory of Environmental Policy*, Cambridge University Press, Cambridge

Bejranonda, S., Hitzhusen, F .J. and Hite, D. (1999) 'Agricultural sedimentation impacts on lakeside property values'. *Agricultural and Resource Economics Review*, vol 28, pp208–218

Bollman, R. A. and Bryden, J. M. (eds) (1997) *Rural Employment: An International Perspective*, CAB International, Wallingford

Bonnieux, F., Rainelli, P. and Vermersch, D. (1998) 'Estimating the supply of environmental benefits by agriculture: A French case study'. *Environmental and Resource Economics*, vol 11, pp135–153

Brouwer R. (1999) *Market Integration of Agricultural Externalities: A Rapid Assessment Across EU Countries*, European Environment Agency, Copenhagen

Buzby, J. C., Roberts, T. and Allos, B. M. (1997) 'Estimated Annual Costs of *Campylobacter*-associated Guillain-Barré Syndrome'. Agricultural Economic Report No 756, USDA, Economic Research Service, Washington, DC

Cabe, R. and Johnson, S. R. (1990) 'Natural resource accounting systems and environmental policy modeling'. *Journal of Soil and Water Conservation*, vol 45, pp533–539

Cavigelli, M. A., Deming, S. R., Probyn, L. K. and Harwood, R. R. (eds) (1998) *Michigan Field Crop Ecology: Managing Biological Processes for Productivity and Environmental Quality.* Extension Bulletin E-2646, Michigan State University, East Lansing, Michigan

Centers for Disease Control and Prevention (1996) *Surveillance for Waterborne-Disease Outbreaks: United States, 1993–1994.* 45(SS-1). Centers for Disease Control and Prevention, Atlanta, Georgia. Available at www.cdc.gov/epo/mmwr/preview/mmwrhtml/00040818.htm

Centers for Disease Control and Prevention (2002) 'Disease information – Foodborne infections'. Available at www.cdc.gov/ncidod/dbmd/diseaseinfo/foodborneinfections_t.htm

Chicago Climate Exchange (2004) www.chicagoclimatex.com/

Clark, E. H. II, Haverkamp, J. A. and Chapman, W. (1985) *Eroding Soils: The Off-farm Impacts*, The Conservation Foundation, Washington, DC

Clean Water Network, the Izaak Walton League of America, and the Natural Resources Defense Council (2000) *Spills and Kills: Manure Pollution and America's Livestock Feedlots*, Clean Water Network, Washington, DC, available at www.cwn.org

Colacicco, D., Osborn, T. and Alt, K. (1989) 'Economic damage from soil erosion', *Journal of Soil and Water Conservation*, vol 44, pp35–39

Cole, D. C., Carpio, F. and León, N. (2000) 'Economic burden of illness from pesticide poisonings in highland Ecuador'. *Pan American Journal of Public Health*, vol 8, pp196–201

Commission of the European Communities, International Monetary Fund, Organisation for Economic Co-operation and Development, United Nations and World Bank (1993) *System of National Accounts*, available at http://unstats.un.org/unsd/sna1993/introduction.asp

Conway, G. R. and Pretty, J. N. (1991) *Unwelcome Harvest: Agriculture and Pollution*. Earthscan, London

Council for Agricultural Science and Technology (1994) *Foodborne Pathogens: Risks and Consequences*, Task Force Report No 122, Council for Agricultural Science and Technology, Washington, DC

Crissman, C. C., Cole, D. C. and Carpio, F. (1994) 'Pesticide use and farm worker health in Ecuadorian potato production'. *American Journal of Agricultural Economics*, vol 76, pp593–597

Crosson, P. (1986) 'Soil erosion and policy issues', in Phipps, T., Crosson, P. and Price, K. (eds) *Agriculture and the Environment*, Resources for the Future, Washington, DC, pp35–73

Crowder, B. M. (1987) 'Economic costs of reservoir sedimentation: A regional approach to estimating cropland erosion damage'. *Journal of Soil and Water Conservation*, vol 42, pp94–197

Crutchfield, S. R., Cooper, J. C. and Hellerstein, D. (1997) *Benefits of Safer Drinking Water: The Value of Nitrate Reduction*, Agricultural Economic Report No 752, USDA, Economic Research Service, Washington DC

Evans, R. (1996) *Soil Erosion and Its Impact in England and Wales*, Friends of the Earth Trust, London

Faeth, P. and Repetto, R. (1991) *Paying the Farm Bill: US Agricultural Policy and the Transition to Sustainable Agriculture*, World Resources Institute, Washington, DC

Farber, S. C., Costanza, R. and Wilson, M. A. (2002) 'Economic and ecological concepts for valuing ecosystem services'. *Ecological Economics*, vol 41, pp375–392

Federal Emergency Management Agency (2002) 'Resources for parents and teachers'. Available at www.fema.gov/kids/98wdgen.htm

Flora, J. L., Hodne, C. J., Goudy, W., Osterberg, D., Kliebenstein, J., Thu, K. M. and Marquez, S. P. (2002) 'Social and community impacts. In Iowa State University and The University of Iowa Study Group', *Iowa Concentrated Animal Feeding Operations Air Quality Study*, Environmental Health Sciences Research Center, Iowa City, Iowa, available at www.public-health.uiowa.edu/ehsrc/CAFOstudy.htm

Food and Agriculture Organization of the United Nations (1996) *A System of Economic Accounts for Food and Agriculture*. Available at www.fao.org/docrep/W0010E/W0010E00.htm

Food and Drug Administration (2002) 'Budget information'. Available at www.fda.gov/oc/oms/ofm/budget/2002/CJ2002/HTML/CFSAN.htm

Freeman, A. M., III (1982) *Air and Water Pollution Control: A Benefit–Cost Assessment*, John Wiley and Sons, New York

Freeman, A. M., III (1998) 'On valuing the services and functions of ecosystems', in Freeman, A. M. (ed) *The Economic Approach to Environmental Policy: The Selected Essays of A. Myrick Freeman III*, Edward Elgar, Cheltenham, UK and Northampton, Massachusetts

Gianessi, L. P. and Marcelli, M. B. (2000) *Pesticide Use in US Crop Production: 1997*. National Center for Food and Agricultural Policy, Washington, DC, available at www.ncfap.org/ncfap/nationalsummary1997.pdf

Graf, W. L. (1993) 'Landscapes, commodities, and ecosystems: The relationship between policy and science for American rivers', in Water Science and Technology Board, National Research Council, *Sustaining our Water Resources*, National Academy Press, Washington, DC

Hanley, N., Shogren, J. F. and White, B. (1997) *Environmental Economics in Theory and Practice*, Oxford University Press, New York and Oxford

Hartridge, O. and Pearce, D. W. (2001) *Is UK Agriculture Sustainable? Environmentally Adjusted Economic Accounts for UK Agriculture*, CSERGE-Economics, University College London, London

Holmes, T. (1988) 'The offsite impact of soil erosion on the water treatment industry'. *Land Economics*, vol 64, pp356–366

Hrubovcak, J., LeBlanc, M. and Eakin, B. K. (2000) 'Agriculture, natural resources and environmental accounting'. *Environmental and Resource Economics*, vol 17, pp145–162

Iowa State University and The University of Iowa Study Group (2002) *Iowa Concentrated Animal Feeding Operations Air Quality Study*, Environmental Health Sciences Research Center, Iowa City, Iowa: Environmental Health Sciences Research Center, available at www.public-health.uiowa.edu/ehsrc/CAFOstudy.htm

Juranek, D. (1995) '*Cryptosporidiosis*: Source of infection and guidelines for prevention'. *Clinical Infectious Diseases*, vol 21, pp57–61

Lawrence, G. B., Goolsby, D. A. and Battaglin, W. A. (1999) *Atmospheric Deposition of Nitrogen in the Mississippi River Basin*. Proceedings of the US Geological Survey Toxic Substances Hydrology Program technical meeting, Charleston, South Carolina, 8–12 March. Available at http://toxics.usgs.gov/pubs/wri99-4018/Volume2/sectionC/2413_Lawrence/index.html

le Goffe, P. (2000) 'Hedonic pricing of agriculture and forestry externalities'. *Environmental and Resource Economics*, vol 15, pp397–401

Mead, P. S., Slutsker, L., Dietz, V., McCaig, L. F., Bresee, J. S., Shapiro, C., Griffin, P. M. and Tauxe, R. V. (1999) *Food-related Illness and Death in the United States-Synopses*, Centers for Disease Control and Prevention, Atlanta, Georgia. Available at www.cdc.gov/ncidod/eid/vol5no5/mead.htm#Figure%201

Mineau, P., Baril, A., Collins, B. T., Duffe, J., Joerman, G. and Luttik, R. (2001) 'Pesticide acute toxicity reference values for birds'. *Reviews of Environmental Contamination and Toxicology*, vol 170, pp13–74

Miranowski, J. A. and Carlson, G. A. (1993) 'Agriculture resource economics: An overview', in Carlson, G. A., Zilberman, D. and Miranowski, J. A. (eds) *Agricultural and Environmental Resource Economics*, Oxford University Press, Oxford

Morris, G. L. and Fan, J. (1998) *Reservoir Sedimentation Handbook: Design and Management of Dams, Reservoirs, and Watersheds for Sustainable Uses*, McGraw-Hill, New York

Morse, R. A. and Calderone, N. W. (2000) 'The value of honey bees as pollinators of U.S. crops in 2000'. Available at http://bee.airoot.com/beeculture/pollination2000/pg1.html

National Coalition Against the Misuse of Pesticides (2002) 'News release: EPA allows use of banned insecticide, deadly to birds, on 10,000 acres, will decide whether to allow the program to go forward this week', 3 July. Available at www.beyondpesticides.org/WATCH-DOG/media/carbofuran_07_03_02.htm

National Research Council, Board on Agriculture, Committee on the Role of Alternative Farming Methods in Modern Production Agriculture (1989) *Alternative Agriculture*, National Academy Press, Washington, DC

Organisation for Economic Co-operation and Development (2001) *Environmental Indicators for Agriculture*. Available at www.oecd.org/dataoecd/0/9/1916629.pdf

Pell, A. N. (1997) 'Manure and microbes: Public and animal health problem?' *Journal of Dairy Science*, vol 80, pp2673–2681

Pesticide Management Education Program at Cornell University (1991) 'News release: Carbofuran phased out under settlement agreement 5/91', 14 May. Available at http://pmep.cce.cornell.edu/profiles/insect-mite/cadusafos-cyromazine/carbofuran/gran-carbo-dec.html

Pimentel, D., Acquay, H., Biltonen, M., Rice, P., Silva, M., Nelson, J., Lipner, V., Giordano, S., Horowitz, A. and D'Arnore M. (1992) 'Environmental and economic costs of pesticide use'. *BioScience*, vol 42, pp750–760

Pimentel, D., Harvey, C., Resosudarmo, P., Sinclair, K., Kurz, D., McNair, M., Crist, S., Shpritz, L., Fitton, L., Saffouri, R. and Blair, R. (1995) 'Environmental and economic costs of soil erosion and conservation benefits'. *Science*, vol 267, pp1117–1123

Piot-Lepetit, I., Vermersch, D. and Weaver, R. D. (1997) 'Agriculture's environmental externalities: DEA evidence for French agriculture'. *Applied Economics*, vol 29, pp331–338

Pretty, J. N., Brett, C., Gee, D., Hine, R. E., Mason, C. F., Morison, J. I. L., Raven, H., Rayment, M. D. and van der Bijl, G. (2000) 'An assessment of the total external costs of UK agriculture'. *Agricultural Systems*, vol 65, pp113–136

Pretty, J. N., Mason, C. F., Nedwell, D. B., Hine, R. E., Leaf, S. and Dils, R. (2003) 'Environmental costs of freshwater eutrophication in England and Wales'. *Environmental Science and Technology*, vol 37, pp201–208

Ribaudo, M. O. (1989) *Water Quality Benefits from the Conservation Reserve Program.* Agricultural Economic Report No 606, USDA, Economic Research Service, Washington, DC

Rola, A. and Pingali, P. (1993) *Pesticides, Rice Productivity, and Farmers' Health: An Economic Assessment*, International Rice Research Institute, Manila

Samuelson, P. A. and Nordhaus, W. D. (1995) *Economics.* McGraw-Hill, New York

Schou, J. S. (1996) 'Indirect regulation of externalities: The case of Danish agriculture'. *European Environment*, vol 6, pp162–167

Smith, V. K. (1992) 'Environmental costing for agriculture: Will it be standard fare in the Farm Bill of 2000?' *American Journal of Agricultural Economics*, vol 74, pp1076–1088

Southwick, E. E. and Southwick, L., Jr (1992) 'Estimating the economic value of honeybees (*Hymenoptera Apidae*) as agricultural pollinators in the United States'. *Economic Entomology*, vol 85, pp621–633

Steiner, R. A., McLaughlin, L., Faeth, P. and Janke, R. R. (1995) 'Incorporating externality costs into productivity measures: A case study using US agriculture', in Barnett, V., Payne, R. and Steiner, R. (eds) *Agricultural sustainability: Economic, Environmental and Statistical Considerations,* John Wiley and Sons, New York

Thorne, P. S. (2002) 'Air quality issues', in Iowa State University and The University of Iowa Study Group. *Iowa Concentrated Animal Feeding Operations Air Quality Study*, Environmental Health Sciences Research Center, Iowa City, Iowa. Available at www.public-health.uiowa.edu/ehsrc/CAFOstudy.htm

Tiezzi, S. (1999) 'External effects of agricultural production in Italy and environmental accounting'. *Environmental and Resource Economics*, vol 13, pp459–472

Titus, J. G. (1992) 'The costs of climate change to the United States', in Majumda, S. K., Kalkstein, L. S., Yarnal, B., Miller, E. W. and Rosenfeld L. M. (eds) *Global Climate Change: Implications, Challenges, and Mitigation Measures*, Pennsylvania Academy of Sciences, East Stroudsburg, Pennsylvania. Available at http://yosemite.epa.gov/oar/globalwarming.nsf/content/ResourceCenterPublicationsSLR_US_Costs.html

United Nations, European Commission, International Monetary Fund, Organisation for Economic Co-operation and Development and World Bank (2003) *Handbook of National Accounting: Integrated Environmental and Economic Accounting.* Available at http://unstats.un.org/unsd/environment/seea2003.htm

US Army Corp of Engineers, Navigation Data Center, Dredging Program (2003) 'FY2002 analysis of dredging costs'. Available at www.iwr.usace.army.mil/ndc/dredge/dredge.htm

US Department of Agriculture, Animal and Plant Health Inspection Service (1994) *Cryptosporidium and Giardia in Beef Calves.* National Animal Health Monitoring System report, USDA, Washington, DC

US Department of Agriculture, Natural Resources Conservation Service (1995) *RCA III, Sedimentation in Irrigation Water Bodies, Reservoirs, Canals, and Ditches.* Working Paper No 5. USDA, Washington, DC. Available at www.nrcs.usda.gov/technical/land/pubs/wp05text.html

US Department of Agriculture (2000a) 'Cleaning up our act: Food safety is everybody's business'. Available at www.reeusda.gov/success/impact00/safefood.htm

US Department of Agriculture, Economic Research Service (2000b) 'Land use. In Economic Research Service', *Agricultural Resources and Environmental Indicators*, USDA, Washington, DC

US Department of Agriculture, Natural Resources Conservation Service (2000c) *National Resources Inventory, 1997*, USDA, Washington, DC. Available at www.nrcs.usda.gov/technical/land/meta/m5112.html

US Department of Agriculture, Economic Research Service (2000d) 'Water quality impacts of agriculture', in *Agricultural Resources and Environmental Indicators*, USDA, Washington DC

US Department of Agriculture, Natural Resources Conservation Service (2000e) *Waterborne Pathogen Information Sheet – Principal Pathogens of Concern, Cryptosporidium and Giardia.* USDA, Washington, DC. Available at http://wvlc.uwaterloo.ca/biology447/modules/module8/SludgeDisposal/Pathogen_Information_Sheet-Cryptosporidium_and_Giardia.pdf

US Department of Agriculture, Economic Research Service (2001a) 'Briefing room – Government food safety policies: Features'. Available at www.ers.usda.gov/briefing/FoodSafetyPolicy/features.htm

US Department of Agriculture, Natural Resources Conservation Service (2001b) 'National resources inventory highlights'. Available at www.nrcs.usda.gov/technical/land/pubs/97highlights.pdf

US Department of Agriculture, Economic Research Service (2001c) 'Research emphasis – Food safety: Features'. Available at www.ers.usda.gov/Emphases/SafeFood/features.htm

US Department of Agriculture (2002a) 'Action plan'. Available at www.nps.ars.usda.gov

US Department of Agriculture (2002b) 'FY2003 budget summary'. Available at www.usda.gov/agency/obpa/Budget-Summary/2003

US Department of Agriculture, Economic Research Service (2004) 'Farm income data'. Available at www.ers.usda.gov/Data/FarmIncome/finfidmu.htm

US Environmental Protection Agency (1997a) *Drinking Water Infrastructure Needs Survey: First Report to Congress.* EPA 812-R-97-001, USEPA, Washington DC

US Environmental Protection Agency, Office of Wetlands, Oceans and Watersheds (1997b) *Managing Nonpoint Source Pollution from Agriculture.* Pointer No 6. EPA841-F-96-004F, USEPA, Washington, DC. Available at www.epa.gov/OWOW/NPS/facts/point6.htm

US Environmental Protection Agency (1997c) 'National primary drinking water regulations: Interim enhanced surface water treatment rule notice of data availability-proposed rule'. *Federal Register* November 3, 59486-59557, Office of the Federal Register, National Archives and Records Administration, Washington, DC

US Environmental Protection Agency (1997d) 'Potential of chemicals to affect the endocrine system'. Available at www.epa.gov/pesticides/factsheets/3file.htm

US Environmental Protection Agency, Office of Water (1998a) *Interim Enhanced Surface Water Treatment Rule.* EPA 815-F-99-009, USEPA, Washington, DC. Available at www.epa.gov/safewater/mdbp/ieswtr.html

US Environmental Protection Agency, Office of Water (1998b) *Small System Compliance Technology List for the Non-microbial Contaminants Regulated before 1996.* EPA 815-R-98-002, USEPA, Washington, DC. Available at www.epa.gov/safewater/standard/tlstnm.pdf

US Environmental Protection Agency, Biological and Economic Analysis Division, Office of Pesticide Programs (1999a) *Pesticide Industry Sales and Usage: 1996 and 1997 Market Estimates Report.* 733-R-99-001, USEPA, Washington, DC. Available at www.epa.gov/oppbead1/pestsales/

US Environmental Protection Agency, Office of Atmospheric Programs (1999b) *Regulatory Impact Analysis for the Final Section 126 Petition Rule*, USEPA, Washington, DC. Available at www.epa.gov/ttn/ecas/regdata/126fn0.pdf

US Environmental Protection Agency, Office of Wetlands, Oceans and Watersheds (2001a) *Nonpoint Source Pollution: The Nation's Largest Water Quality Problem.* Pointer No 1. EPA841-F-96-004A, USEPA, Washington, DC. Available at www.epa.gov/OWOW/NPS/facts/point1.htm

US Environmental Protection Agency (2001b) 'Pesticides and food: Health problems pesticides may pose'. Available at www.epa.gov/pesticides/food/risks.htm

US Environmental Protection Agency, Office of Water (2001c) 'Where does my drinking water come from?' Available at www.epa.gov/OGWDW/wot/wheredoes.html

US Environmental Protection Agency (2002a) 'Food Quality Protection Act (FQPA) background'. Available at www.epa.gov/opppsps1/fqpa/backgrnd.htm

US Environmental Protection Agency, Clean Air Market Programs (2002b) *Nitrogen: Multiple and Regional Impacts*. EPA-430-R-01-006. USEPA, Washington, DC. Available at www.epa.gov/airmarkets/articles/nitrogen.pdf

US Environmental Protection Agency (2002c) 'Pesticide effects'. Available at www.epa.gov/ebt-pages/pestpesticideeffects.html

US Environmental Protection Agency (2002d) 'Summary of the EPA's budget, FY2003'. Available at www.epa.gov/ocfo/budget/2003/2003bib.pdf

US Environmental Protection Agency and US Department of Energy (2003a) *Fuel Economy Guide – Model 2004*. DOE/EE-0283, USEPA/DOE, Washington DC. Available at www.fueleconomy.gov/feg/FEG2004.pdf

US Environmental Protection Agency (2003b) *Inventory of US Greenhouse Gas Emissions and Sinks: 1990–2001*. EPA 430-R-03-004, USEPA, Washington, DC. Available at http://yosemite.epa.gov/oar/globalwarming.nsf/content/ResourceCenterPublicationsGHGEmissionsUSEmissionsInventory2003.html

US Geological Survey (1998) *Estimated Use of Water in the United States in 1995*. Circular 1200, Government Printing Office, Washington, DC. Available at http://water.usgs.gov/watuse/pdf1995/html/

Vaughan, W. J. and Russell, C. S. (1982) *Freshwater Recreational Fishing – The National Benefits of Water Pollution Control*. Prepared for Resources for the Future, Johns Hopkins University Press, Baltimore, Maryland

Vitousek, P. M., Aber, J. D., Howarth, R. W. and Likens, G. E. (1997) 'Human alternation of the global nitrogen cycle – Sources and consequences'. *Ecological Applications*, vol 7, pp737–750

Waibel, H. and Fleischer, G. (1998) *Kosten und Nutzen des chemischen Pflanzenschutzes in der deutschen Landwirtschaft aus gesamtwirtschaftlicher Sicht* [Social Costs and Benefits of Chemical Pesticide Use in German Agriculture]. Wissenschaftsverlag Vauk, Kiel. Available at www.ifgb.uni-hannover.de/ppp/publications.htm

Watson, W. A., Litovitz, T. L., Rodgers, G. C., Klein-Schwartz, W., Youniss, J., Rutherford-Rose, S., Borys, D. and May, M. E. (2003) '2002 Annual report of the American Association of Poison Control Centers Toxic Exposure Surveillance System'. *The American Journal of Emergency Medicine*, vol 21, pp353–421, available at www.aapcc.org

Zilberman, D. and Marra, M. (1993) 'Agricultural externalities', in Carlson, G. A., Zilberman, D. and Miranowski, J. A. (eds) *Agricultural and Environmental Resource Economics*, Oxford University Press, Oxford

10

From Pesticides to People: Improving Ecosystem Health in the Northern Andes

Stephen Sherwood, Donald Cole, Charles Crissman and
Myriam Paredes

Introduction

Since the early 1990s, a number of national and international organizations have been working with communities in Carchi, Ecuador's northernmost province, on projects to assess the role and effects of pesticide use in potato production and to reduce its adverse impacts. These are National Institute of Agricultural Research from Ecuador (INIAP), CIP (International Potato Center), Montana State University (US), McMaster University and University of Toronto (Canada), Wageningen University (the Netherlands), and the FAO's Global IPM Facility.

These projects have provided quantitative assessments of community-wide pesticide use and its adverse effects. Through system modelling and implementation of different alternatives, we have demonstrated the effectiveness of different methods to lessen pesticide dependency and thereby improve ecosystem health. Meanwhile, the principal approach to risk reduction of the national pesticide industry continues to be farmer education through 'Safe Use' campaigns, despite the safe use of highly toxic chemicals under the social and environmental conditions of developing countries being an unreachable ideal. These conflicting perspectives and the continued systematic poisoning of many rural people in Carchi have motivated a call for international action (Sherwood et al, 2002).

The project members have worked with interested stakeholders to inform the policy debate on pesticide use at both the provincial and national levels. Our position has evolved to include the reduction of pesticide exposure risk through a combination of hazard removal (in particular, the elimination of highly toxic pesticides from the market), the development of alternative practices and ecological education. The experience reported here has led us to conclude that more knowledge-based and socially oriented interventions are needed. These must be aimed at political changes for enabling new farmer learning and organizational capacity, differentiated markets and increased participation of the most affected parties in policy formulation and implementation. Such measures

Note: Reprinted from *The Pesticide Detox* by Pretty, J. (ed), copyright © (2004) Earthscan, London

involve issues of power that must be squarely faced in order to foster continued transformation of potato production in the Andes towards sustainability.

Potato farming in Carchi

The highland region of Carchi is part of a very productive agricultural region, the Andean highlands throughout Northern Ecuador, Colombia and Venezuela. Situated near the equator, the region receives adequate sunlight throughout the year which, coupled with evenly distributed rainfall, means that farmers can continuously cultivate their land. As a result, the province is one of Ecuador's most important producers of staple foods, with farmers producing nearly 40 per cent of the national potato crop on only 25 per cent of the area dedicated to potato (Herrera, 1999).

Carchi is a good example of the spread of industrialized agricultural technologies in the Americas during the Green Revolution that began in the 1960s. A combination of traditional sharecropping, land reform, market access and high value crops provided the basis for rural economic development (Barsky, 1984). Furthermore, as a result of new revenues from the oil boom of the 1970s, the Ecuadorian government improved transportation and communication infrastructure in Carchi, and the emerging agricultural products industry was quick to capitalize on the availability of new markets. A typical small farm in Carchi is owned by an individual farm household and consists of several separate, scattered plots with an average area of about six hectares (Barrera et al, 1998).

Not surprisingly, agricultural modernization underwent a local transformation. In Carchi, mechanized, agrochemical and market oriented production technologies are mixed with traditional practices, such as sharecropping arrangements, payments in kind, or planting in *wachu rozado* (a pre-Colombian limited tillage system) (Paredes, 2001). Over the last half-century, farming in Carchi has evolved towards a market oriented potato-pasture system dependent on external inputs. Between 1954 and 1974 potato production increased by about 40 per cent and worker productivity by 33 per cent (Barsky, 1984). Until recently, the potato growing area in the province continued to increase, and yields have grown from about 12 tonnes per hectare (t ha^{-1}) in 1974 to about 21 today, a remarkable three times the national average (Crissman et al, 1998).

To confront high price variability in potato (by factors of 5 to 20 in recent years), farmers have applied a strategy of playing the 'lottery', which involves continual production while gambling for high prices at harvest to recover overall investment. Nevertheless, the dollarization of the Ecuadorian Sucre in 2000 led to triple digit inflation and over 200 per cent increase in agricultural labour and input costs over three years (World Development Index, 2003). Meanwhile, open trade with neighbouring Colombia and Peru has permitted the import of cheaper commodities. As a result of a trend towards increased input costs and lower potato prices, in 2003 Carchense farmers responded by decreasing area planted in potato from about 15,000ha in previous years to less than 7000ha. It remains to be seen how farmers ultimately will compensate for the loss of competitiveness brought about by dollarization.

Carchi farmers of today rely on insecticides to control the tuber-boring larva of the Andean weevil (*Premnotrypes vorax*) and a variety of foliage damaging insects. They also rely on fungicides to control late blight (*Phytophtera infestans*). One economic study of pesticides in potato production in Carchi confirmed that farmers used the products efficiently (Crissman et al, 1994), and later attempts during the 1990s by an environmental NGO to produce pesticide-free potatoes in Carchi failed (Frolich et al, 1999). After 40 years inorganic fertilizers and pesticides appear to have become an essential part of the social and environmental fabric of the region (Paredes, 2001).

Pesticide use and returns

Our 1990s study of pesticide use found that farmers applied 38 different commercial fungicide formulations (Crissman et al, 1998). Among the fungicides used, there were 24 active ingredients. The class of dithiocarbamate contact-type fungicides were the most popular among Carchi farmers, with mancozeb contributing more than 80 per cent by weight of all fungicide active ingredients used. The dithiocarbamate family of fungicides has recently been under scrutiny in the Northern Andes due to suspected reproductive (Restrepo et al, 1990) and mutagenic effects in human cells (Paz-y-Mino et al, 2002). Similar concerns have been raised in Europe and the United States (USEPA, 1992; Lander et al, 2000).

Farmers use three of the four main groups of insecticides in 28 different commercial products. Although organochlorine insecticides can be found in Ecuador, farmers in Carchi did not use them. The carbamate group was represented only by carbofuran, but this was the single most heavily used insecticide – exclusively for control of the Andean weevil. Carbofuran was used in its liquid formulation, even though it is restricted in the US and Europe due to the ease of absorption of the liquid and the high acute toxicity of its active ingredient. Another 18 different active ingredients from the organophosphate and pyrethroid groups were employed to control foliage pests, though only four were used on more than 10 percent of plots. Here the OP methamidophos, also restricted in the US due to its high acute toxicity, was the clear favourite. Carbofuran and methamidophos, both classified as highly toxic (1b) insecticides by the World Health Organization (WHO), respectively made up 47 per cent and 43 per cent of all insecticides used (by weight of active ingredient applied). In sum, 90 per cent of the insecticides applied in Carchi were highly toxic. A later survey by Barrera et al (1998) found no significant shifts in the products used by farmers.

Most insecticides and fungicides come as liquids or wettable powders and are applied by mixing with water and using a backpack sprayer. Given the costs associated with spraying, farmers usually combine several products together in mixtures known locally as cocktails, applying all on a single pass through the field. On average, each parcel receives more than seven applications with 2.5 insecticides and/or fungicides in each application (Crissman et al, 1998). Some farmers reported as many as seven products in a single concoction. On many occasions different commercial products were mixed containing the same active ingredient or different active ingredients intended for the same type of control. Women and very young children typically did not apply pesticides: among the 2250 applications that we documented, women made only four.

Product and application costs together account for about one-third of all production costs among the small and medium producers in the region. The benefit to yields (and revenues) from using pesticides exceeded the additional costs of using them (including only direct production costs such as inputs and labour but not the costs of externalities). Nevertheless, Crissman et al (1998) found that farmers lost money in four of ten harvests, largely due to potato price fluctuations and price increases in industrial technologies, particularly mechanized land preparation, fertilizers, and pesticides, that combined can represent 60 per cent of overall production outlays. Unforeseen ecological consequences on natural pest control mechanisms, in particular parasitoids and predators in the case of insect pests and selective pressure on *Phytophtora infestans* in the case of disease, raises further questions about the real returns on pesticides (Frolich et al, 1999). As we shall see, long-term profitability of pesticide use is even more questionable when associated human health costs to applicators and their families are taken into account.

Pesticide exposure and health effects

Based on survey, observational and interview data, the majority of pesticides are bought by commercial names. Only a small minority of farmers reported receiving information on pesticide hazards and safe practices from vendors (Espinosa et al, 2003). Pesticide storage is usually relatively brief (days to weeks) but occurs close to farmhouses because of fear of robbery. Farmers usually mix pesticides in large barrels without gloves, resulting in considerable dermal exposure (Merino and Cole, 2003). Farmers and, on larger farms, day labourers apply pesticides using backpack sprayers on hilly terrain. Few use personal protective equipment for a variety of reasons, including social pressure (eg masculinity has become tied to the ability to withstand pesticide intoxications), and the limited availability and high cost of equipment. As a result, pesticide exposure is high. During pesticide applications, most farmers wet their skin, in particular the back (73 per cent of respondents) and hands (87 per cent) (Espinosa et al, 2003). Field exposure trials using patch-monitoring techniques showed that considerable dermal deposition occurred on legs during foliage applications on mature crops (Cole et al, 1998a, b). Other studies have shown that additional field exposure occurs in the field during snack and meal breaks, when hand washing rarely occurs (Paredes, 2001).

Family members are also exposed to pesticides in their households and in their work through a multitude of contamination pathways. Excess mixed product may be applied to other tuber crops, thrown away with containers in the field, or applied around the house. Clothing worn during application is often stored and used repeatedly before washing. Contaminated clothing is usually washed in the same area as family clothing, though in a separate wash. Extent of personal wash up varies but is usually insufficient to remove all active ingredients from both the hands of the applicator and the equipment. Separate locked storage facilities for application equipment and clothing are also uncommon. Swab methods have found pesticide residues on a variety of household surfaces and farm family clothing (Merino and Cole, 2003).

Pesticide poisonings in Carchi are among the highest recorded in developing countries (Cole et al, 2000). In active poisoning surveillance, though there were some suicides and accidental exposures, most reported poisonings were of applicators. While the extensive use of fungicides causes dermatitis, conjunctivitis and associated skin problems (Cole et al, 1997a), we focused our attention on neurobehavioral disorders caused by highly toxic methamidophos and carbofuran. The results were startling.

The health team applied WHO recommended battery of tests to determine the effects on peripheral and central nervous system functions (Cole et al, 1997b, 1998a). The results showed high proportions of the at-risk population affected, both farmers and their family members. Average scores for farm members were a standard deviation below the control sample, the non-pesticide population from the town. Over 60 per cent of rural people were affected and women, although not commonly active in field agriculture', were nearly as affected as field workers. Alarmingly, both Mera-Orcés (2000) and Paredes (2001) found that poisonings and deaths among young children were common in rural communities.

Contamination resulted in considerable health impacts that ranged from sub-clinical neurotoxicity (Cole et al, 1997a, 1998a), poisonings with and without treatment (Crissman et al, 1994) to hospitalizations and deaths (Cole et al, 2000). In summary, human health effects included poisonings (at a rate of 171/100,000 rural population), dermatitis (48 per cent of applicators), pigmentation disorders (25 per cent of applicators), and neurotoxicity (peripheral nerve damage, abnormal deep tendon reflexes and coordination difficulties). Mortality due to pesticide poisoning is among the highest reported anywhere in the world (21/100,000 rural population). These health impacts were predominantly in peri-urban and rural settings. This high incidence of poisoning may not be because the situation is particularly bad in Carchi, but because researchers sought systematically to record and document it.

Acute pesticide poisonings led to significant financial burden on individual families and the public health system (Cole et al, 2000). At the then current exchange rates, median costs associated with pesticide poisonings were estimated as follows: public health care direct costs of US$9.85/case; private health costs of $8.33/case; and lost time indirect costs for about six worker days of $8.33/agricultural worker. All of these were over five times the daily agricultural wage of about $1.50 at the time (1992). Antle et al (1998a) showed that the use of some products adversely affects farmer decision-making capacity to a level that would justify worker disability payments in other countries. Neither group of researchers included financial valuation of the deaths associated with pesticide poisonings nor the effects of pesticides on quality of life, both of which would substantially increase overall economic burden of illness estimates.

A myth: The highly toxics can be safely used

Following the research results, limitations in the pesticide industry's safe use of pesticides (SUP) campaign became apparent. In a letter to the research team, the Ecuadorian Association for the Protection of Crops and Animal Health (APCSA, now called Crop Life Ecuador) noted that an important assumption of SUP was that exposure occurred because of 'a lack of awareness concerning the safe use and handling of [pesticide] products'. Although our Carchi survey showed a low percentage of women in farm families had received any training on pesticides (14 per cent), most male farmers (86 per cent)

Table 10.1 *Hierarchy of controls for reducing pesticide exposure*

Most effective
1 Eliminate more highly toxic products, eg carbofuran and methamidophos
2 Substitute less toxic, equally effective alternatives
3 Reduce use through improved equipment eg low volume spray nozzles
4 Isolate people from the hazard, eg locked separate pesticide storage
5 Label products and train applicators in safe handling
6 Promote use of personal protection equipment
7 Institute administrative controls, eg rotating applicators
Least effective

Source: adapted from Plog et al, 1996

had received some training on pesticide safety practices. Furthermore, labels are supposed to be an important part of the 'hazard communication process' of salesmen. Yet our work in Carchi indicated that farm members often could not decipher the complex warnings and instructions provided on most pesticide labels.

Although 87 per cent of the population in our project area was functionally literate, over 90 per cent could not explain the meaning of the coloured bands on pesticide containers indicating pesticide toxicity. Most believed that toxicity was best ascertained through the odour of products, potentially important for organophosphates with sulphur groups but not generalizable to all products that are impregnated by formulators for marketing purposes. Hence even the universal, seemingly simple toxicity warning system of coloured bands on labels has not entered the local knowledge system. If industry is seriously concerned about informing farmers of the toxicity of its products, it should better match warning approaches to current perceptions of risk, such as considering using toxicity-related odour indicators.

In addition, the SUP campaign's focus on pesticides and personal protective equipment (PPE) is misguided. Farmers regard PPE as uncomfortable and 'suffocating' in humid warm weather, leading to the classic problem of compliance associated with individually oriented exposure reduction approaches (Murray and Taylor, 2000). Examination of the components of the classic industrial hygiene hierarchy of controls (Table 10.1) shows PPE to be among the least effective controls and suggests that the industry strategy of prioritizing PPE is similar to locking the stable after the horse has bolted. Our research has shown the ineffectiveness of product labelling (point 5). Isolation (point 4) is difficult in open environments such as field agriculture where farming infrastructure and housing are closely connected and some contamination of the household is virtually inevitable, particularly in poorer households. Priority should be given to other more effective strategies of exposure reduction, beginning with point 1: eliminating the most toxic products from the work and living environments. Likewise, this is the highest priority of the IPPM 2015 initiative.

INIAP, the Ecuadorian agricultural research institute, is prepared to declare that alternative technologies exist for the Andean weevil and foliage pests and that highly toxic pesticides are not necessary for potato production and other highland crops in Ecuador (Gustavo Vera, INIAP Director General of Research, pers comm). Meanwhile, pesticide industry representatives have privately acknowledged that they understand that highly toxic pesticides eventually will need to be removed from the market. Nevertheless, the Ecuadorian Plant and Animal Health Service (SESA) and Crop Life Ecuador have taken the position that they will continue to support the distribution and sale of WHO Class I products in Ecuador until the products are no longer profitable or that it is no longer politically viable to do so.

One seven-year study by Novartis (now Syngenta) found that SUP interventions in Latin America, Africa and Asia were expensive and largely ineffective, particularly with smallholders (Atkin and Leisinger, 2000). The authors argue that 'the economics of using pesticides appeared to be more important to [small farmers] than the possible health risks' (p121). The most highly toxic products are the cheapest on the market in Carchi, largely because the patents on these early generation products have expired, permitting free access to chemical formulas and competition, and because farmers have come to accept the personal costs associated with poisonings.

Policies and trade-offs

Pesticide use in agricultural production conveys the benefit of reducing losses due to pests and disease. That same use, however, can cause adverse environmental and health impacts. Previously, we cited a study that showed that pesticide use by farmers was efficient from a narrow farm production perspective. Nevertheless, that study examined pesticide use solely from the perspective of reducing crop losses. If the adverse health and environmental effects were also included in the analysis, the results would be different. Integrated assessment is a method to solve this analytical problem. The Carchi research team devised an innovative approach to integrated assessment called the Trade-off Analysis (TOA) method (Antle et al, 1998b; Stoorvogel et al, 2004).

The TOA method is an interactive process to define, analyse and interpret results relevant to policy analysis. At its heart is a set of linked economic, biophysical and health models inside a user shell called the TOA Model. Based on actual dynamic data sets from the field, we used simulations in the TOA method to examine policy options for reducing pesticide exposure in Carchi.

The policy options we explored were a combination of taxes or subsidies on pesticides, price increases or declines in potatoes, technology changes with IPM, and the use of personal protective equipment. We examined the results in terms of farm income, leaching of pesticides to groundwater and health risks from pesticide exposure. Normally, policy and technology changes produce trade-offs – as one factor improves, the other factor worsens. Our analysis of pesticide taxes and potato price changes produced such a result. As taxes decrease and potato prices increase, farmers plant more of their farm with potatoes and tend to use more pesticide per hectare. Thus a scenario of pesticide subsidies and potato price increases produce growth in income and increases in groundwater contamination and health risks from pesticide exposure.

With the addition of technology change to these price changes, the integrated analysis produced by the TOA Model showed that a combination of IPM and protective clothing could produce a win–win outcome throughout the range of price changes:

neurobehavioral impairment and environmental contamination decreased while agricultural incomes increased or held steady (Antle et al, 1998c; Crissman et al, 2003).

Transforming awareness and practice: The experience of EcoSalud

The unexpected severity of pesticide related health problems and the potential to promote win–win solutions motivated the research team to search for ways to identify and break the pervasive cycle of exposure for the at-risk population in Carchi. The EcoSystem Approaches to Human Health Program of IDRC (www.idrc.ca/ecohealth) offered that opportunity through support to a project called EcoSalud (*salud* means health in Spanish). The Eco-System Approaches to Human Health Program was established on the understanding that ecosystem management affects human health in multiple ways and that a holistic, gender sensitive, participatory approach to identification and remediation of the problem is the most effective manner to achieve improvements (Forget and Lebel, 2001).

The EcoSalud project in Carchi was essentially an impact assessment project designed to contribute directly to ecosystem improvements through the agricultural research process. The aims were to improve the welfare of the direct beneficiaries through enhanced neurobehavioral function brought about by reduced pesticide exposure, and to improve the wellbeing of indirect beneficiaries through farming innovation. The project design called for before-and-after measurements of a sample population that changed its behaviour as result of the intervention. Consistent with IDRC's EcoSystem Health paradigm, the intervention was designed to be gender sensitive and increasingly farmer- and community-led.

EcoSalud started by informing members of three rural communities of past research results on pesticide exposure and health impacts. To illustrate pesticide exposure pathways, we used a non-toxic fluorescent powder that glowed under ultraviolet light as a tracer (Fenske et al, 1986). Working with volunteers in each community, we added the tracer powder to the liquid in backpack sprayers and asked farmers to apply as normally. At night we returned with ultraviolet lights and video cameras to identify the exposure pathways. During video presentations, community members were astonished to see the tracer not only on the hands and face of applicators, but also on young children who played in fields after pesticide applications. We also found traces on clothing and throughout the house, such as around wash areas, on beds and even on the kitchen table. Perhaps more than other activities, the participatory tracer study inspired people to take action themselves.

People, in particular mothers, began to speak out at community meetings. The terms *el remedio* (the treatment) and *el veneno* (the poison) were often used interchangeably when referring to pesticides. Spouses explained that the need to buy food and pay for their children's' education when work options were limited led to an acceptance of the seemingly less important risks of pesticides. They explained that applicators often prided themselves on their ability to withstand exposure to pesticides. As one young girl recounted (in Paredes, 2001):

> One time, my sister Nancy came home very pale and said that she thought she had been poisoned. I remembered that the pesticide company agricultural engineers had spoken about this, so I washed her with lots of soap on her back, arms and face. She said she felt dizzy, so I helped her vomit. After this she became more resistant to pesticides and now she can even apply pesticides with our father.

Despite stories such as this, many women became concerned about the health impacts of pesticides on their families. During one workshop, a women's group asked for disposable cameras to document pesticide abuse. Children were sent to spy on their fathers and brothers and take photos of them handling pesticides carelessly or washing sprayers in creeks. Their presentations led to lively discussions. The results of individual family studies showed that poisonings caused chronic ill-health for men and their spouses, and ultimately jeopardized household financial and social stability. Concern about the overall family vulnerability was apparent during community meetings, when women exchanged harsh words with their husbands over their agricultural practices that resulted in personal and household exposure to toxic chemicals. The men responded that they could not grow crops without pesticides and that the safer products were the most expensive. Communities called for help.

INIAP's researchers and extensionists in Carchi had gained considerable experience with farmer participatory methodologies for technology development, including community-led varietal development of late blight disease resistant potatoes. We know that such approaches can play an important role in enabling farmers to acquire new knowledge, skills and attitudes needed for improving their agriculture. INIAP built on existing relationships with Carchi communities to run Farmer Field Schools (FFSs), a methodology recently introduced to the Andes. In part, FFSs attempt to strengthen the position of farmers to counterbalance the messages from pesticide salespeople. As one FFS graduate said (in Paredes, 2001):

> Prior to the Field School coming here, we used to go to the pesticide shops to ask what we should apply for a problem. Then the shopkeepers wanted to sell us the pesticides that they could not sell to others, and they even changed the expiry date of the old products. Now we know what we need and we do not accept what the shopkeepers want to give us.

FFS have sought to challenge the most common of IPM paradigms that centres on pesticide applications based on economical thresholds and transfer of single element technologies within a framework of continuing pesticide use (Gallagher, 2000). In contrast, FFS programmes propose group environmental learning on the principles of crop health and ecosystem management as an alternative to reliance on curative measures to control pests. As a FFS graduate in Carchi noted (in Paredes, 2001):

> When we talk about the insects [in the FFS] we learn that with the pesticides we kill everything, and I always make a joke about inviting all the good insects to come out of the field before we apply pesticides. Of course, it is a poison, and we kill everything. We destroy nature when we do not have another option for producing potatoes.

In practice, the FFS methodology has broadened technical content beyond common understanding of IPM to a more holistic approach for improving plant and soil health. The FFS methodology adapts to the diverse practical crop needs of farmers, be they production, storage or commercialization. FFS ultimately aspires to catalyse the innovative capacity of farmers, as exemplified by how a graduate has improved cut foliage insect traps tested in his FFS (in Paredes, 2001):

> I always put out the traps for the Andean weevil, even if I plant 100 [bags of seed] because it decreases the number of adults. It is advantageous because we do not need to buy much of that poison Furadan. But I do use them differently. After ploughing, I transplant live potato plants from another field, then I do not need to change the dead plants every eight days.

In an iterative fashion, FFS participants conduct learning experiments on comparative (conventional vs IPM) small plots (about 2500m²) to fill knowledge gaps and to identify opportunities for reducing external inputs while improving production and overall productivity. After two seasons, initial evaluation results in three communities were impressive. Through the use of alternative technologies, such as Andean weevil traps, late blight resistant potato varieties, specific and low toxicity pesticides, and careful monitoring before spraying, farmers were able to decrease pesticide sprays from 12 in conventional plots to seven in IPM plots while maintaining or increasing production (Barrera et al, 2001). The amount of active ingredient of fungicide applied for late blight decreased by 50 per cent, while insecticides used for the Andean weevil and leafminer fly (*Liriomysa quadrata*), that had commonly received the highly toxic carbofuran and methamidophos, decreased by 75 per cent and 40 per cent, respectively.

Average yields for both conventional and IPM plots were unchanged at about 19 t ha⁻¹ but net returns increased as farmers were spending less on pesticides. FFS participants identified how to maintain the same level of potato production with half the outlays in pesticides and fertilizers, decreasing the production costs from about US$104 to $80 per tonne. Because of the number of farmers involved in FFS test plots, it was difficult to assess labour demands in the economic analysis. Nonetheless, farmers felt that the increased time for scouting and using certain alternative technologies, such as the insect traps, would be compensated by decreased pesticide application costs, not to mention decreased medical care visits. A recent ex-post study that INIAP will publish in early 2004 has confirmed this trend at the level of individual farms of FFS graduates in Carchi (Barrera et al, 2004).

In addition to the intensive six month FFS experience, EcoSalud staff visited individual households to discuss pesticide safety strategies such as improved storage of pesticides, PPE, use of low volume nozzles that achieve better coverage with less pesticide, and more consistent hygienic practices. Based on widespread disinterest in PPE, we were surprised when participants began to request help in finding high quality personal protective equipment, that they said was unavailable at the dozens of local agrochemical vendors. EcoSalud staff found high quality PPE (mask, gloves, overalls and pants) through health and safety companies in the capital city, costing $34 per set, the equivalent of over a week's labour at the time. The project agreed to grant interest free, two month credit towards the purchase price to those interested in buying the gear. Remarkably, 46 of the 66 participating families in three communities purchased complete packages of equipment. A number of farmers rented their equipment to others in the community in order to recuperate costs. Follow-up health studies are not complete, but anecdotal evidence is promising. As the wife of one FFS graduate who previously complained of severe headaches and tunnel vision due to extensive use of carbofuran and metamidophos said:

> Carlos no longer has headaches after working in the fields. He used to return home [from applying pesticides] and could hardly keep his eyes open from the pain. After the Field School and buying the protective equipment, he is a far easier person to live with (farm family, Santa Martha de Cuba, pers comm).

Complementary projects have supported follow-up activities in Northern Ecuador and elsewhere, including the production of FFS training materials (Pumisacho and Sherwood, 2000; Sherwood and Pumisacho, in press), the training of nearly 100 FFS facilitators in Carchi and nearby Imbabura, the transition of FFS to small-enterprise production groups and the establishment of farmer-to-farmer organization and capacity-building. Concurrently, over 250 facilitators have been trained nationwide and hundreds of FFS have been completed. Recently, Ecuador's Ministry of Agriculture decided to include FFS as an integral part of its burgeoning national Food Security Program. Furthermore, in part due to the successful experience in Carchi, FFS methodology has subsequently spread to Peru, Bolivia, and Colombia as well as El Salvador, Honduras and Nicaragua, where over 1500 FFS had been conducted by mid 2003 (LEISA, 2003).

Conclusions

Much conventional thinking in agricultural development places emphasis on scientific understanding, technology transfer, farming practice transformation and market linkages as the means to better futures. Consequently, the focus of research and interventions tends to be on the crops, the bugs and the pesticides, rather than the people who design, choose and manage practices. Recent experiences of rural development and community health, however, argue for a different approach (see for example, Norgaard, 1994; Uphoff et al, 1998; Latour, 1998; Röling, 2000). Of course, technologies can play an important role in enabling change, but the root causes of the ecosystem crisis such as in Carchi appear to be fundamentally conceptual and social in nature, that is, people sourced and dependent.

There is a general need for organizing agriculture around the development opportunities found in the field and in communities (van der Ploeg, 1994). Experience with people-centred and discovery-based approaches has shown promise at local levels, but ultimately such approaches do not address structural power issues behind complex, multi-stakeholder, socio-environmental issues, such as pesticide sales, spread and use.

The search for innovative practice less dependant on agrochemical markets needs to focus on the diversity of farming and the socio-technical networks that enable more socially and ecologically viable alternatives. Progress in this area would require a new degree of political commitment from governments to support localized farming diversity and the change of preconceived, externally designed interventions towards more flexible, locally driven initiatives. In addition, local organizations representing the most affected people must aim to influence policy formulation and implementation.

Our modern explanations are ultimately embedded in subtle mechanisms of social control that can lead to destructive human activity. The social and ecosystem crises common to modernity, evident in the people-pest-pesticide crises in the Northern Andes, are not just a question of knowledge, technology, resource use and distribution, or access to markets. Experience in Carchi demonstrates that approaches to science, technology, and society are value-laden and rooted in power relationships among the diverse actors – such as farmers, researchers, industry representatives and government officials – that can drive farming practice inconsistent with public interest and the integrity of ecosystems. Solutions will only be successful if they break with past thinking and more effectively empower communities and broader civil society to mobilize enlightened activity for more socially and environmentally acceptable outcomes.

References

Antle, J. M., Capalbo, S. M. and Crissman, C. C. (1998a) 'Tradeoffs in policy analysis: Conceptual foundations and disciplinary integration' in Crissman, C. C., Antle, J. M. and Capalbo, S. M. (eds) *Economic, Environmental and Health Tradeoffs in Agriculture: Pesticides and the Sustainability of Andean Potato Production*. International Potato Center, Lima, Kluwer Academic Press, Boston, pp21–40

Antle, J. M., Capalbo, S. M., Cole, D. C., Crissman, C. C. and Wagenet, R. J. (1998b) 'Integrated simulation model and analysis of economic, environmental and health tradeoffs in the Carchi potato-pasture production system' in Crissman, C. C., Antle, J. M. and Capalbo, S. M. (eds) *Quantifying Tradeoffs in the Environment, Health and Sustainable Agriculture: Pesticide Use in the Andes*. International Potato Center, Lima, Kluwer Academic Press, Boston, pp243–268

Antle, J. M., Cole, D. C. and Crissman, C. C. (1998c) 'Further evidence on pesticides, productivity and farmer health: Potato production in Ecuador', *Agricultural Economics*, vol 18, pp199–207

Atkin, J. and Leisinger, K. M. (eds) (2000) *Safe and Effective Use of Crop Protection Products in Developing Countries*. CABI Publishing, London

Barrera, V. H., Norton, G. and Ortiz, O. (1998) *Manejo de las principales plagas y enfermedades de la papa por los agricultores en la provincia del Carchi, Ecuador*. INIAP, Quito, Ecuador

Barrera, V., Escudero, L., Norton, G. and Sherwood, S. (2001) *Validación y difusión de modelos de manejo integrado de plagas y enfermedades en el cultivo de papa: Una experiencia de capacitación participativa en la provincia de Carchi, Ecuador*. Revista INIAP, Quito, Ecuador, pp16, 26–28

Barsky, O. (1984) *Acumulación Campesina en el Ecuador: Los productores de papa del Carchi. Colección de Investigaciones*, No 1. Facultad Latinoamericana de Ciencias Sociales, Quito, Ecuador

Cole, D. C., Carpio, F., Julian, J. and León, N. (1997a) 'Dermatitis in Ecuadorian farm workers', *Contact Dermatitis – (Environmental and Occupational Dermatitis)*, vol 37, pp1–8

Cole, D. C., Carpio, F., Julian, J., León, N., Carbotte, R. and De Almeida, H. (1997b) 'Neurobehavioral outcomes among farm and non-farm rural Ecuadorians', *Neurotoxicology and Teratology*, vol 19, pp277–286

Cole, D. C., Carpio, F., Julian, J. and León, N. (1998a) 'Assessment of peripheral nerve function in an Ecuadorean rural population exposed to pesticides', *Journal of Toxicology and Environmental Health*, vol 55, pp77–91

Cole, D. C., Carpio, F., Julian, J. and León, N. (1998b) 'Health impacts of pesticide use in Carchi farm populations', in Crissman, C. C., Antle, J. M. and Capalbo, S. M. (eds) *Economic, Environmental and Health Tradeoffs in Agriculture: Pesticides and the Sustainability of Andean Potato Production*, CIP (International Potato Center), Lima, Peru and Kluwer Academic Publishers, Dordrecht /Boston/London, pp209–230

Cole, D. C., Carpio, F. and León, N. (2000) 'Economic burden of illness from pesticide poisonings in highland Ecuador', *Pan American Review of Public Health*, vol 8, pp196–201

Crissman, C. C., Antle, J. M. and Capalbo, S. M. (eds) (1998) *Economic, Environmental and Health Tradeoffs in Agriculture: Pesticides and the Sustainability of Andean Potato Production*. International Potato Center, Lima and Kluwer Academic Press, Boston, 280pp

Crissman, C. C., Cole, D. C. and Carpio, F. (1994) 'Pesticide use and farm worker health in Ecuadorian potato production', *American Journal of Agricultural Economics*, vol 76, pp593–597

Crissman, C. C., Yanggen, D., Antle, J., Cole, D., Stoorvogel, J., Barrera, V. H., Espinosa, P. and Bowen, W. (2003) 'Relaciones de intercambio existentes entre agricultura, medio ambiente y salud humana con el uso de plaguicidas', in Yanggen, D., Crissman, C. C. and Espinosa, P. (eds) *Plaguicidas: Impactos en producción, salud y medioambiente en Carchi, Ecuador*, CIP, INIAP, Ediciones Abya Yala, Quito, Ecuador, pp146–162

Espinosa, P., Crissman, C. C., Mera-Orcés, V., Paredes, M. and Basantes, L. (2003) 'Conocimientos, actitudes y practicas de manejo de plaguicidas por familias productoras de papa en Carchi', in Yanggen, D., Crissman, C. C. and Espinosa, P. (eds) *Los Plaguicidas. Impactos en producción, salud y medio ambiente en Carchi, Ecuador*, CIP, INIAP, Ediciones Abya-Yala, Quito, Ecuador, pp25–48

Fenske, R., Wong, S., Leffingwell, J. and Spear, R. (1986) 'A video imaging technique for assessing dermal exposure II. Fluorescent tracer testing', *American Industrial Hygiene Association Journal*, vol 47, pp771–775

Forget, G. and Lebel, J. (2001) 'An ecosystem approach to human health', *International Journal of Occupational and Environmental Health*, vol 7, ppS1–S38

Frolich, L. M., Sherwood, S., Hemphil, A. and Guevara, E. (2000) 'Eco-papas: Through potato conservation towards agroecology', *ILEA Newsletter*, December, pp44–45

Gallagher, K. D. (2000) 'Community study programmes for integrated production and pest management: Farmer Field Schools', in FAO, *Human Resources in Agricultural and Rural Development*, FAO, Rome, pp60–67

Herrera, M. (1999) *Estudio del subsector de la papa en Ecuador*. INIAP PNRT-Papa. Quito, Ecuador

Lander, B. F., Knudsen, L. E., Gamborg, M. O., Jarventaus, H. and Norppa, H. (2000) 'Chromosome aberrations in pesticide-exposed greenhouse workers', *Scandinavian Journal of Work, Environment & Health*, vol 26, pp436–442

Latour, B. (1998) 'To modernize or to ecologize? That's the question', in Castree, N. and Willems-Braun, B. (eds) *Remaking Reality: Nature at the Millennium*, Routledge, London and New York, pp221–242

LEISA (2003) Aprendiendo con las ECAs. LEISA: Revista de Agroecología, *Junio*, vol 14, p87

Mera-Orcés, V. (2000) *Agroecosystems Management, Social Practices and Health: A Case Study on Pesticide Use and Gender in the Ecuadorian Highlands*. A Technical Report to the IDRC. Canadian-CGIAR Ecosystem Approaches to Human Health Training Awards with a particular focus on gender

Merino, R. and Cole, D. C. (2003) 'Presencia de plaguicidas en el trabajo agrícola, en los productos de consumo, y en el hogar' in Yanggen, D., Crissman, C. C. and Espinosa, P. (eds) *Los Plaguicidas. Impactos en producción, salud y medio ambiente en Carchi, Ecuador*, CIP, INIAP, Ediciones Abya-Yala, Quito, Ecuador, pp71–93

Murray, D. L. and Taylor, P. L. (2000) 'Claim no easy victories: Evaluating the pesticide industry's global safe use campaign', *World Development*, vol 28, pp1735–1749

Norgaard, R. (1994) *History in Development Betrayed. The end of the progress and a coevolutionary revisioning of the future*, Routledge Press, London, pp280

Paredes, M. (2001) 'We are like the fingers of the same hand: Peasants' heterogeneity at the interface with technology and project intervention in Carchi, Ecuador'. MSc thesis. Wageningen University, Wageningen, the Netherlands

Paz-y-Mino, C., Bustamente, G., Sanchez, M. E. and Leone, P. E. (2002) 'Cytogenetic monitoring in a population occupationally exposed to pesticides in Ecuador', *Environmental Health Perspectives*, vol 110, pp1077–1080

Plog, B. A., Niland, J., Quinlan, P. J. and Plogg, H. (eds) (1996) *Fundamentals of Industrial Hygiene*, 4th edition, Ithaca, NY, National Safety Council

Pumisacho, M. and Sherwood, S. (eds) (2000) *Herramientas de Aprendizaje para Facilitadores. Manejo Integrado del Cultivo de Papa*, INIAP and CIP, Quito, Ecuador

Röling, N. (2000) 'Gateway to the global garden: Beta-gamma science for dealing with ecological rationality'. Eighth annual Hopper Lecture. University of Guelph, Canada, 24 October, www.uoguelph.ca/cip

Sherwood, S. and Pumisacho, M. (in press) *Guia Metodológica de Escuelas de Campo de Agricultores*, INIAP-CIP-FAO-WN

Sherwood, S., Crissman, C. and Cole, D. (2002) *Pesticide Exposure And Poisonings in the Northern Andes: A Call for International Action*, spring edition, Pesticide Action Network, UK

Stoorvogel, J. J., Antle, J. M., Crissman, C. C. and Bowen, W. (2004) 'The tradeoff analysis model: Integrated bio-physical and economic modeling of agricultural production systems', *Agricultural Systems*, vol 80, pp43–66

Uphoff, N. Esman, M. J. and Krishna, A. (1998) *Reasons for success: Learning from Instructive Experiences in Rural Development*, Kumarian Press, West Hartford, Connecticut

USEPA (1992) 'Ethylene bisdithiocarbamates (EBDCs); Notice of intent to cancel and conclusion of Special Review', *Federal Register*, vol 57, pp7434–7539

van der Ploeg, J. D. (1994) 'Styles of farming: An introductory note on concepts and methodology', in van der Ploeg, J. D. and Long, A. (eds) *Born from Within: Practice and Perspectives of Endogenous Rural Development*, Van Gorcum, Assen, pp7–31

World Development Index (2003) World Bank, Washington, DC, an on-line database available at www.worldbank.org [accessed 6 January 2004]

11

Agroecology and Agroecosystems

Stephen R. Gliessman

Agriculture is more than an economic activity designed to produce a crop or to make as large a profit as possible on the farm. A farmer can no longer pay attention to the objectives and goals for his or her farm only and expect to adequately deal with the concerns of long-term sustainability. Discussions about sustainable agriculture must go far beyond what happens within the fences of any individual farm. Farming is now viewed as a much larger system with many interacting parts, including environmental, economic and social components (Gliessman, 2001; Flora, 2001). It is the complex interaction and balance among all of these parts that has brought us together to discuss sustainability, to determine how to move toward this broader goal, and to learn how an agroecological perspective focused on sustainable agroecosystems is a way to achieve these long-term objectives.

Much of modern agriculture has lost the balance needed for long-term sustainability (Kimbrell, 2002). With their excessive dependence on fossil fuels and external inputs, most industrialized agroecosystems are overusing and degrading the soil, water, genetic and cultural resources upon which agriculture has always relied. Problems in sustaining agriculture's natural resource foundation can only be masked for so long by modern practices and high input technologies. In a sense, as we borrow ever-increasing amounts of water and fossil fuel resources from future generations, the negative impacts on farms and farming communities will continue to become more evident. The conversion to sustainable agroecosystems must become our goal (Gliessman, 2001).

In an attempt to clarify my own thinking about agroecosystems, I often think of agriculture as a stream, and farms are different points along that stream. When we think of an individual farm as a 'pool' in a calm eddy at some bend in the stream's flow, we can imagine how many things 'flow' into a farm, and we also expect that many things flow out of it as well. As a farmer, I work hard to keep my pool in the stream (my farm) clean and productive. I try to be as careful as possible in terms of how I care for the soil, which crops I plant, how I control pests and diseases, and how I market my harvest. Back in the days when there were fewer farms, fewer people to feed, and smaller demands on farmers and farmland, I could keep my farm in pretty good shape. I could keep my pool in the stream pretty clean and did not have to worry very much about what was going on 'downstream' from my farm.

Note: Reprinted from *Agroecosystem Analysis* by Richert, D. and Francis, C. (eds), Agronomy Monograph Series, copyright © (2004), with permission from the American Society of Agronomy.

But such a strategy has become much more difficult today. I find that I have less and less control over what comes into my pool. I face a variety of 'upstream impacts' that in combination can threaten the sustainability of my farm. These include the inputs into my farm that either I purchase or which arrive from the surrounding area. They include labour availability and cost, market access for what I produce, legislated policies that determine how much water I use, pesticides I apply, or how I care for my animals – not to mention the vagaries of the weather! My pool can become quickly muddied.

I must also increasingly consider how the way I take care of my pool can have 'downstream effects' in the stream below. Soil erosion and groundwater depletion can negatively affect other farms than my own. Inappropriate or inefficient use of pesticides and fertilizers can contaminate the water and air, as well as leave potentially harmful residues on the food that my family and others will consume. How well I do on my farm is reflected in the viability of rural farm economies, our local community, and society broadly. Key indicators are the losses of farmland to other activities and the loss of family farms in general. Both upstream and downstream factors are linked in complex ways, often beyond my control, and they impinge upon the sustainability of my farm.

The agroecology perspective

The agroecosystem

Any definition of sustainable agriculture must include how we examine the production system as an agroecosystem. We need to look at the entire system, the entire stream in the above analogy. This definition must move beyond the narrow view of agriculture that focuses primarily on the development of practices or technologies designed to increase yields and improve profit margins. These practices and technologies must be evaluated on their contributions to the overall sustainability of the farm system. The new technologies have little hope of contributing to sustainability unless the longer-term, more complex impacts of the entire agricultural system are included in the evaluation. The agricultural system is an important component of the larger food system (Francis et al, 2003).

A primary foundation of agroecology is the concept of the ecosystem, defined as a functional system of complementary relations between living organisms and their environment, delimited by arbitrarily chosen boundaries, which in space and time appears to maintain a steady yet dynamic equilibrium (Odum, 1996; Gliessman, 1998). Such an equilibrium can be considered to be sustainable in a definitive sense. A well-developed, mature natural ecosystem is relatively stable, self sustaining, recovers from disturbance, adapts to change, and is able to maintain productivity through using energy inputs of solar radiation alone. When we expand the ecosystem concept to agriculture and consider farm systems as agroecosystems, we have a basis for looking beyond a primary focus on traditional and easily measured system outputs (yield or economic return). We can instead look at the complex set of biological, physical, chemical, ecological and cultural interactions determining the processes that permit us to achieve and sustain yields.

Agroecosystems are often more difficult to study than natural ecosystems because they are complicated by human management, which alters normal ecosystem structures and functions. There is no disputing the fact that for any agroecosystem to be fully sustainable, a broad series of interacting ecological, economic and social factors and processes must be taken into account. Still, ecological sustainability is the building block upon which other elements of sustainability depend.

An agroecosystem is created when human manipulation and alteration of an ecosystem take place for the purpose of establishing agricultural production. This introduces several changes in the structure and function of the natural ecosystem (Figure 11.1) and resulting changes in a number of key system-level qualities. These qualities are often referred to as the emergent qualities or properties of systems, qualities that manifest themselves once all of the component parts of the system are organized. These same qualities can also serve as indicators of agroecosystem sustainability (Gliessman, 2001). Four key emergent qualities of ecosystems and how they are altered as they are converted to agroecosystems are discussed in the following sections.

Energy flow

Energy flows through a natural ecosystem as a result of complex sets of trophic interactions, with certain amounts being dissipated at different stages along the food chain, and with the greatest amount of energy within the system ultimately moving along the detritus pathway (Odum, 1971). Annual production of the system can be calculated in terms of net primary productivity or biomass, each component with its corresponding energy content. Energy flow in agroecosystems is altered greatly by human interference (Rappaport, 1971; Pimentel and Pimentel, 1997). Although solar radiation is obviously the major source of energy, many inputs are derived from human-manufactured sources and are most often not self sustaining. Agroecosystems too often become through-flow systems, with a high level of fossil fuel input and considerable energy directed out of the system at the time of each harvest. Biomass is not allowed to otherwise accumulate within the system or contribute to driving important internal ecosystem processes (eg organic detritus returned to the soil serving as an energy source for micro-organisms that are essential for efficient nutrient cycling). For sustainability to be attained, renewable sources of energy must be maximized, and energy must be supplied to fuel the essential internal trophic interactions needed to maintain other ecosystem functions.

Nutrient cycling

Small amounts of nutrients continually enter an ecosystem through several hydrogeochemical processes. Through complex sets of interconnected cycles, these nutrients then circulate within the ecosystem, where they are most often bound in organic matter (Borman and Likens, 1967). Biological components of each system become very important in determining how efficiently nutrients move, ensuring that minimal amounts are lost from the system. In a mature ecosystem, these small losses are replaced by local inputs, maintaining a nutrient balance. Biomass productivity in natural ecosystems is linked very closely to the annual rates at which nutrients are able to be recycled. In an agroecosystem, recycling of nutrients can be minimal, and considerable quantities are lost from the system with the harvest or as a result of leaching or erosion due to a great reduction in permanent biomass levels held within the system (Tivy, 1990). The frequent exposure of bare soil between crop plants during the season, or in open fields between cropping seasons, creates 'leaks' of nutrients from the system. Modern agriculture has come to rely heavily upon nutrient inputs derived or obtained from petroleum-based sources to replace these losses. Sustainability requires that these leaks be reduced to a minimum and recycling mechanisms be reintroduced and strengthened. Ultimately, human societies need to find ways to return nutrients consumed in agricultural products back to the fields, the agroecosystems that consumed and produced them in the first place.

Population regulating mechanisms

Through a complex combination of biotic interactions and limits set by the availability of physical resources, population levels of the various organisms are controlled, and thus eventually link to and determine the productivity of an ecosystem. Selection through time tends toward the establishment of the most complex structure biologically possible within the limits set by the environment, permitting the establishment of diverse trophic interactions and niche diversification. Due to human-directed genetic selection and domestication, as well as the overall simplification of agroecosystems (i.e. the loss of niche diversity and a reduction in trophic interactions), populations of crop plants or animals are rarely self reproducing or self regulating. Human inputs in the form of seed or control agents, often dependent on large energy subsidies, determine population sizes. Biological diversity is reduced, natural pest control systems are disrupted, and many niches or micro-habitats are left unoccupied. The danger of catastrophic pest or disease outbreak is high, often despite the availability of intensive human interference and inputs. A focus on sustainability requires the reintroduction of the diverse structures and species relationships that permit the functioning of natural control and regulation mechanisms. We must learn to work with and profit from diversity, rather than focus on agroecosystem simplification.

Dynamic equilibrium

The species richness or diversity of mature ecosystems permits a degree of resistance to all but very damaging perturbations. In many cases, periodic disturbances ensure the highest diversity, and even highest productivity (Connell, 1978). System stability is not a steady state, but rather a dynamic and highly fluctuating one that permits ecosystem recovery following disturbance. This promotes the establishment of an ecological equilibrium that functions on the basis of sustained resource use which the ecosystem can maintain indefinitely and which can even shift if the environment changes. At the same time, rarely do we witness what might be considered large-scale disease outbreaks in healthy, balanced ecosystems. With a reduction of natural structural and functional diversity, much of the resilience of the system is lost, and constant human-derived external inputs must be maintained. An overemphasis on maximizing harvest outputs upsets the former equilibrium and leads to a dependence on outside interference. To reintegrate sustainability, the emergent qualities of system resistance and resiliency must once again play a determining role in agroecosystem design and management.

We need to be able to analyse both the immediate and future impacts of agroecosystem design and management so we can identify the key areas in each system on which to focus the search for alternatives or solutions to problems. We must learn to be more competent in our agroecological analysis in order to avoid problems or negative changes before they occur, rather than struggling to reverse the problems after they have been created. The agroecological approach provides us one such alternative (Altieri, 1995; Gliessman, 1998).

Applying agroecology

The process of understanding agroecosystem sustainability has its foundations in two kinds of ecosystems: natural ecosystems and traditional (also known as local or indigenous) agroecosystems. Both provide ample evidence of having passed the test of time in terms of long-term productive ability, but each offers a different knowledge base from which to understand this ability. Natural ecosystems are reference systems for understanding the ecological basis for sustainability in a particular location. Traditional agroecosystems provide many examples of how a culture and its local environment have coevolved with time through processes that balance the needs of people, expressed as ecological, technological, and socio-economic factors. Agroecology, defined as the application of ecological concepts and principles to the design and management of sustainable agroecosystems (Gliessman, 1998), draws on both to become a research approach that can be applied to converting unsustainable and conventional agroecosystems into sustainable ones.

Natural ecosystems reflect a long period of evolution in the use of local resources and adaptation to local ecological conditions. They have each become complex sets of plants and animals that coinhabit a given environment, and as a result, provide extremely useful information for the design of more locally adapted agroecosystems. As I have suggested (Gliessman, 1998), 'the greater the structural and functional similarity of an agroecosystem to the natural ecosystems in its biogeographical region, the greater the likelihood that the agroecosystem will be sustainable'. If this suggestion holds true, natural ecosystem structures and functions can be used as benchmarks or threshold values for more sustainable systems. Scientists have begun to explore how an understanding of natural ecosystems can be used to guide our search for sustainable agroecosystems that respect and protect the environment and natural resources (Soule and Piper, 1992; Jackson and Jackson, 2002).

Traditional and indigenous agroecosystems are different from conventional systems in that they developed originally in times or places where inputs other than human labour and local resources were generally not available or desirable to the local people. Production takes place in ways that demonstrate people's concerns about long-term sustainability of the system, rather than solely maximizing output and profit. Traditional systems continue to be important as the primary sources of food production for a large part of the populations of many developing countries, while at the same time maintaining their foundations in ecological knowledge (Wilken, 1988; Altieri, 1990). This reality demonstrates their importance for the development of sustainable agroecosystems. This is especially true today when so many modern conventional agroecosystems have caused severe degradation of their ecological foundations, as socioe-conomic factors have become the predominant forces in the food system (Altieri, 1990). Many traditional agroecosystems are actually very sophisticated examples of the application of ecological knowledge, and can serve as the starting point for the conversion to more sustainable agroecosystems in the future. The traditional Mesoamerican intercrop of corn *(Zea mays* L.), bean, and squash is a well-known cropping system where higher yields in the mixtures come about due to a complex of interactions among components of the agroecosystem (Amador and Gliessman, 1990). Examples of such interactions range from the increased presence of beneficial insects due to attractive microclimates and a greater abundance of pollen and nectar sources (Letourneau, 1986), to biologically fixed nitrogen being made available to corn through mycorrhizal fungi connections with roots of bean (Bethlenfalvay et al, 1991).

How can agroecology link our understanding of natural ecosystem structure and function with the knowledge inherent in traditional agroecosystems? On the one hand, the knowledge of place that comes from understanding local ecology is an essential foundation. Another is the local experience with farming that has its roots in many generations of living and working within the limits of that place. We put both of these approaches together when we work with farmers going through the transition process to more environmentally sound management practices, and thus realize the potential for contributing to long-term sustainability. This transition is already occurring. Many farmers, despite the heavy economic pressure on agriculture, are in the process of converting their farms to more sustainable design and management (National Research Council, 1989; OAC/SCOAR, 2003). In California the dramatic increase in organic acreage for a range of crops has been based largely on farmer innovation (Swezey and Broome, 2000). It is incumbent that agroecologists play an important role in contributing to this conversion process.

Converting an agroecosystem to a more sustainable design is a complex process. It is not just the adoption of a new practice or a new technology. There are no silver bullets. Instead, this conversion uses the agroecological approach described above. The farm is perceived as part of a larger system of interacting parts, an agroecosystem. We must focus on redesigning that system in order to promote the functioning of an entire range of different ecological processes (Gliessman, 1998). In a study of the conversion of conventional strawberries (*Fragaria Ananassa* Rozier) to organic management, several changes were observed (Gliessman et al, 1996). As the use of synthetic chemical inputs was reduced or eliminated and recycling was emphasized, agroecosystem structure and function changed as well. A range of processes and relationships began to transform, beginning with improvement in basic soil structure, an increase in soil organic matter content, and greater diversity and activity of beneficial soil biota. Major changes began to occur in the activity and relationships among weed, insect and pathogen populations, and in the functioning of natural control mechanisms. For example, predatory mites gradually replaced the use of synthetic acaricides for the control of two-spotted spider mites (*Tetranychus urticae* Koch), the most common arthropod pest in strawberries in California.

Ultimately, nutrient dynamics and cycling, energy use efficiency, and overall agroecosystem productivity are affected. Changes may be required in day-to-day management of the farm, planning, marketing and even philosophy. The specific needs of each agroecosystem will vary, but the principles for conversion listed in Table 11.1 can serve as general guidelines for working through the transition. It is the role of the agroecologist to help the farmer measure and monitor these changes during the conversion period in order to guide, adjust and evaluate the conversion process. Such an approach provides an essential framework for determining the requirements for and indicators of sustainable agroecosystem design and management.

Comparing ecosystems and agroecosystems

The key to developing sustainability is building a strong ecological foundation under the agroecosystem, using the ecosystem knowledge inherent to agroecology as discussed above. This foundation then serves as the framework for producing the sustainable

Table 11.1 *Guiding principles for the process of conversion to sustainable agroecosystems design and management*

- Shift from through-flow nutrient management to recycling of nutrients, with increased dependence on natural processes, such as biological N fixation and mycorrhizal relationships
- Use renewable sources of energy instead of non-renewable sources
- Eliminate the use of non-renewable off-farm human inputs that have the potential to harm the environment or the health of farmers, farm workers or consumers
- When materials must be added to the system, use naturally occurring materials instead of synthetic, manufactured inputs
- Manage pests, diseases and weeds instead of 'controlling' them
- Reestablish the biological relationships that can occur naturally on the farm instead of reducing and simplifying them
- Make more appropriate matches between cropping patterns and the productive potential and physical limitations of the farm landscape
- Use a strategy of adapting the biological and genetic potential of agricultural plant and animal species to the ecological conditions of the farm rather than modifying the farm to meet the needs of the crops and animals
- Value most highly the overall health of the agroecosystem rather than the outcome of a particular crop system or season
- Emphasize conservation of soil, water, energy and biological resources
- Incorporate the idea of long-term sustainability into overall agroecosystem design and management

Source: modified from Gliessman, 1998

harvests needed by humans. In order to maintain sustainable harvests, though, human management is a requirement. Agroecosystems are not self sustaining, but rely on natural processes for maintenance of their productivity. An agroecosystem's resemblance to natural ecosystems allows the system to be sustained, in spite of the long-term human removal of biomass, without large subsidies of non-renewable energy and without detrimental effects on the surrounding environment.

Table 11.2 compares natural ecosystems with three types of agroecosystems in terms of several ecological criteria. Traditional agroecosystems most closely resemble natural ecosystems, since they most often are focused on the use of locally available and renewable resources, local use of agricultural products, and the return of biomass to the farming system. Sustainable agroecosystems are very similar in many properties, but they are more dissimilar in others because of the probable focus on export of harvest to distant markets, the need to purchase a significant part of their nutrients externally, and the much stronger impact of market systems on agroecosystem diversity and management. Compared with conventional systems, sustainable agroecosystems have somewhat lower and more variable yields due to the weather variation that occurs from year to year. Such reductions in yields can be more than offset, from the perspective of sustainability, through the advantages gained in reduced dependence on external inputs, more reliance on natural controls of pests, and reduced negative off-farm impacts of farming activities.

Table 11.2 *Emergent properties of natural ecosystems, traditional agroecosystems, conventional agroecosystems and sustainable agroecosystems. Agroecosystem properties are most applicable to the farm scale and for the short- to medium-term time frame*

Emergent ecological property	Natural ecosystem	Agroecosystem type		
		Traditional	Conventional	Sustainable
Productivity (process)	medium	medium	low/med	med/high
Species diversity	high	med/high	low	medium
Structural diversity	high	med/high	low	medium
Functional diversity	high	med/high	low	med/high
Output stability	medium	high	low/med	high
Biomass accumulation	high	high	low	med/high
Nutrient recycling	high	high	low	high
Tropic relationships	high	high	low	med/high
Natural population regulation	high	high	low	med/high
Resistance	high	high	low	medium
Resilience	high	high	low	medium
Dependence on external human inputs	low	low	high	medium
Autonomy	high	high	low	high
Human displacement of ecological processes	low	low	high	low/med
Sustainability	high	med/high	low	high

Source: Modified from Odum (1984), Conway (1985), Altieri (1995), and Gliessman (1998)
Note: Agroecosystem properties are most applicable to the farm scale and for the short- to medium-term time frame.

Future perspectives

Problems in agriculture create the pressures for the changes that will bring about a sustainable agriculture. However, it is one thing to express the need for sustainability, and quite another to actually quantify it and bring about the changes that are required. Designing and managing sustainable agroecosystems, as an approach, is in its formative stages. Initially it builds upon the fields of ecology and agricultural science and is emerging as the science of agroecology. This combination can play an important role in developing the understanding necessary for a transition to sustainable agriculture.

But sustainable agriculture is more. It takes on a cultural perspective as the concept expands to include humans and their impacts on agricultural environments. Agricultural systems are a result of the coevolution that occurs between culture and environment, and a sustainable agriculture values the human as well as the ecological components.

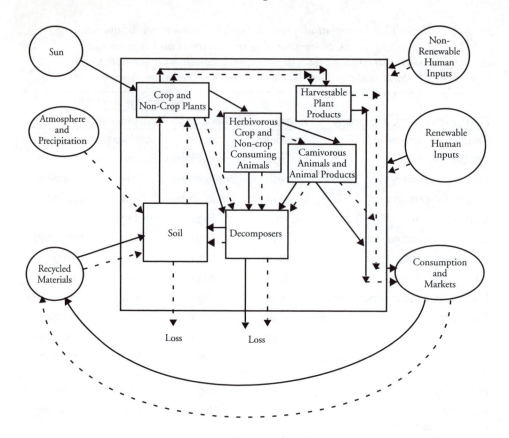

Note: Solid lines are energy flow, and dotted lines are nutrient cycles. This model assumes that nutrients and leftover energy are returned to the agroecosystem as reusable materials, and that the use of non-renewable human inputs is minimized.

Figure 11.1 *Functional and structural components of an ecosystem converted to a sustainable agroecosystem*

Our small pool in the stream becomes the focal point for changing how we do agriculture, but that change must occur in the context of the human societies within which agriculture is practiced, the whole stream in our analogy.

All agricultural systems can no longer be viewed as strictly production activities driven primarily by economic pressures. We need to reestablish an awareness of the strong ecological foundation upon which agriculture originally developed and ultimately depends. Too little importance has been given to the 'downstream' effects that are manifest off the farm, either by surrounding natural ecosystems or by human communities. We need an interdisciplinary basis upon which to evaluate these impacts.

In the broader context of sustainability, we must study the environmental background of the agroecosystem, as well as the complex of processes involved in the maintenance of long-term productivity. We must first establish the ecological basis of sustainability in terms of resource use and conservation, including soil, water, genetic resources and air quality. Then we must examine the interactions among the many organisms of the agroecosystem, beginning with interactions at the individual species level and culminating at the ecosystem level as our understanding of the dynamics of the entire system is revealed.

Our understanding of ecosystem-level processes should then integrate the multiple aspects of the social, economic and political systems within which agroecosystems function, making them even more complex systems. Such an integration of ecosystem and social system knowledge about agricultural processes will not only lead to a reduction in synthetic inputs used for maintaining productivity; it will also permit the evaluation of such qualities of agroecosystems as the long-term effects of different input-output strategies, the importance of the environmental services provided by agricultural landscapes, and the relationship between economic and ecological components of sustainable agroecosystem management. By properly selecting and understanding the 'upstream' inputs into agriculture, we can be assured that what we send 'downstream' will promote a sustainable future.

Acknowledgements

The author is extremely grateful for support provided by the Alfred Heller Endowed Chair for Agroecology. An early version of this chapter was rigorously discussed in a Kellogg Foundation National Fellowship Forum, and the valuable input from the Fellows in Group VI is warmly acknowledged. Detailed edits from Diane Rickerl, Chuck Francis, and an anonymous reviewer are also much appreciated. Joji Muramoto graciously assisted with the formatting of Figure 11.1.

References

Altieri, M. A. (1990) 'Why study traditional agriculture?' in Carroll, R. C. et al (eds) *Agroecology*, McGraw-Hill, New York, pp551–564

Altieri, M. A. (1995) *Agroecology: The scientific Basis of Alternative Agriculture*. 2nd ed, Westview Press, Boulder, Colorado

Amador, M. F. and Gliessman, S. R. (1990) 'An ecological approach to reducing external inputs through the use of intercropping', in Gliessman, S. R. (ed) *Agroecology: Researching the ecological basis for sustainable agriculture*, Springer-Verlag, New York, pp146–154

Bethlenfalvay, G. J., Reyes-Solis, M. G., Cametyand R. and Ferrera-Cerrato, S. B. (1991) 'Nutrient transfer between the root zones, of soybean and maize plants connected by a common mycorrhizal inoculum', *Physiol. Plant*, vol 82, pp423–432

Borman, F. H. and Likens, G. E. (1967) 'Nutrient cycles', *Science*, vol 155, pp424–429

Connell, J. H. (1978) 'Diversity in tropical rain forests and coral reefs', *Science*, vol 199, pp1302–1310

Conway, G. R. (1985) 'Agroecosystem analysis', *Agric. Admin*, vol 20, pp31–55

Flora, C. (ed) (2001) 'Interactions between agroecosystems and rural communities', *Advances in Agroecology*, CRC Press, Boca Raton, Florida

Francis, C., Lieblein, G., Gliessman, S. T., Breland, A., Creamer, N., Harwood, R., Salomonsson, L., Helenius, J., Rickerl, D., Salvador, R., Wiendehoeft, M., Simmons, S., Allen, P., Altieri, M. Porter Flora, J. and Poincelot, R. (2003) 'Agroecology: The ecology of food systems', *Journal of Sustainable Agriculture*, vol 22, pp99–119

Gliessman, S. R. (1998) *Agroecology: Ecological Processes in Sustainable Agriculture*, Lewis/CRC Press, Boca Raton, Florida

Gliessman, S. R. (ed) (2001) 'Agroecosystem sustainability: Toward practical strategies', *Advances in Agroecology*, CRC Press, Boca Raton, Florida

Gliessman, S. R., Werner, M. R., Swezey, S., Caswell, E., Cochran, J. and Rosado-May, F. (1996) 'Conversion to organic strawberry management changes ecological processes', *Californian Agriculture*, vol 50, pp24–31

Jackson, D. L. and Jackson, L. L. (2002) *The Farm as Natural Habitat*, Island Press, Washington, DC

Kimbrell, A. (ed) (2002) *Fatal Harvest: The Tragedy Of Industrial Agriculture*, Island Press, Washington, DC

Letourneau, D. K. (1986) 'Associational resistance in squash monoculture and polycultures in tropical Mexico', *Environmental Entomology*, vol 15, pp285–292

National Research Council (1989) *Alternative Agriculture*, National Academic Press, Washington, DC

Odum, E. P. (1971) *Fundamentals of Ecology*, W.B. Saunders, Philadelphia, Pennsylvania

Odum, E. P. (1984) 'Properties of agroecosystems', in Lowrance, R. et al. (eds) *Agricultural Ecosystems: Unifying Concepts*, John Wiley & Sons, New York, pp5–12

Odum, E. P. (1996) *Ecology: Bridging Science and Society*, Sinauer Associates Inc., Sunderland, Massachusetts

Organic Agriculture Consortium (OAC)/Scientific Congress on Organic Agriculture Research (SCOAR) (2003) *Organic Agricultural Information*. Econ. Res. Serv. Issues Center, Washington, DC, Available at www.organicaginfo.org

Pimentel, D. and Pimentel, M. (ed) (1997) *Food, Energy and Society*. 2nd ed, University Press of Colorado, Niwot

Rappaport, R. A. (1971) 'The flow of energy in an agricultural society', *Scientific American*, vol 224, 117–132

Soule, J. D., and Piper, J. K. (1992) *Farming in Nature's Image*, Island Press, Washington, DC

Swezey, S. L. and Broome, J. (2000) 'Growth predicted in biologically integrated and organic farming', *Californian Agriculture*, vol 54, pp26–35

Tivy, J. (1990) *Agricultural Ecology*, Longman Scientific and Technical, London

Wilken, G. C. (1988) *Good Farmers: Traditional Agricultural Resource Management in Mexico and Central America*, University of California Press, Berkeley.

12

The Doubly Green Revolution

Gordon Conway

Creative and ambitious measures must be taken to shatter the deeply ingrained uncritical and dependent attitude towards pesticides which prevails at all levels in developing countries, from ministries to the smallest farms. (Patricia Matteson, Kevin Gallagher and Peter Kenmore, 'Extension of integrated pest management for planthoppers in Asian irrigated rice: empowering the user' [1].

Pests, pathogens and weeds are the most visible of threats to sustainable food production.[2] Just how much crop and livestock loss they cause is largely guesswork; estimates range between 10 and 40 per cent. But in some situations the potential losses can be considerably higher. Much depends on the nature of the crop: where a premium is placed on the quality of the harvested product – for example, cotton, or fruits or vegetables – even a small pest or pathogen population can cause the farmer serious financial loss. Grain crops are not in this category, but the intensity of cultivation of the new varieties encourages severe pest and pathogen attack, on occasion resulting in total destruction of the crop.

Since World War II, the common approach to pest, pathogen and weed problems has been to spray crops with pesticides (insecticides, nematocides, fungicides, bactericides and herbicides). Quite apart from the hazards they pose to human health and wildlife, they are frequently costly and inefficient. This has been especially true of the modern insecticides: they have to be repeatedly sprayed if control is to be maintained, insect pests commonly become resistant to them and, as ecological research has shown, they can make the problem worse by killing off the natural enemies – the parasites and predators – which normally control pests.[3]

I first encountered the problems pesticides can cause when I was working as an ecologist in north Borneo (later the state of Sabah, Malaysia) in 1961. Cocoa was a recently introduced crop being grown in large, partial clearings in the primary forest. When I arrived the crop was being devastated by pests: cocoa loopers and bagworms were stripping off all the leaves, cossid borers were destroying the branches, ring-bark borers were killing whole trees, and a pest new to science, the bee bug, was damaging the cocoa pods.[4] At the time the cocoa fields were being heavily and repeatedly sprayed with insecticides, sometimes consisting of cocktails of organochlorines, such as DDT and dieldrin. They were having little effect. On the contrary, I believed they were making the

Note: Reprinted from *The Doubly Green Revolution* by Conway, G., copyright © (1997) with permission from the author and Penguin Books

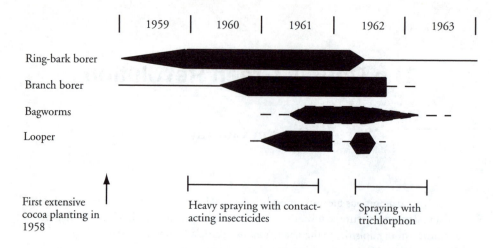

Figure 12.1 *The control of cocoa pests in North Borneo, 1961*[5]

situation worse. In their natural forest home the pest species were probably being controlled by a variety of natural enemies and, it seemed to me, the problem was being caused by the pesticides, which, being unselective in their action, were killing off the natural enemies.

At my recommendation all spraying was stopped. Two of the pests, the branch borer and the cocoa looper, soon came under control by parasitic wasps. The bagworms continued to cause damage and they were controlled by use of a highly selective pesticide before eventually being naturally controlled by a parasitic fly. The ring-bark borer was largely eliminated by destroying a secondary forest tree that had remained in the fields and was the borer's natural host. Very selective spraying kept the bee bug in check. Within a year all the major pests were being satisfactorily controlled, and this has persisted to the present day (Figure 12.1).

Since the 1960s the broad-spectrum organochlorine insecticides have been replaced by more selective compounds, which also tend to be less damaging to wildlife and human health.[6] Increasingly stringent regulations in the developed countries have forced manufacturers to engage in exhaustive safety and environmental testing, both in the laboratory and in natural field conditions.[7] New pesticides were discovered in the past by a largely random process of screening thousands of synthetic compounds. Now, with the greater understanding conferred by modern cellular and molecular biology, chemical companies have begun to search for tailor-made pesticides. One group are compounds that mimic the effects of juvenile hormones in insects, disrupting the transition from one life cycle stage of an insect to another, for example preventing caterpillars from becoming moths. They are valuable because they often only affect one species of insect. Another successful group of insecticides is based on the bacterium, *Bacillus thuringiensis*. When ingested by a caterpillar feeding on a sprayed leaf, a toxic protein is released which paralyses the caterpillar's gut and mouthparts, causing it to die. Since natural parasites and predators do not feed on the sprayed leaves they are unaffected. This property of *Bacillus thuringiensis* has been exploited in genetic engineering.

There is also growing interest in natural plant compounds that have been tradition-ally used by farmers for pest control. They include custard apple, turmeric, croton oil tree, Simson weed, castor oil, ryania and chilli pepper.[8] The pyrethroids, based on the compound pyrethrum found in chrysanthemum plants, are effective against certain pests and are very safe. One of the best-known sources of a natural insecticide is the neem tree, which has been used against rice pests in India for centuries.[9] The bitter com-pound, azardirachtin, contained in the seed acts as an anti-feedant, making crops unpal-atable to pests. It does not harm birds or mammals or beneficial insects such as honey bees. Unfortunately, it degrades fairly rapidly in sunlight. An effective formulation which prevents azardirachtin from degrading is on the market, but is considerably more costly than the natural product.[10]

An alternative, or a complement, to using selective pesticides is to encourage the natural enemies of pest directly. Sometimes, although rarely, this can be spectacularly suc-cessful. A recent example is the biological control of the cassava mealybug in Africa.[11] The mealybug first appeared in the Congo and Zaire in 1973, but soon spread across a wide belt of central Africa, from Mozambique to Senegal, producing yield losses of up to 80 per cent. Cassava originated in South America and a search was made there for the mealybug's natural enemies. A parasitic wasp was found in Paraguay and released in Nigeria in 1981. The results were dramatic, with yield increases of up to 2–5 t ha^{-1}; overall benefits are estimated in terms of billions of dollars. Plant pathogens can also be controlled by their 'enemies', organisms which act as antagonists.[12] In the US, commer-cially available *Agrobacterium radiobacter* (K84) produces an antibiotic which prevents the growth of the pathogen causing crown galls.[13] Biological control has also sometimes been effective against weeds, by releasing herbivorous insects – such as leaf-eating bee-tles. But more promising is the use of the phenomenon of 'allelopathy'. Certain plants release compounds which are harmful to weeds. One approach may be to introduce the allelopathic genes into crop plants; another may be to synthesize the toxic allelopathic compounds.[14]

Often natural enemies of pests can be encouraged by creating a more diverse agroe-cosystem. There are very few pest problems in Javanese home gardens. The diversity of plants in each garden encourages a diversity of insects which, in turn, supports a large population of general predators – spiders, ants, assassin bugs – that keep potential pests under control. Sometimes even growing a mixture of two crops is enough.[15] In the Phil-ippines, intercropping of maize and peanuts helps to control the maize-stem borer. The predator is a spider which, as an adult, feeds on the stem borer caterpillars. But the young spiders feed on springtails and these they find in the leaf litter under the peanut plants. The simple intercrop is sufficient to create a complex and beneficial food web. Often the mechanism of control is subtle. Aromatic odours from the intercropping of cabbages and tomatoes repel the diamondback moth; the shading effect of mung bean or sweet potato grown with maize reduces weed growth; and the liberation and spread of pathogen inocula can be reduced by growing cowpeas with maize.[16]

The move towards large areas of monoculture has been one of the reasons why pest and disease outbreaks have grown in the wake of the Green Revolution.[17] There have been other factors. Rice-stem borer and sheath blight attacks have increased as a result of higher nitrogen applications and leaf disease is more prevalent in the microclimate created by the densely leaved, short-strawed wheats and rices (although, it should be noted, fertilizers increase resistance to rice tungro virus, while irrigation reduces losses to rice blast).[18] The narrow genetic stock of the new varieties has also been a contributory factor, as has the misuse of pesticides.

Pests and pathogens, in common with all organisms, have the ability to adapt, through natural selection, to new situations. Michael Loevinsohn, working in the Philippines, has shown the remarkable capacity of rice pests to evolve in response to variation in the timing of rice cultivation. Within a few years genetically different populations of pests have arisen, each adapted to rice-cropping patterns separated by only a few kilometres. At Mapalad, at the base of the Sierra Madre Mountains in Luzon, where a single rain-fed crop is grown, populations of the yellow stem borer had a shorter generation time than did populations at Zaragoza, 10km away in the centre of the irrigated plain where two crops are grown. They also laid more eggs and had a lower survival rate. Planting is carried out more or less at the same time at Mapalad and the crop matures uniformly; in these conditions there is a selective advantage for pests that mature quickly and increase rapidly in numbers. By contrast, under irrigated double-cropping the planting is asynchronous and the pests are more heavily attacked by natural enemies – predators and parasites. In these circumstances it is advantageous, for the pests, to mature more slowly but have a higher survival rate.

Not surprisingly, pest and pathogen populations responded very rapidly to the continuous cropping of the new varieties of wheat and rice. The first ten years of double-cropping of rice at the Independent Research Institute (IRRI) in the Philippines resulted in dramatic growth in pest populations (Figure 12.2). The numbers increased directly because of the introduction of a dry-season rice crop; but there were more pests and more damage on the wet season crop as well. Thirteen per cent of the wet season crop was lost under single-cropping, but this rose to 33 per cent when double-cropping was introduced.[19] Under triple-cropping the numbers and damage were even higher. Only where there is a break in cultivation, such as a fallow, or where a cereal is alternated with a dissimilar crop, are pests and diseases held in check.

Pests and pathogens are also capable of evolving rapid resistance to threats and adverse circumstances, in particular to the use of pesticides.[20] By the mid-1980s some 450 pest species in the world were resistant to one or more insecticides and about 150 fungi and bacteria were resistant to or tolerated fungicides. Nearly 50 weed species were resistant to herbicides. Several important insect pests are resistant to all the major classes of insecticides: the diamondback moth, a pest of cabbages and other crucifers, is resistant in Malaysia not only to the older organochlorines and carbamates but also to the newer organophosphates and pyrethroids.[21]

Pests and pathogens are equally adept at evolving ways of overcoming the defences which naturally occur in crop plants or are bred into them by plant breeders. In 1950 a new race of wheat-stem rust suddenly exploded in the US and southern Canada, and was carried by high winds into Mexico.[22] This was only the first of a series of epidemics. New races continued to arrive and by 1960 a group of virulent races had almost completely replaced the existing, relatively weak forms. The wheat-breeding programme was able to keep pace with this changing pattern of disease but only by virtue of having new resistant varieties quickly available when each change in race occurred. A similar situation arose when new biotypes of the brown planthopper (BPH) suddenly appeared on rice in Southeast Asia in the 1970s and 1980s.

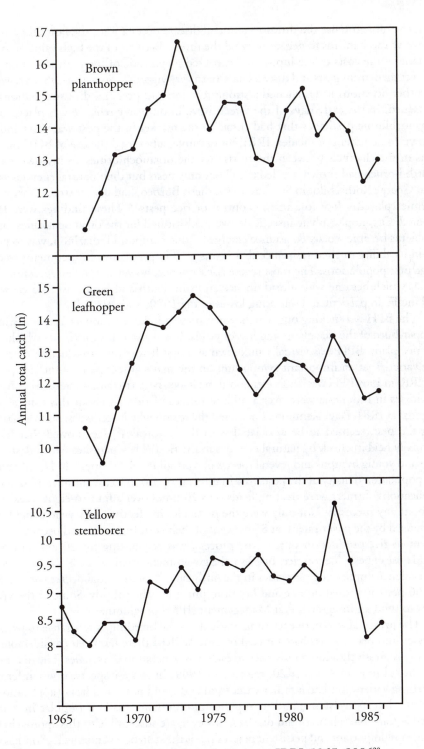

Figure 12.2 *Numbers of three rice pests at IRRI, 1965–1986*[23]

The more intense the threat to pest and disease populations, the faster they are able to evolve mechanisms to negate or avoid the threat. Because of the high value of the new varieties – the costs of the inputs and expected magnitude of the returns – attempts to protect them from pests and diseases have been very vigorous. Sometimes this has worked; on other occasions it has elicited a strong reaction and pest and disease situations have worsened. In the early days of the Green Revolution heavy reliance was placed on the organochlorine pesticides that had come on the market in the post-war years and were known to be potent pesticides. IRRI, for example, advocated the use of BHC placed in cans in the channels between the fields. Yet the organochlorines, as my own work in North Borneo had shown, tended to kill not only pests but their natural enemies as well. And George Rothschild, in Sarawak in southern Borneo, had demonstrated that natural enemies played a key role in the control of rice pests.[24] These findings were largely ignored. Organophosphate insecticides were substituted for the organochlorines, but the problems became worse. Resistance emerged – one compound, diazinon, was so powerful in its ability to elicit resistance that it was used in pest-control experiments to create large pest populations. The most severe consequence, however, was the epidemic of the BPH, which became widespread on rice crops in South and Southeast Asia causing in Indonesia, in particular, devastating losses in the 1970s and 1980s.[25]

The BPH is a sucking bug which, when present in large numbers, causes distinctive hopper-burn of the rice plants and loss of yields. It will also transmit viruses which attack the rice plant. BPH was virtually unknown as a pest before the introduction of the new rice varieties; at an important symposium on the major insect pests of the rice plant at the IRRI in 1964 BPH was hardly mentioned. Yet it soon began to cause severe damage. By 1977 the losses in Indonesia were over a million tonnes of rice. At about this time a young scientist at IRRI, Peter Kenmore, uncovered the reason why insecticides, far from controlling the pest, seemed to be associated with the outbreaks.[26] He showed that BPH is normally held in check by natural enemies in the ricefields – parasites which destroy the eggs and young nymphs and several species of wolf spider planthopper. In North Sumatra the population density of the pests rose in direct proportion to the number of insecticide applications; farmers were treating fields 6 to 20 times over a four-to-eight-week period without any success.[27] Not only were the pesticides ineffective, they were being heavily subsidized by the government, at 85 per cent of their cost. Indonesia had become self-sufficient in rice production but, in the process, was accounting for 20 per cent of the world's use of pesticides on rice. In 1986, the government acted on the basis of the mounting evidence implicating pesticides in the outbreaks. A presidential decree banned 57 of the 66 pesticides used on rice and began to phase out the subsidy. Some of the savings went to fund an Integrated Pest Management (IPM) programme.

The principal alternative to using pesticides – the breeding of resistance to pests and diseases into plants – has had a mixed record. At IRRI there has been considerable success in progressively adding resistance to each new generation of varieties. The first variety with broad resistance was IR20, released in 1969. In some respects, it was inferior to preceding varieties; it had a rather weak stem and could not stand heavy applications of fertilizer, but its resistance and its superior grain quality made it popular in Asia for over 15 years. Not all pests and diseases, however, are susceptible to this approach. One of the most important and persistent pests of rice is the stem borer and much effort has been devoted to producing resistant varieties, but so far with little success. The more serious defect of this strategy is the likelihood of a breakdown in resistance and the need for breeders always to keep one step ahead. IR20 eventually fell from favour because it was susceptible to the brown planthopper. The Indonesian BPH outbreak of 1977 was initially tackled by introducing new resistant rice variety, IR26, but within three seasons it had failed and losses in 1979 were again very severe. Next to be introduced was IR36. It was more successful, and was rapidly adopted. By 1984 Indonesia had become self-sufficient in rice production. But the resistance of IR36 was also short-lived. By 1986 the planthoppers had exploded to the levels of 1977, threatening over 50 per cent of Java's riceland. Losses in 1986/7 were estimated to be nearly US$400 million.[28] One explanation of this sequence that BPH exists in a number of different biotypes or races:

- The original resistance in the variety IR26 had been to biotype 1;
- Then biotype 2 emerged, to which IR36 was resistant. It kept its resistance from 1977 to 1982;
- Then in 1983 a new biotype (biotype 3) invaded Sumatra and attacked IR36;
- IR56 was then introduced and was resistant to all the biotypes.[29]

However, Peter Kenmore and his colleagues believe this is too simplistic an explanation.[30] In their experience, planthopper populations are extremely variable and can rapidly evolve to local circumstances (as Michael Loevinsohn has also shown for the rice-stem borer, see above). They argue that the heavy pesticide spraying accelerates the adaptation of BPH to new rice varieties. In their view plant-breeding approaches are only sustainable if they are a part of an integral strategy.

The approach plant breeders have taken in attempting to create resistant cereal varieties has been to seek out and introduce single, major resistant genes. These effectively confer immunity against one or, at most, a few strains or races of the pest or disease.[31] It has proven to be a very effective strategy. One variety, the Mexipak wheat, Sonalika, developed in Pakistan from the Mexican varieties, stood up to major rust attack in South Asia for over 20 years. But it can be risky if there is a likelihood of a new virulent pest or disease arising. Sonalika's resistance to rust is now breaking down and new races of rice blast are producing repeated breakdown of resistance in the new rice varieties.[32] Resistance to blast usually breaks down in two to three years. This is not a problem if new sources of resistance can be found quickly. When IR20, originally resistant to tungro virus, was wiped out by a new strain in the Philippines in 1972, a new resistant variety, IR26, was produced by IRRI within a year.

The alternative strategy is to build up a combination of genes each of which contributes only a partial degree of resistance. It is slower to achieve; accumulating the necessary genes from different parents can take 10 to 12 years.[33] However, because the pest or pathogen is being resisted in a number of ways and is rarely completely controlled, there is less likelihood of a new, more virulent strain emerging. As experience over the past 30 years has shown, the greater the apparent success in achieving pest or disease control in the short term, the greater the likelihood of a serious breakdown. In the longer term it is better to live with low levels of pest attack, utilizing a diversity of approaches to keep on top of the problem.

Pest and pathogen control has been, and to some extent still is, a hit-and-miss affair.[34] Usually the first response is to try a pesticide; if this does not work or causes further problems, a different pesticide or an alternative method is tried. And so the process continues. Often what works for one pest on a crop will not work for another, or may actually make the other pest problem worse. Professionals in crop protection have long recognized the problem and since the 1950s have been developing a systematic approach to pest control that goes under the name of IPM.[35] It looks at each crop and pest situation *as a whole* and then devises a programme that integrates the various control methods in the light of all the factors present. As practised today it combines modern technology, the application of synthetic, yet selective, pesticides and the engineering of pest resistance, with natural methods of control, including agronomic practices and the use of natural predators and parasites. As I demonstrated in one of the first applications of IPM in the developing countries, the control of cocoa pests in Sabah, the outcome is sustainable, efficient pest control that is often cheaper than the conventional use of pesticides.

A recent, highly successful example is IPM developed for the brown planthopper and other rice pests in Indonesia. Under the programme, farmers are trained to recognize and regularly monitor the pests and their natural enemies. They then use simple, yet effective, rules to determine the minimum necessary use of pesticides. The outcome is a reduction in the average number of sprayings from over four to under one per season, while yields have grown from 6 to nearly 7.5 tonnes per hectare. Since 1986 rice production has increased 15 per cent while pesticide use has declined 60 per cent, saving $120 million a year in subsidies. The total economic benefit to 1990 was estimated to be over $1 billion.[36] The farmers' health has improved, and a not insignificant benefit has been the return of fish to the ricefields.

In many ways, pest control is like a multidimensional game of chess. We pit ourselves against a variety of pests, drawing on a range of methods of control; the pests respond by evolving new defences. As we have learnt, it is not an unequal contest – there is rarely a final check-mate. Sustainable pest control depends on developing new strategies and tactics, in a continuing game. A useful way to envisage the contest is to characterize pests in terms of different evolutionary strategies. I have suggested there are three such strategies (recognizing, of course, that these are not conscious strategies – they have evolved through the process of natural selection):[37]

1   *Opportunist pests:* these are the invaders, moving from place to place, multiplying rapidly, attacking many kinds of crop and causing enormous damage because of their numbers. They include locusts and army worms, and diseases such as rust.
2   *Specialist pests:* present most of the time, with low rates of increase, causing losses because they attack a very valuable part of the plant or transmit a disease – they include the rhinoceros beetle that eats the heart of the coconut tree and the green leafhopper that carries tungro disease of rice.
3   *Intermediate pests:* these lie between the two other types, but are distinguished by being controlled by natural enemies. They include the BPH.

On the cocoa in North Borneo in the early 1960s, the bee bug and the ring-bark borer were specialists, the branch borer and cocoa looper intermediates and the bagworms were opportunists.

On our side we have, essentially, four methods of control at our disposal:

1 *Pesticide control,* the application of chemical compounds to directly kill or deter pests.
2 *Biological control,* the utilization of natural enemies, either by augmenting those already present or by introducing them from other regions or countries.
3 *Cultural control,* the use of agricultural or other practices to change adversely the habitat of the pest.
4 *Plant and animal resistance,* the breeding of animals and crop plants for resistance to pests.

Which control we use depends on the pests present and their strategies. The best approach is to identify one or more key pests that need tackling first. They are usually intermediate pests and need to be specifically targeted to ensure that their natural enemies are able to work effectively. For them, biological control is the appropriate strategy. Pesticides may have to be used against opportunists (eg the cocoa bagworms), but they should be selective, particularly if intermediate pests are present. Specialists such as the cocoa ring-bark borer can be controlled by manipulation of the crop environment. Box 12.1 provides a suggested matching between control and pest strategies.

Over the past 40 years, IPM has grown into a sophisticated approach to pest control and has had a number of notable successes.[39] Savings have often been considerable. In Madagascar, a programme based on cultural control, plant resistance and moderate herbicide use has dispensed with a very costly aerial spraying programme covering 60,000 hectares of riceland.[40] Jules Pretty's review of IPM in the developing countries identified several programmes where annual savings are in the range of $1 – $10 million.[41] But IPM has not been as widely adopted as might be expected. Part of the reason is that, despite its grounding in ecological principles, it has remained until recently a traditional top-down approach in its implementation. IPM programmes have been worked out by specialists and then instructions passed on to farmers.

IPM is a more complex process than one relying on a regular calendar of spraying. Farmers, it is often believed, cannot understand some of the technicalities involved. However, in recent years, this view has been effectively challenged. In Zamorano in Honduras, training programmes at the Escuela Agricola Paramericana have been discovering what farmers do and do not know about pest control.[42] They know a great deal about bees, but are unaware of the existence of solitary wasps that prey on insects, or of parasitic wasps that, as larvae, live inside other insects. They are very knowledgeable about many aspects of the ear rot disease of maize, but not how it reproduces. They are aware that pesticides are toxic, but equate this with the smell of the pesticide, and take few precautions when they spray. In the training course they look at fungi under the microscope, they watch parasitoids emerge from pests and, in the field, observe wasps and ants preying on pests. A most rewarding result has been the farmers' readiness to experiment with their new-found understanding, integrating it with their traditional knowledge. One farmer intercropped amaranth among his vegetables to encourage predators; another placed his box of stored potatoes on an ants' nest; a third took parasite cocoons from his farm to a neighbour's farm.

The most extensive involvement of farmers in IPM has been the Indonesian rice programme which I outlined earlier.[43] By 1993 over 100,000 farmers had attended

Farmer Field Schools where they used simple Agroecosystem Analysis diagrams to understand and discuss the relationships between various pests and the rice crop. The life histories of pests and their predators and parasites are explained using an 'insect zoo' and dyes are placed in knapsack sprayers to demonstrate where the insecticide sprays end up. The schools themselves have become the basis of farmer IPM groups where farmers continue to meet to discuss their problems and to organize village-wide monitoring of pests and predator populations. In 1990 an outbreak of white-stem borer threatened to undermine the success of the programme, but the calls to revert to spraying were successfully resisted. Through the schools, farmers were taught to recognize the egg masses of the stem borers and in a massive campaign searched for and destroyed them. Only a handful of ricefields were infested a year later.

IPM in Indonesia has thus become institutionalized and hence sustainable. Since 1990 some 20 per cent of the farmer training has been paid for by the farmers themselves. Observers are convinced this accounts for the very considerable savings on pesticide applications and the attainment of higher yields. As one graduate put it, 'After following the field school I have peace of mind. Because I now know how to investigate, I am not panicked any more into using pesticides as soon as I discover some pest damage symptoms'.[44] This approach is now being extended to farmers in eight other countries of Asia.

Box 12.1 *Control and pest strategies*[38]			
Pest/control strategy	Opportunists	Intermediates	Specialists
Pesticides	Based on forecasting	Selective	Targeted, based on monitoring
Biological control		Introduction or enhancement of natural enemies	
Cultural control	Cultivation, rotations, timing of planting		Destruction of alternative hosts
Resistance	Polygenic		Monogenic

Notes

1   P. C. Matteson, K. D. Gallagher, and P. E. Kenmore (1992) 'Extension of integrated pest management for planthoppers in Asian irrigated rice: empowering the user', in R. F. Denno and T. J. Perfect (eds), *Ecology and Management of Planthoppers*, Chapman & Hall, London, pp656–685.
2   'Pests' include insects, mites, nematodes and vertebrate pests such as rats and Quelea birds. 'Pathogens' cause diseases and include fungi, bacteria, viruses and, in the case of livestock, various protozoa and worms. 'Weeds' are any plants that adversely compete with crop plants.

3   G. R. Conway (1971) 'Better methods of pest control', in W. W. Murdoch (ed) *Environment: Resources, Pollution and Society*, Sinauer Assoc., Inc., Stanford, California D. Dent (1991) *Insect Pest Management*, Wallingford, UK, CAB International.

4   G. R. Conway (1972) 'Ecological aspects of pest control in Malaysia', in J. Farvar and J. Milton (eds) *The Careless Technology: Ecological Aspects of International Development*, Natural History Press, Doubleday & Co., Garden City, New York, pp467–488; D. Dent (1987) 'Man versus pests', in R. M. May (ed) *Theoretical Ecology: Principles and Applications*, 2nd edn, Blackwell Scientific, Oxford, pp356–386.

5   Conway (1972) op cit.

6   G. R. Conway and J. N. Pretty (1991) *Unwelcome Harvest: Agriculture and Pollution*, Earthscan, London; J. N. Pretty (1995) *Regenerating Agriculture: Policies and Practice for Sustainability and Self-reliance*, Earthscan, London.

7   N. O. Crosland (1989) 'Laboratory to experiment', Proceedings of the Vth International Congress of Toxicology, July 1989, Brighton, UK, pp184–192.

8   Pretty, op cit.

9   R. C. Saxena (1987) 'Antifeedants in tropical pest management', *Insect Science and its Applications*, vol 8, pp731–736.

10 FAO (1993) *Harvesting Nature's Diversity*, Food and Agriculture Organizations, Rome, Italy

11 P. Neuenschwander and H. R. Herren (1988) 'Biological control of the cassava mealybug *Phenacoccus manihoti*, by the exotic parasitoid *Epidinocarsis lopezi* in Africa', *Philosophical Transactions of the Royal Society of London*, B, 318, 319–333; A. Kiss and F. Meerman (1991) *Integrated Pest Management in African Agriculture*, Washington, DC, World Bank (Technical Paper 142, African Technical Department Series).

12 R. Campbell (1989) *Biological Control of Microbial Plant Pathogens*, Cambridge University Press, Cambridge.

13 M. A. Altieri (1995) *Agroecology: the Science of Sustainable Agriculture,* 2nd edition, Boulder, Colorado, Westview Press, London, Intermediate Technology.

14 Altieri, op cit.

15 Conway (1971) op cit.

16 Altieri, op cit.

17 R. F. Smith (1972) 'The impact of the Green Revolution on plant protection in tropical and subtropical areas', *Bulletin of the Entomological Society of America*, vol 18, pp7–14.

18 E. E. Saari and R. Wilcoxson (1974) 'Plant and disease situation of high-yielding dwarf wheats in Asia and Africa', *Annual Review of Phytopathology*, vol 12, pp49–68; IRRI (1985) Proceedings of the Second Upland Rice Conference, International Rice Research Institute, Los Banos, Philippines.

19 M. E. Loevinsohn, J. A. Litsinger and E. A. Heinrichs (1988) 'Rice insect pests and agricultural change', in M. K. Harris and C. E. Rogers (eds) *The Entomology of Indigenous and Naturalized Systems in Agriculture*, Westview Press, Boulder, Colorado, pp161–182.

20 M. Dover and B. Croft (1984) *Getting Tough: Public Policy and the Management of Pesticide Resistance*, World Resources Institute, Washington, DC; G. P. Georghiou (1985) 'The magnitude of the problem', in National Research Council, *Pesticide Resistance: Strategies and Tactics for Management*, Washington DC.

Committee on Strategies for the Management of Pesticide Resistant Pest Populations, Board of Agriculture, National Research Council, National Academy Press

21 K. I. Sudderuddin (1979) 'Insecticide resistance in agricultural pests with special reference to Malaysia', in Proceedings of MAPPS Seminar, 1–2 March, 1979, Kuala Lumpur, Malaysia, pp138–148.

22 H. Hanson, N. E. Borlaug and R. G. Anderson (1982) *Wheat in the Third World*, Westview Press, Boulder, Colorado.

23 Annual catches of moths at light traps. – Loevinsohn (1994) op cit.

24 G. H. L. Rothschild (1971) 'The biology and ecology of rice-stem borers in Sarawak (Malaysian Borneo)', *Journal of Applied Ecology*, vol 8, pp287–322.

25 P. Kenmore (1991a) 'Getting policies right, keeping policies right: Indonesia's Integrated Pest Management policy, production and environment', paper presented at the Asia Region and Private Enterprise Environment and Agriculture Officers' Conference, Sri Lanka; P. Kenmoe (1991b) *How Rice Farmers Clean Up the Environment, Conserve Biodiversity, Raise More Food, Make Higher Profits: Indonesia's IPM – a Model for Asia*, Food and Agriculture Organization, Manila; K. D. Gallagher, P. E. Kenmore and K. Sogawa (1994) 'Judicial use of insecticides deter planthopper outbreaks and extend the life of resistant varieties in Southeast Asian rice', in Denno and Perfect, op cit, pp599–614; R. Stone (1992) 'Researchers score victory over pesticides and pests – in Asia', *Science*, vol 256, p5057.

26 P. E. Kenmore (1980) 'Ecology and outbreaks of a tropical pest of the green revolution, the brown planthopper, *Nilaparvata lugens*. Stahl.', PhD thesis, University of California, Berkeley, California; R. Norgaard (1988) 'The biological control of cassaya mealybug in Africa', *American Journal of Agricultural Economics*, vol 70, pp366–371.

27 P. E. Kenmore (1986) *Status Report on Integrated Pest Control in Rice in Indonesia with Special Reference to Conservation of Natural Enemies and the Rice Brown Planthopper (Nilaparvata lugens)*, Food and Agriculture Organization, Jakarta, Indonesia.

28 E. B. Barbier (1987) 'Natural resources policy and economic framework', in J. Tarrant et al (eds) *Natural Resources and Environmental Management in Indonesia*, Annex 1, United States Agency for International Development (USAID), Jakarta, Indonesia

29 R. W. Herdt and C. Capule (1983) *Adoption, Spread, and Production Impact of Modern Rice Varieties in Asia*, International Rice Research Institute, Los Banos, the Philippines.

30 Matteson et al, op cit.

31 N. W. Simmonds (1981) *Principles of Crop Improvement*, Longman, Harlow, UK. J. E. Vanderplank (1982) *Host – Pathogen Interactions in Plant Disease*, NY Academic Press, New York.

32 E. E Saari (1985) 'South and South-east Asian region', in CIMMYT, *Report on Wheat Improvement*, 1983, International Maize and Wheat Improvement Center; Mexico, DF, S. H. Ou, (1977) 'Genetic defence of rice against disease', in P. R. Day (ed) *The Genetic Basis of Epidemics in Agriculture. Annals of the New York Academy of Science*, pp275–286.

33 G. S. Khush (1992) 'Selecting rice for simply inherited resistances', in H. T. Stalker and J. P. Murphy (eds) Plant Breeding in the 1990s: Proceedings of the Symposium on Plant Breeding in the 1990s, CAB International, Wallingford, UK, pp303–322.

34 Conway (1971) op cit.
35 M. L. Flint and R. van den Bosch (1981) *Introduction to Integrated Pest Management*, Plenum Press, New York; J. R. Cate and M. K. Hinkle (1994) *Integrated Pest Management: The Path of a Paradigm*, National Audubon Society, Washington, DC.
36 Kenmore (1991a, 1991b) op. cit.
37 Conway (1971) op cit.
38 ibid.
39 L. A. Thrupp (ed) (1996) *New Partnerships for Sustainable Agriculture*, World Resources Institute, Washington, DC.
40 A. Von Hildebrand (1993) 'Integrated pest management in rice: the case of the paddy fields in the region of Lake Alaotra', paper presented at the East/Central/Southern Africa Integrated Pest Management Implementation Workshop, Harare, Zimbabwe, 19–24 April, 1993.
41 Pretty, op cit.
42 J. W. Bentley, G. Rodrigues and A. Gonzalez (1993) 'Science and the people: Honduran campesinos and natural pest control inventions', in D. Buckles (ed) *Gorras y Sombreros: Caminos hacia la Colaboracion entre Tecnicos y Campesionosia*, El Department of Crop Protection, El Zamarano, Honduras'; J. W. Bentley, G. Rodrigues and A. Gonzalez (1994) 'Stimulating farmer experiments in non-chemical pest control in Central America', in I. Scoones and J. Thompson (eds) *Beyond Farmer First: Rural People's Knowledge, Agricultural Research and Extension Practice*. Intermediate Technology, London, pp147–150.
43 Pretty, op cit; Kenmore (1991) op cit; Matteson et al, op cit; Y. Winarto (1994) 'Encouraging knowledge exchange: integrated pest management in Indonesia', in Scoones and Thompson, op cit, pp150–154.
44 E. van der Fliert (1993) *Integrated Pest Management: Farmer Field Schools Generate Sustainable Practices,* Wageningen, the Netherlands, Wageningen Agricultural University, (WAU Paper 93–3).

Part 3

Social Perspectives

Introduction to Part 3:
Social Perspectives

Jules Pretty

Part 3 of the *Reader in Sustainable Agriculture* contains five articles on social perspectives for sustainable agriculture by Robert Chambers, Jules Pretty, Richard Bawden, Niels Röling and Kevin Gallagher and co-authors. Many of the technological changes known to be necessary to make progress towards sustainability require collective social action as a prerequisite – and this in turn requires individuals to act and think differently.

In the first article, drawn from the seminal 1989 book *Farmer First*, Robert Chambers discusses the reversals necessary to put farmers' knowledge and capacities at the heart of agricultural transformations. For decades, agricultural research and extension institutions have used a transfer-of-technology mode of working, with farmers and their communities simply as recipients of technologies and practices developed on research stations. A farmer-first approach requires professionals to adopt different attitudes and behaviour, becoming, for example, convenors, catalysts, advisers, travel agents and supporters of farmers' own analyses, choices and experiments. The complex, diverse and risk-prone environments of many developing country contexts are not well suited to homogenous technologies, however effective they have been on research stations. They require that professionals reverse past practices, and encourage farmers to conduct their own analyses and experiments, thus adapting and fitting technologies to their own situations. Such reversals of 'normal practice' also require institutional change, with policies and institutions needing to facilitate such efforts. As Chambers said, the stakes are high, and a decade and a half after this chapter was written, they remain disturbingly high for millions of people and their environments.

In the second article, Richard Bawden describes the remarkable experience of Hawkesbury College in Australia, and its quarter century journey of institutional change towards action research. Their aim was to create reflective practitioners, in which theory and practice inform one another. Praxis is the emergent property of experiential learning processes by which we try to make sense of the world. Finding out and taking action are thus linked together. This paper also sets out key principles for soft systems learning (in contrast with hard systems approaches), and indicates how researched systems need to evolve through researching systems to critical systemic discourse. These principles are then put into practice through the redesign and continued adaptation of the curriculum and learning at Hawkesbury (later to be part of the University of Western Sydney). Even after having achieved so much, Bawden indicates that there is a very long way to go.

In the third article, Jules Pretty sets out the importance of social capital in the collective management of natural resources. The term social capital captures the idea that social bonds and norms are important for people and communities. It emerged as a term following detailed analyses of the effects of social cohesion on regional incomes, civil society and life expectancy. As social capital lowers the transaction costs of working together, it facilitates cooperation. People have the confidence to invest in collective activities, knowing that others will also do so. They are also less likely to engage in unfettered private actions with negative outcomes, such as resource degradation. Four features are important: relations of trust; reciprocity and exchanges; common rules, norms and sanctions; connectedness in networks and groups.

Collective resource management programmes that seek to build trust, develop new norms, and help form groups have become increasingly common, and are variously described by the terms community-, participatory-, joint-, decentralized- and co-management. They have been effective in several sectors, including watershed, forest, irrigation, pest, wildlife, fishery, farmers' research, and micro-finance management. Since the early 1990s, some 400,000–500,000 new local groups were established in varying environmental and social contexts, mostly evolving to be of similar small size, typically with 20–30 active members, putting total involvement at some 8–15 million households. The majority continue to be successful, and show the inclusive characteristics identified as vital for improving community wellbeing, and evaluations have confirmed that there are positive ecological and economic outcomes, including for watersheds, forests and pest management.

Niels Röling makes the case in the fourth article for the expansion of beta/gamma science to address issues of ecological rationality and agricultural sustainability. This paper is based on the assumption, as Röling puts it, 'that we live, not in epoch of change, but in a change of epoch'. Though we have built economies and technologies that have transformed the lives of most, allowing many to escape from poverty and misery, we have not done a good job of managing the Earth. The paper iterates the importance of a constructivist perspective for mobilizing the reflexivity needed for change, and provides the theoretical underpinning for the human predicament for having to address coherence and correspondence. The challenge, says Röling, is 'not in dealing with land, but in how people use it'.

The paper presents a model for knowledge based action, showing how organisms, including humans, can learn or adapt a coherent set of elements of cognition in order that we might change circumstances. The idea drives the change. At the same time, it is our thinking that brings forth a world. And what does it look like – a society geared to generating wealth and satisfaction, in which we think we control many of the factors that cause suffering? Ask the question, and we may not get the answer we want – such as of a Chinese women spreading night soil on her land, a picture of tradition it appears, until she reveals that she is fed up with this stupid and dirty job, and would rather be in Paris. The agricultural treadmill appears to give benefits, but the costs are enormous. The paper draws these theses together by indicating the implications for interdisciplinarity, through the domains of using instruments and incentives, and continuous learning (and co-learning).

In the final article on social perspectives, Kevin Gallagher and co-authors describe the ecological and social basis for integrated pest management in rice agroecosystems in Asia. The chapter, drawn from the book *The Pesticide Detox*, first sets out the specific ecological basis of rice fields, and why continued pesticide applications have not resulted in cheap and effective pest control. Pesticides may kill pests in the short term, but they also eliminate natural enemies that exert good ecological control over pests. The idea behind integrated pest management (IPM) is to put technologies into the hands of farmers and communities, so that they learn to farm with low to zero use of pesticides, yet also do not suffer pest losses. Many tens of thousands of Farmer Field Schools have been held throughout Asia, and these have been highly effective at increasing farmers' own capabilities and knowledge for ecological management of rice fields.

Many countries are now reporting large reductions in pesticide use. In Vietnam, one million farmers have cut pesticide use from more than three sprays to one per season; in Sri Lanka, 55,000 farmers have reduced use from three to a half per season; and in Indonesia, one million farmers have cut use from three sprays to one per season. In no case has reduced pesticide use led to lower rice yields. Amongst these are reports that many farmers are now able to grow rice entirely without pesticides: a quarter of field school trained farmers in Indonesia, a fifth to a third in the Mekong Delta of Vietnam, and three-quarters in parts of the Philippines.

13

Reversals, Institutions and Change

Robert Chambers

Farmer first and TOT

The new behaviours and attitudes presented by the contributors to this book conflict with much normal professionalism and with much normal bureaucracy. Normal professional training and values are deeply embedded in the transfer of technology (TOT) mode, with scientists deciding research priorities, generating technology and passing it to extension agents to transfer to farmers. Normal bureaucracy is hierarchical and centralizes, standardizes and simplifies. When the two combine, as they do in large organizations, whether agricultural universities, international agricultural research centres, or national agricultural research systems (NARSs), they have an impressive capacity to reproduce themselves and to resist change.

But to serve well the resource-poor farm families of the third – complex, diverse and risk-prone – agriculture with which much of this book has been concerned, requires these 'normal' tendencies to be reversed: for farmers' analysis to be the basis of most research priorities, for farmers to experiment and evaluate, for scientists to learn from and with them; and for research and services to farmers to be decentralized, differentiated and versatile.

The difficulty of effecting major changes and reversals in large organizations underlines the importance of seeing what changes of behaviour and attitude are required, what institutional conditions are necessary for them to be sustained and spread, and how these might be achieved. To do this, we need to outline in more detail the contrast between TOT and the farmer-first approach and methods represented in this book (see Table 13.1).

With farmer first, the main objective is not to transfer known technology, but to empower farmers to learn, adapt and do better; analysis is not by outsiders – scientists, extensionists or non-governmental organization (NGO) workers – on their own but by farmers and by farmers assisted by outsiders; the primary location for research and development is not the experiment station, laboratory or greenhouse, necessary though they are for some purposes, but farmers' fields and conditions; what is transferred by outsiders

Note: Reprinted from *Farmer First: Farmer Innovation and Agricultural Research* by Chambers, R., Paley, A. and Thrupp, L. A. (eds), copyright © (1989), with permission from IT Publications, London

Table 13.1 *Transfer of technology and farmer-first compared*

	TOT	FF
Main objective	Transfer technology	Empower farmers
Analysis of needs and priorities by	Outsiders	Farmers assisted by outsiders
Primary R&D location	Experiment station, laboratory, greenhouse	Farmers' fields and conditions
Transferred by outsiders to farmers	Precepts Messages Package of practices	Principles Methods Basket of choices
The 'menu'	Fixed	A la carte

to farmers is not precepts but principles, not messages but methods, not a package of practices to be adopted but a basket of choices from which to select. The menu, in short, is not fixed or table d'hote, but a la carte and the menu itself is a response to farmers' needs articulated by them. All this demands changes in activities and roles.

Farmer-first activities and roles

Contributions to this chapter show farmers carrying out or participating in various activities which in the TOT mode are conducted only by scientists. Three of these, again and again, are analysis, choice and experiment. To support farmers in these activities generates and requires new roles for outsiders (Table 13.2):

Table 13.2 *Roles generated for outsiders by farmers' activities*

Farmers' activities	New roles for outsiders
analysis	convenor, catalyst, adviser
choice	searcher, supplier, travel agent
experiment	supporter, consultant

What these activities and roles entail can be illustrated by contributions to this chapter, supported by other sources.

(i) Analysis. Analysis by farmers takes many forms and can be promoted in many ways, involving outsiders to different degrees. In the examples in this chapter an outsider has often played a role, whether as questioner, convenor of a group, stimulator of discussion, or catalyst whose presence speeds up the process.

Analysis can be part of or generated by the use of a method. Some examples are:

- open interview and iterative group conversations (Floquet, 1989);
- ethnohistory and ethnobiography (the biography of a crop, or of a person's experience of a crop, an historical analysis of the experience of a community, etc) (Rocheleau et al, 1985; Box, 1987);
- inspection and discussion: visiting trial sites, observing innovations, field days, and visits by farmers to research stations when they observe and discuss (Norman et al, 1982; Ashby et al, 1987);
- visual aids to analysis: seasonal or other diagramming (Conway, 1987), aerial photographs (Carson, 1987) systems diagramming on a board (Lightfood et al, 1987), other uses of diagrams with and by farmers and communities (Kabutha and Ford, 1988; McCracken, 1988) and drawing maps (Gupta, 1987a, b);
- eliciting clients' criteria and preferences, where individuals or groups (women, men, farmers etc) articulate their reasons for preferences, and then rank items according to them (Ashby et al, 1987; Chambers, 1988); key questions and approaches to questioning: 'ways in' or 'points of entry' such as 'What would a desirable variety look like to you?' (Ashby et al, 1987, p27), 'What would you like your landscape to look like in the future?' (Rocheleau, 1987) 'When you were a boy, what was the oldest variety of (a particular crop) that you knew about?' and 'Comparing agriculture practiced at the time of your father and grandfather with the agriculture practiced by you today, what are the major changes that have occurred?' (Gubbels, 1988);
- contrast analysis, where groups or individuals are asked to explain the contrasting conditions or behaviour of others, thus setting a frame of reference before analysing their own (Gupta, 1987);
- sequences of meetings and visits (Rocheleau et al, 1985, Mathema and Galt, 1987, Norman et al, 1985, Lightfoot et al, 1987, Repulda et al, 1985);
- innovator workshops where farmer innovators meet to discuss their new practices (Abedin and Haque, 1987; Ashby et al, 1987).

The role of the outsider is to elicit, encourage, facilitate and promote analysis by farmers, providing where necessary the stimulus, the occasion and the incentive for meetings and discussions. The outsider can take part, but does not dominate. Farmers' own analysis, criteria and priorities come first. Requests are generated for outsiders to search for what farmers want and need, and to provide them with choices or ideas for experiments to solve a problem or exploit an opportunity (Lightfoot et al, 1987, Repulda et al, 1987).

(ii) Choice. Choice by farmers is prominent in the farmer-first paradigm. It has two aspects. First, farmers' analysis generates an agenda of requests for information and material. Second, farmers need a range of choice, so that they can pick and choose to suit their conditions, extend their repertoire and enhance their adaptability. Norman et al note 'the technology assessment process in which a wide range of options are presented to a large number of volunteer farmers' (p141). To find and present variety and choices to farmers is largely a task for outsiders. Some examples are:

- providing farmers with varied genetic materials to test and appraise (Maurya et al, 1987, Ashby et a, 1987, Norman et al, 1987);
- planting a variety of lines or species, to be followed by 'wait-and-see and pick-and-choose';
- issuing mini-kits of seeds and fertilizers to farmers for them to try out in various combinations;
- requiring nurseries, as with forestry in Kenya, to plant and provide a range of species, including a preponderance of indigenous species;
- transferring genetic material between regions, countries and continents, especially of non-cereal plants (multi-purpose trees, shrubs, grasses, vining plants, root crops etc) and livestock;
- transferring indigenous technical knowledge and practices between farmers in different regions;
- enabling farmers to travel, visit, see and learn for themselves the farming practices of others.

The role of the outsider, whether scientist, extensionist, or NGO worker, is to search for and supply the species, varieties, treatments, cultural practices, scientific principles or combinations of these which fit and meet farmers' requests and needs. It may also be that of travel agent or tour operator, to arrange for farmers to visit research stations, other farmers, or other regions, to learn from other farmers and scientists and to widen their experience and options.

(iii) Experimenting. Experimenting by farmers has long been under-perceived. The professional world has been slow to recognize farmers' experimental inclinations and abilities (but see Johnson, 1972; Richards, 1985; Rhoades, 1987). Rhoades and Bebbington (1988) have identified three reasons why farmers experiment: to satisfy curiosity; to solve problems; and to adapt technology. As we have seen, their farming is both performance (Richards, 1987) and in a sense a continuous experiment: Hossain and Islam (1985) point out that farmers in Bangladesh are continually changing their cropping patterns (p35) and Juma puts it that 'a farmer is a person who experiments constantly because he is constantly moving into the unknown' (1987, p34).

In the farmer-first approach, it is not packages of technology that are provided to farmers, but genetic material, principles, practices and methods for them to test and use. Genetic material can take many forms and may come from nearby, from other regions, or from other countries or continents. Similarly, principles can originate from different sources: in West Africa, the principle of alley cropping was taken from the research station and was adapted and experimented with by farmers (Sumberg and Okali, 1988); the principle of diffused light to inhibit potato sprouting in store originated with farmers in Kenya and was spread internationally and laterally to other farmers in many countries, who made their own applications with local materials to fit local farm architecture (Rhoades, 1987). Experimental principles and methods suitable for their conditions and needs can also be provided to farmers to improve their investigations and innovations (Bunch 1985, Bunch, 1987).

Farmers' experiments are, then, encouraged and supported by outsiders. This is close to Biggs' (1987) collegiate mode of farmer–scientist *interaction*. Farmers take part in design (Fernandez and Salvatierra, 1987), determine management conditions and implement and evaluate the experiments. They 'own' the experiments and the outsiders provide support and advice.

Evaluation of experiments is also by farmers and continuous. An authoritative World Bank publication (Casley and Kumar, 1987, p116) has pointed out that it is often assumed that illiterate, tradition-bound farmers cannot assess the dynamics of change, but that their knowledge and judgments are in many instances more accurate than those of project staff. One of DM Maurya's criteria for assessing a line given to a farmer to try is whether other farmers ask for seed (pers comm). It is farmers' judgements, interest and adoption that count.

Stimulating, servicing and supporting these farmers' activities – analysis, choice and experiment – requires reversals of normal and expected roles on the part of outsiders, be they scientists, extensionists or workers in NGOs. This does not mean that they have to be purely passive catalysts. It would be as absurd for their ideas and knowledge not to be brought into play, as it has been for those of farmers to be neglected. In raising questions, in providing tools for analysis, in presenting what they already know to be feasible and available choices, and in supporting and advising on farmers' experiments, they have a part to play. But their role is not that of teacher, of the bearer of superior modern technology, of the person who knows what is good for others better than they know for themselves. It is neither the role of traditional agricultural extension, nor that of normal agricultural science. An open, learning process approach is indicated, of a sort encouraged neither by the content of university curricula nor by the hierarchy and style of government bureaucracies.

For these changes and reversals of role to occur on any scale is not easy. It requires resolute changes in institutions, in incentives and in methods and interactions.

Institutional change

Unfortunately, normal bureaucracy tends to centralize, standardize and simplify, and agricultural research and extension are no exceptions. They fit badly, therefore, with the conditions of resource-poor farm families, with their geographical scatter, heterogeneity and complexity within any farm and farm household. In resource-rich areas of industrial and green revolution agriculture, production has been raised through packages, with the environment managed and controlled to fit the genotype. The third agriculture, being complex, diverse and risk-prone, requires the reverse, with searches for genotypes to fit environments. In industrial and green revolution agriculture, higher production has come from intensification of inputs and simplification and standardization of practices; in the third agriculture, it comes more from diversifying enterprises and multiplying linkages. Green revolution agriculture has been convergent, evolving towards common practices; the third agriculture often needs to be divergent, evolving towards a greater variety of differing enterprises and practices.

At first sight, then, the farmer-first approach appears incompatible with normal bureaucracy. But reversals in government research organizations, though difficult to start and to sustain, are not impossible. Some contributors were working in special projects linked with NARSs; others were working in more normal conditions, as with the innovator workshops in Bangladesh (Abedin and Haque, 1987) and the distribution to farmers of advanced lines of rice in India (Maurya et al, 1987).

For the future, to achieve farmer-first reversals in national bureaucracies, especially NARSs, three aspects of management merit special attention: decentralization and resources; search and supply; and incentives.

(i) Decentralization and resources. Central controls need loosening if local actions are to fit diverse conditions. Centralized permissions for expenditures constrain flexibility. Centrally coordinated trials limit discretion and the ability to serve local priorities. When resources such as transport and money for travel are scarce, local discretion and control become more important than ever. The essence of farmer-first approaches is to serve and support local diversity, with a reversal of demands on staff, the demands to come from fanners below more than from seniors above.

Decentralization is difficult in normal bureaucracies. Central accountants fear loss of control over expenditures. Central officials fear loss of power and prestige. Reports are harder to collate and present, and work harder to supervise, when activities are varied. Methods are needed, and perhaps easier now with microcomputers, for valuing local diversity in staff activities in place of counting reported achievements of standard targets. For NARSs, the practical implications are to devolve resources and discretion more to the local level.

Freedom and means for staff to visit and spend time with farmers are crucial. For travel, something can usually be done quite simply. In the joint trek in Nepal, scientists walk together for days (Mathema and Galt, 1987). Foot, bicycle, horse and public transport can, variously, be used. For cost-effectiveness, though, other means of travel can be important, especially when distances are great and environments diverse. Unfortunately, access to transport and permission to use it are frequent problems, though less so with foreign-funded programmes. Travel and allowances can be high-profile privileges for which staff compete, jealously guarded and sparingly allocated by directors of institutes and heads of units. Worse, when revenue shortfalls or national policy reforms force cuts in recurrent budgets, staff are usually protected and it is other votes that suffer. Fuel, vehicles and nights out allowances are favourite victims. In Zambia, the Ministry of Agriculture's vote for petrol and maintenance had been reduced by 1980 to only one fifth of its 1973 level despite an increase in vehicles and staff (ILO, 1981, pxxvi). Scientists can usually work with farmers close to their research stations and residences; but without hassle-free and adequate access to means for travel, it is difficult for them to work regularly and well with others further afield.

(ii) Search and supply. Search is neglected and rarely rewarded as a professional activity. This includes search for farmer-innovators and experimenters, for genetic material, and for principles, practices and technologies, whether locally, regionally, nationally or internationally.

Search is basic for meeting farmers' needs and widening their choices. In complex, diverse and risk-prone agriculture, what farmers want and need often differs from the simplifications of centrally planned priorities. Agricultural research and extension have, for example, a tendency to specialize on single commodities. But farmers' analysis will often specify a non-commodity need, such as multipurpose trees for agroforestry, or a rapidly vining legume to suppress weeds, or a range of vegetable seeds, or means to create, improve and exploit microenvironments, or technology for harvesting water, capturing and concentrating soil, or improving the supply of plant nutrients. As a result of past neglect, the potential for search and supply of such varied material and technologies seems still very large.

Search and supply have institutional implications. These include that grass-roots extension staff and scientists have resources and are rewarded for finding farmers' innovations and experiments and for stimulating and articulating realistic demand from farmers; and an ability of national and international agricultural research systems to respond with supplies of genetic material, principles and methods.

These reversals face two major obstacles. First, extensionists and scientists may not be rewarded for raising problems and making requests. Extensionists seen in the TOT and normal bureaucratic mode are there to pass on messages and packages downwards, not to multiply work for their senior officers by passing varied requests upwards. Second, most NARSs lack capacity to respond to needs and requests articulated by farmers for material or information. In practice, most management information systems are designed to feed information upwards to serve central management, rather than to draw it downwards to serve farmers. Six of the seven management information systems listed in 1987 for agricultural research in the Philippines were for central management; only one, the Research Information Storage and Retrieval System, was to provide information useful at the grass roots, and that was described in the future tense, with the statement that financial support was needed to extend it into the regions. Many NARSs have poor institutional memories for research findings (Kean and Singogo, 1988, p48), and work often has to be repeated because earlier records cannot be found. Few, if any, are yet set up well enough to provide diverse information, genetic material and technologies to meet diverse local demand.

The practical implications are for agricultural research and extension organizations to make three changes: to encourage field staff to search for, support and spread farmers' innovations; to judge and reward staff by the requests they make upwards in response to analysis and demands by farmers; and to develop information and supply systems to respond to those demands.

(iii) Incentives. As with any new paradigm, professionals who innovate in the farmer-first mode risk being marginalized. In the short term, the safest route to promotion will often seem to be work on-station not on-farm; on irrigated agriculture, not rainfed (and least of all on unreliable rainfed); on a single commodity, not complex combinations; on industrial, commercial and major cereal crops not low status subsistence food crops; with quick maturing annuals not slow maturing perennials like shrubs and trees; and with validation through standard experimental design not farmers' adoption. Nor does improving complex, diverse and risk-prone (CDR) farming lend itself to the statistical testing methods taught in textbooks, involving as it often does complex and multiple simultaneous change, for example, agroforestry combined with water harvesting, growing fish with rainfed rice, home gardening with several canopies, or the creation and exploitation of protected microenvironments in semi-arid conditions. More papers can be produced more reliably by using conventional methods on conventional crops in conventional

environments, where there is already a good information base, than by using unconventional methods on unconventional agricultural practices in unconventional environments. Where promotions boards judge candidates only by adherence to standard methods, or numbers of publications, rather than farmers' adoption, then pioneers in farmer-first modes will not do as well as their less innovative colleagues.

The rapid transfer of agricultural research staff poses a further problem especially in sub-Saharan Africa. The costs in lost continuity and effectiveness in formal on-station research are well known. Less well recognized is the way in which rapid turnover reduces incentives for staff to build up relations with farmers, and undermines farmers' confidence in them.

The practical implications of these obstacles are to develop enabling conditions and incentives. The several forms these can take include the following:

- assessing research staff less on publications, and extension staff less on the achievement of targets; and both more on the demands and searches they initiate on behalf of farmers, on farmers' interest and innovation and on adoption and spread of technology;
- rewarding those who pioneer and write about new methods. Until recently, farmer-first research methods were not much the subject of articles in the harder scientific journals, but as the summer 1988 issue of *Experimental Agriculture* (Farrington, 1988) has shown, this is changing. As scientists come to realize that they can publish articles about their methods and experiences, and that these bring national and international recognition, publishing disincentives should not just disappear but be reversed;
- ensuring more continuity for scientists in field posts. This may be difficult for many reasons. Fortunately, where lack of staff continuity is endemic, experimenting farmers and local organizations may be able, more and more, to provide their own continuity;
- networking between farmer-first researchers, providing mutual support and recognition.

The strongest incentive, though, is professional and personal satisfaction. Those who make reversals and changes in directions, and who work collegially with farmers, soon find it intellectually and professionally exciting, enjoyable, and even fun, with the supreme reward of effectively helping farmers to do better. This is the most hopeful aspect. For even if other conditions are adverse, more and more will want to work in the farmer-first mode for the simple and sound reason that it satisfies and succeeds.

Methods and interactions

In themselves, these three things – decentralization and resources, organization for search and supply and providing incentives – are not enough. Much also depends on what is done and how it is done – on the methods available and the quality of interactions.

The need here is to develop further, describe and disseminate farmer-first methods in detail. Just as the aim is to widen choice of practices for resource-poor farmers, so it is to widen choice of methods for scientists and extensionists. Some of these are methods for decentralization, for search and supply and for farmers' experiments; yet others are for interactions between professionals and farmers. Many such methods are now known. Those that are most promising deserve to be evaluated, written up and made accessible through manuals and practical training.

The more important methods to be developed and described include:

- aiding farmers' analysis and learning their agendas;
- getting started with families and communities;
- finding out about agricultural research (for NGOs);
- finding and supporting farmers' experiments;
- convening and assisting groups;
- convening and managing innovator workshops;
- searching, and supplying farmers with what they want and need;
- designing and managing incentives for scientists;
- communicating: farm family and outsider face to face.

This last, concerning the quality of interaction between farmers and scientists, is as crucial as it has been neglected. Most accounts and manuals concentrate on the mechanics of methods, as though rules guarantee results. This is not so. As social anthropologists, sociologists and some psychologists know, and as is only common sense, the quality of the face-to-face relationship can make or mar an interview or discussion; and much depends on mutual respect and rapport.

Good advice is available (Rhoades, 1982; Grandstaff and Grandstaff, 1987) but one may still ask how many scientists and extensionists have a grounding in the significance of non-verbal cues, of seating arrangements, of demeanour and manners and of that respect for and interest in people and what they have to show and say which makes for free and open communication.

Even good manuals and training for farmer-first methods and manners cannot by themselves guarantee good results. After institutions, incentives and interactions, there remains personality. Personal styles and aptitudes differ. The contrast between the closed blueprint approach to development and the open learning process (Korten, 1980) parallels the contrast between TOT and farmer first. Some people are more at home with blueprints, with fixed plans and rules, and with clear ideas of what is expected and what will be officially rewarded. For them, the TOT mode fits better. Others are more at ease with learning processes, with open-ended exploration, with deciding for themselves how to proceed as they go along, and with the reward of knowing in themselves that they have done well. They will be better with the farmer-first mode.

Practical action: Starting and sustaining change

Professionals concerned with agricultural innovation, research and extension – whether they are farmers, or physical, biological or social scientists, and whether they are independent or working in universities, training institutes, government departments or NGOs – will have found in this book many ideas for what they might do. Non-farming agricultural professionals, just like resource-poor farmers, are faced with diversity and complexity, and similarly need a repertoire of methods so that they can be versatile and adaptable.

At a personal level, it is tempting to say that nothing can be done until a whole bureaucratic and professional system changes. Usually, though, there is room for manoeuvre. Some steps can be taken; a start can almost always be made. Even if the start is small and progress slow, it may be the seed of a self-sustaining movement. In the spirit of the learning process approach to development, it is better to start, to do something and to learn on the way, than to wait for better conditions before acting.

In the spirit of pluralism, action can and should start in many places. But not everything can be done at once. There are questions of how and where to start.

Two principles help here. The first is to start where it is easier, simpler and quicker, while weighing the danger of biases against poorer farmers. It is better to start and learn by doing and through mistakes than to wait for perfect conditions. By starting, experience is gained and confidence built up.

The second principle is to change behaviour before attitudes. Preaching about attitudes invites acquiescence without deep change. Action means experience gained and that, more than exhortation, reorients attitudes and habits of thought.

Taking these two principles together, analysis by and with farmers appears the most promising point of entry, followed by search, choice and experiment. A basic question to ask is what farmers would like in their basket of choices. From this question follow demands which reverse the normal top-down flow. Whether a department of agriculture, a university, an NGO or combinations of these can handle such requests can then be put to the test. Activities and roles then have to change. Procedures to accept and handle demands are required. Information systems for management from below have to be created and made to work. Subsequently, other elements of the paradigm become active, with testing and experiments by farmers and consultative support by others.

Finally, for professionals to innovate by working in the farmer-first mode demands vision and leadership on the part of those with power and responsibility. These include senior officials in capital cities, vice-chancellors and deans, directors of research stations, leaders of teams and senior staff in regional, provincial and district headquarters, as well as in aid agencies and NGOs. Leaders can act like normal professionals and normal bureaucrats who simplify, standardize and stifle; or they can break out and encourage and support initiative and change, providing resources and room for manoeuvre for those under their management who have the aptitude and will to work in new participatory ways; and they can reorganize departments, procedures and management information systems so that searches can be made to meet farmers' demands and fill their basket with choices.

Alliances and mutual support also help. Those who see or sense the potential will do well to seek out and support like-minded fellow professionals in their own and other organizations. Shared ideas and experiences speed up learning. If those in this book provide stimulus and encouragement, they will have served their purpose. And if the new paradigm fulfills its promise, and is accepted and practiced much more in the 1990s and the 21st century, then those who take risks now to support, develop and spread it will not have acted in vain.

For the stakes are high. Over a billion people are supported by the third agriculture. The challenge is to enable many of the poorer among them to secure better and more sustainable livelihoods from their complex, diverse and risk-prone farming when normal agricultural research has so largely failed. This book points to new potentials. It shows that reversals in the farmer-first mode can be effective for farmers and exciting for professionals. A quiet revolution has already started, but it is scattered and still small-scale. Which countries, institutions and individuals will now lead remains to be seen. Change depends on personal decisions and action. Those who now explore the frontiers of farmer participation cannot expect Nobel prizes, or be confident of early recognition or promotion; but they will be joining a vanguard. Their rewards, more surely, will be the exhilaration of pioneering, the satisfaction of seeing innovations spread and the knowledge that through their work, poor farm families are being truly served.

References

An asterisk (*) indicates a paper presented at the IDS Workshop on Farmers and Agricultural Research: Complementary Methods, held at the Institute of Development Studies, University of Sussex, UK, 26–31 July, 1987. Only the short title of such papers is given, together with a reference to any published version. Where papers have not been formally published, copies can be obtained from: Institute of Development Studies, University of Sussex, Brighton BN1 9RE, UK, or from Overseas Development Institute, Regent's College, Inner Circle, Regent's Park, London NW2 4NS, UK.

*Abedin, M. Z. and Haque, M. F. (1987) 'Learning from farmer innovations and innovator workshops', IDS Workshop

*Ashby, J. A., Quiros, C. A. and Rivera, Y. M. (1987) 'Farmer participation in on-farm varietal trials', IDS Workshop, available as ODI Agricultural Administration (Research and Extension) Network *Discussion Paper 22*, December 1987

Biggs, S. D. (1987) 'Interactions between resource-poor farmers and scientists in agricultural research', School of Development Studies, University of East Anglia, Discussion Paper for OFCOR research group; ISNAR study on Organization and Management of On-farm Research

*Box, L. (1987) 'Experimenting cultivators', IDS Workshop, available in full as 'Experimenting cultivators: A methodology for adaptive agricultural research', ODI Agricultural Administration (Research and Extension) Network *Discussion Paper 23*, December 1987

Bunch, R. (1985) *Two Ears of Corn: A Guide to People-centered Agricultural Improvement*, World Neighbors, Oklahoma City

*Bunch, R. (1987) 'Small farmer research', IDS Workshop

Carson, B. (1987) 'Appraisal of rural resources using aerial photography: An example from a remote hill region in Nepal', in KKU 1987, Proceedings of the 1985 Conference on Rapid Rural Appraisal, pp174–190

Casley, D. and Kumar, K. (1987) *Project Monitoring and Evaluation in Agriculture*, Johns Hopkins University Press, Baltimore

*Conway, G. R. (1987) 'Diagrams for farmers', IDS Workshop

Farrington, J. (ed) (1988) *Experimental Agriculture*, vol 24, part 3, with 'Farmer participatory research: Editorial introduction', pp269–279

*Fernandez, M. E. and Salvatierra, H. (1987) 'Design and implementation of participatory technology validation in highland communities of Peru', IDS Workshop, see also Farming Systems Research Symposium, Kansas State University, Manhattan, Kansas, 5–8 October, 1986

Floquet, A. (1989) 'Conservation of soil fertility by peasant farmers in Atlantic Province, Benin', in Kotschi, J. (ed) *Ecofarming Practices for Tropical Smallholdings – Research and Development in Technical Cooperation*, GTZ, Eschborn

Grandstaff, S. W. and Grandstaff, T. B. (1987) 'Semi-structuring interviewing by multidisciplinary reams in RRA', in KKU 1987, Proceedings of the 1985 Conference on Rapid Rural Appraisal, pp129–143

Gubbels, P. (1988) 'Peasant farmer agricultural self-development: The Wold Neighbors experience in West Africa', *ILEIA Newsletter*, vol 4, no 3, pp11–14

*Gupta, A. K. (1987a) 'Organizing the poor client responsive research system: Can tail wag the dog?', IDS Workshop

*Gupta. A, K. (1987b) 'Scientific perceptions of farmers' innovations', IDS Workshop

*Hossain, S. M. A. and Islam, M. T. (1985) *A Report on the First Workshop of Innovative Rice Farmers*, GTI Publication no 56, Bangladesh Agricultural University, Mymensingh

ILO (1981) *Zambia: Basic Needs in an Economy under Pressure*, International Labour Office, Jobs and Skills Programme for Africa, Addis Ababa

Johnson, A. W. (1972) 'Individuality and experimentation in traditional agriculture', *Human Ecology*, vol 1, pp448–459

*Juma, C. (1987) 'Ecological complexity and agricultural innovation: The use of indigenous genetic resources in Bungoma, Kenya', IDS Workshop

Kabutha, C. and Ford, R. (1988) 'Using RRA to formulate a village resource management plan, Mbusanyi, Kenya', *RRA Notes*, vol 2, IIED, London

Kean, S. A. and Singogo, L. P. (1988) *Zambia: Organization and Management of the Adaptive Research Planning Team (ARPT), Research Branch, Ministry of Agriculture and Water Development*, OFCOR Case Study no 1, ISNAR, the Netherlands, May

Korton, D. C. (1980) 'Community organisation and rural development: A learning process approach', *Public Administration Review*, vol 40, pp480–510

*Lightfoot, C., de Guia, O., Jr, Aliman, A. and Ocado, F. (1987) 'Letting farmers decide in on-farm research', IDS Workshop

McCracken, J. A. (1988) *Participatory Rapid Appraisal in Gujarat: A Trial Model for the Aga Khan Support Programme (Kenya)*, IIED, London

*Mathema, S. B. and Galt, D. (1987) 'The Samuhik Bhraman process in Nepal: A multidisciplinary group activity to approach farmers', IDS Workshop

*Norman, D. W., Baker, D., Heinrich, G., Jonas, G., Maskiera, S. and Worman, F. (1987) 'Farmer groups for technology development: Experiences from Botswana', IDS Workshop

Norman, D. W., Simmons, E. B. and Hays, H. M. (1982) *Farming Systems in the Nigerian Savanna: Research and Strategies for Development*, Westview Press, Boulder

*Repulda, R. T., Quero, F., Ayaso, R., Guia, O. de and Lightfoot, C. (1987) 'Doing research with resource-poor farmers: FSDP-EV perspectives and programmes', IDS Workshop

Rhoades, R. E. (1982) *The Art of the Informal Agricultural Survey*, International Potato Centre, Lima, Peru

*Rhoades, R. E. (1987) 'The role of farmers in the creation and continuing development of agri-technology and systems', IDS Workshop, also available as 'Farmers and Experimentation', ODI Agricultural Adminstration (Research and Extension) Network *Discussion Paper 21*, December

Rhoades, R. E. and Bebbington, A. (1988) 'Farmers who experiment: An untapped resource for agricultural research and development', paper presented at the International Congress on Plant Physiology, New Delhi, 15–20 February

Richards, P. (1985) *Indigenous Agricultural Revolution*, Hutchinson, London, Westview Press, Boulder

*Richards, P. (1987) 'Agriculture as a performance', IDS Workshop

*Rocheleau, D. (1987) 'The user perspective and the agroforestry research and action agenda', IDS Workshop, in Gholz, H. (ed) *Agroforestry*, Martinus Nijhoff, Dordrecht, the Netherlands, September 1987

Rocheleau, D. Khasiala, P., Munyao, M., Mutiso, M., Opala, E., Wanjohi, B. and Wanjuagna, A. (1985) 'Women's use of off-farm lands: Implications for agroforestry research', project report to the Ford Foundation, ICRAF, Nairobi, mimeo

Sumberg, J. and Okali, C. (1988) 'Farmers, on-farm research and the development of new technology', *Experimental Agriculture*, vol 24, pp333–342

14

The Hawkesbury Experience: Tales from a Road Less Travelled

Richard Bawden

In the course of the exponentially growing forces of the modernization process, hazards and potential threats have been unleashed to an extent previously unknown.

We are concerned no longer exclusively with making nature useful, or with releasing mankind from traditional constraints, but also and essentially with problems of techno-development itself. Modernization is itself becoming reflexive; it is becoming its own theme.

Ulrich Beck (1992) Risk Society: Towards a New Modernity

But at a time when the threat of total annihilation no longer seems to be an abstract possibility but the most imminent and real potentiality, it becomes all the more important to try again and again to foster and nurture those forms of communal life in which dialogue, conversation, *phronesis,* practical discourse, and judgment are concretely embodied in our everyday practices.

Richard Bernstein (1983) Beyond Objectivism and Relativism: Science,
Hermeneutics and Praxis

The transfer of allegiance from one paradigm to another paradigm is a conversion experience that cannot be forced.

Thomas Kuhn (1962) The Structure of Scientific Revolutions

The most important feature of the systems approach is that it is committed to ascertaining not simply that the decision maker's choices lead to his desired ends, but whether they lead to ends that are ethically defensible.

C. West Churchman (1971) The Design of Inquiring Systems

Introduction

A quarter of century has now passed since the meeting at which the entire Agriculture School of 50-or so 'aggies' at Hawkesbury Agriculture College, that I had just been appointed to 'lead', took four decisions that would mark the beginnings of a radical approach to curriculum reform (later to embrace the whole issue of 'rural transformation') that has persisted virtually to this day.

1   We would 'go back to basics' and start the whole process of curriculum reform afresh with the exploration of two deceptively 'simple' questions:

   a   What did we mean when we spoke of agriculture? (especially in the context of what role it played in the development of rural Australia – past and present),

   b   What did we mean when we spoke of education? (especially in the context of learning and development, and thence curriculum and pedagogy).

2   We would focus our attention particularly on the identification of those competencies that we believed could/should be needed by the next generation of agricultural/rural practitioners to meet future challenges as well as deal better with the present.

3   We would 'situate' our explorations and discussions within a context of:

   a   An appreciation of the complexity of both current and recent historical changes not just in Australian agriculture, but in the communities and environments in which rural people lived and worked, especially in New South Wales.

   b   A critical concern about the nature of the existing services that were available to rural people, especially higher education, extension, and research, the policies they reflected, and the paradigm that 'drove' them all.

   c   An acute awareness of the dynamic circumstances that were characterizing higher education in Australia at that time, especially in agriculture.

4   We would approach our tasks as a team of 'action researchers' or 'community of action learners', creating and participating in an overarching action research programme as well as in specific projects in the development of our curricula, of ourselves, of our organization, and of the networks in which we become embedded. In the process we collaborated closely with each other, as well as with various other actors who were concerned with rural conditions in Australia and/or with curriculum reform in higher education per se.

In addition to the usual concerns about *theory* and *practice* in agriculture that engage the attention of all agricultural educators, we also committed ourselves, as *action researchers* to including concerns about the *theory and practice of practice itself* (predating what Donald Schon (1987) would later explore in his book *Educating the Reflective Practitioner*). We co-opted the word *praxis* to express the way practitioners use theories (of both types) to inform the practical actions that they take and reflections on those actions to trigger the search for new theories. We thus came to accept *praxis* as a human 'property' that

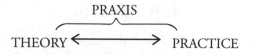

Figure 14.1 *Praxis as an emergent property*

'emerges' through the constant everyday interplay between theory (or understanding) and practice (or action), or as the dynamic synthesis between the *inner world of abstract, concepts* and the *outer world of concrete experiences.*

The foundations of our praxis

The work of David Kolb (1976) would prove especially invaluable to us in helping us to 'operationalize' this crucial inter-relationship through the process of *experiential learning* that he presented as a cycle involving 'four adaptive learning modes': *concrete experiences, reflective observations, abstract conceptualizations* and *active experimentations*. While we would make a number of very significant changes to Kolb's basic model as our work progressed, the concepts that he presented remain foundational to the Hawkesbury *praxial* endeavours to this day. In an early adaptation, we converted the 'cycle' to a 'flux': between 'finding out' (and the generation of theoretical or propositional knowledge) and 'taking action' (and the generation of practical knowledge) as that was what we saw ourselves actually doing in the development of our *praxis* as experiential action researchers – *with experiential knowledge emergent*. As the model in Figure 14.2 then suggests, learning/researching combines a continuing 'recursive' flux between four essential activities: (i) finding out in the concrete (*observing*), (ii) finding out in the abstract (*thinking*), (iii) taking action in the abstract (*designing*), and (iv) taking action in the concrete (*acting*).

After a while, we would come to appreciate the systemic significance of inter-relating these four 'domains' as we began to explore in some detail Churchman's (1971) powerful notion of the inquiring system and Checkland's evocative account of soft systems methodology in his book *Systems Thinking, Systems Practice* (Checkland, 1981).

Praxis is the emergent property of the (experiential learning) process through which we are continually trying to make sense out of the world that we are experiencing about us, as the basis for making adaptations both to ourselves and to 'it' in order to 'find a better fit'. We transform our experiences into knowledge and knowledge into knowledgeable action – with both 'conditioned' by (i) the beliefs that we hold about knowledge itself and (ii) about the nature of reality and (iii) by normative values that we hold. At heart, then, our praxis is our basic way *of being* human. *To be is to learn* – to live the constant dialectical flux between 'finding out' and 'taking (adaptive) action': where 'theory' (propositional knowledge) can be regarded as the key emergent output of the former, and 'practice' (practical knowledge) the output of the latter. Changing the way we take actions in the world is highly interdependent with changing the way that we perceive and value that world – innovation is thus grounded in re-perception.

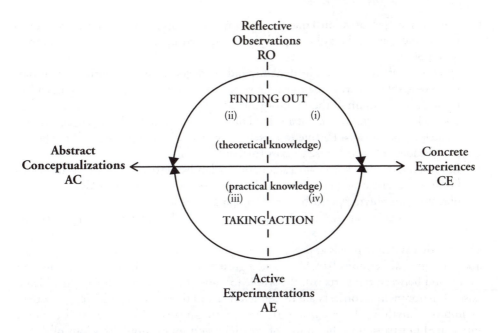

Figure 14.2 *Learning as the flux between finding out and taking action (adapted after Kolb)*

Systems principles and practices are vehicles for re-perceiving and just as the experiential process has provided one of the key foundations of the Hawkesbury work, *systemics* (theories and practices from the systems sciences), has provided another. Because of its significance, it is relevant at this point to digress a little further from the tale, with a brief explication of the systems idea of wholeness through dynamic inter-connectedness – which is born of a philosophy of holism:

- Systems are 'bounded', coherent whole entities that have properties that are unpredictably different from the sums of the component parts that are embedded within them, as well as from the environment in which they themselves are embedded. Systems are essentially regarded as *systems of systems* (*holons*), as both the component parts ('lower order' or sub-systems) (*ss*) of any system (*S*) in which they are embedded, and the 'environment' in which the systems themselves are embedded, are also systems ('higher order' or supra-systems) (*SS*).
- While this [*ss-S-SS*] hierarchy *(holarchy)* can obviously be explored, conceptually at least to an infinity of levels, there are distinct practical and conceptual advantages in dealing only with the three levels of organization here identified.
- The unique properties that 'emerge' at each of the levels of the *holarchy* are assumed to be functions of dynamic inter-relational tensions (a) within the component sub-systems (*ss-ss*), (b) between the sub-systems and the system (*ss-S-ss*) as a whole, (c) within the system (*S-S*), (d) between the system and its supra-system (*S-SS-S*), (e) within the supra-system (*SS-SS*) and (f), between the supra-system and the sub-systems (*SS-ss-SS*).

- Any changes (perturbations) that occur in any part of this holarchy can trigger changes in any other part, and may be amplified or attenuated and be instantaneous or delayed in response.
- Systems are therefore typically dynamic and characteristically in continual co-evolution/co-adaptation with the dynamic supra-systems in which they are embedded, through the medium of their self-organizing abilities.
- For a self-organizing system (at any level of organization) to be able to engage in such co-adaptation, it must be able to match the variety of the supra-system in which it is embedded, or it will cease to exist (mal-adaptation).
- Information feedback is vital to adaptation: adaptive systems may seek 'static' equilibrium – attenuating perturbations through *negative feedback* – or 'dynamic equilibrium' – amplifying perturbations through *positive feedback*.
- Systems in states that are 'far from equilibrium' are said to exist at the 'edge of chaos'.

Those with a systemic praxis approach the world both with a profound sense of wholeness – in three dimensions [*ss.S.SS*] – and a great sensitivity to the inter-connectedness within and between the parts (*ss*), the whole (*S*), and the environment (*SS*) in which it exists. Environmental contexts are therefore crucial to those who would design systems to 'improve' anything. To *systemists* furthermore, the surprise of 'emergence' is not only anticipated but trusted as the source of creativity and innovation! 'Tensions of difference' are therefore encouraged and nurtured, while it is appreciated that as emergence can have both positive and negative impacts on events and phenomena in the world, there has to be a consistent concern for the implications of any actions that are designed to be taken, in the pursuit of development.

In synthesis:

Because of the 'nested embeddedness', observers can identify whichever level within a holarchy, as the 'system' that they wish to study, and then identify both supra-system and relevant sub-systems, as they judge appropriate. These are constructions after all – although as we were to come to learn, there are significant ontological and epistemological issues that relate to the 'realness' of systems in the world!

I can now use these two ' images' of the experiential process and of 'systemics' in synthesis, to illustrate the hypothesis, that, in retrospect, one can recognize three different phases of development at our School through changes in our essential conceptual and methodological foci – (1) *researched systems*, (2) *researching systems* and (3) *critical systemic discourse*. Each phase was also characterized by our activities, by prevailing theory sets and worldviews, and by our collective *praxis*.

- *Phase 1:* Researched systems which we developed our *praxis* as facilitators of the learning of undergraduate students (UGS) through the *study of agro ecosystems*.
- *Phase 2:* Researching Systems in which we developed our *praxis* as facilitators of the learning of UGS, of graduate students (GS), and of project collaborators (PCS), through the *practical application of systems methodologies* to 'real world' issues.
- *Phase 3:* Critical Systemic Discourse in which we developed our *praxis* as facilitators of 'critical learning systems' UGS, PGS, PCS and a wide spectrum of other stake-

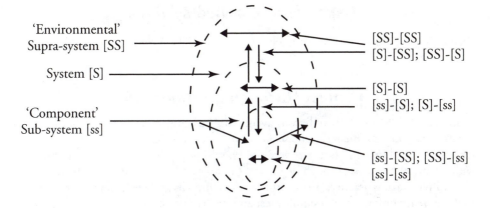

Figure 14.3 *'System of Systems': A coherent network of embedded inter-connections*

holders in agriculture and rural affairs, through *systemic discourse* across an extensive 'actor' network.

Observations on a changing praxis

A few points need to be emphasized before I attempt to describe some of the details of each phase and what led specifically to the transition from one phase to the next. Each change was multi-causal and 'emerged' essentiality from systemic interactions between:

- the dynamic contexts in which we were operating – both external to the School and within its internal milieu;
- the dynamic profile of the faculty through retirements/resignations, secondments and recruitments;
- the dynamic composition of the student body with respect both to the 'type' of student who enrolled and the spectrum of courses and programs that were on offer;
- the dynamic growth and development of the networks in which the faculty and students 'engaged' beyond the School; and most significantly
- the synthesis of ever-changing practical experiences and ever-developing theoretical understandings of the faculty, the students, project collaborators and a wide range of 'stakeholders' with an interest in agriculture, rural community development, rural environment restitution etc.

It is also important to stress that each phase transition resulted not in the total rejection of what had preceded it, but in the *'embeddedment'* of many of its elements and characteristics. Typically 'the seeds for change would be sown' long before the 'transition' actually became apparent with the introduction of new programs, or changes in the way matters were organized, or the *collective praxis*.

Phase 1: researched systems

Facilitating the learning of undergraduate students (UGS) through the study of agro-ecosystems.

Context (*External environment*)

In the late 1970s, rural Australia was awash with urban generated *blame* that was triggering very considerable resentment out in the countryside. Fault was being found in virtually everything that farmers were doing: selling produce that was contaminated with chemical biocides; through degrading the soils and landscape through cultivation and animal grazing practices; being cruel to their animals and neglecting their 'rights' to welfare; being unfair to their employees, and neglecting their rights to adequate well-being; being positively hostile to aboriginal claims for their rights to land title.

The fault-finding was not all one way however: for their part, farmers were blaming law makers for abandoning them in the increasingly dire economic times of high inflation rates, high interest rates, and low commodity prices; scientists for following their own research agendas that seemed to be of little help in providing alternative technologies for husbandry practices that clearly had 'bad' environmental and/or social 'side-effects', or alternative strategies to help them deal with the pervasive drought that refused to break; economists for their exclusive focus on productivity gain and their unfettered embrace for market liberalization and structural adjustment; the financial institutions for foreclosing on their loans and forcing so many among them to first fire their helpers and then abandon their own farms; other rural businesses for closing up shop in the face of a pervasive rural recession marked by high rural unemployment and low disposable incomes; state bureaucracies for reducing health, education and technical extension services and doing little to deal with rising crime and violence; the 'environmentalist greenies' for 'overstating' the environmental impacts of farming practices; and the media for exacerbating everything!

The irony and tragedy both, were that the vast majority of Australian farmers were acutely aware that in their quest to become more productive, they were paradoxically creating conditions that would make further increases in productivity increasingly difficult to achieve in the future. The more they sought security through the intensification of their practices, so the more vulnerable they risked becoming, through the environmentally and socially destructive impacts of many of those practices.

The College was also having to deal with the internal turbulence of a new identity. In 1975 the institution had gained its independence from the state Department of Agriculture, of which it had been a part since its establishment in 1891, becoming an autonomous, but still publicly funded, College of Advanced Education (CAE) – itself only a relatively new institutional category in Australia and regarded by most beyond them, as somewhat inferior, through its essential emphasis on teaching, to the university sector which offered four year undergraduate degrees, research graduate programmes and had strong research profiles.

Following the transformation to CAE status, Hawkesbury had been structurally re-organized into three academic units: the Schools of Agriculture, of Food Science and Technology, and of Management and Human Development for which Heads were then sought by open recruitment. This was not a universally supported move – and nor was my appointment as Head of Agriculture, hailing as I did from the university sector. The School of Agriculture itself was structurally organized into five departments – four of which offered technocentric majors within the six-semester diploma programme in Applied Science (agronomy and crop sciences, animal husbandry, intensive animal production, and agricultural engineering) with the fifth (agricultural biology) being a service department. A 'no-specialization' major was also offered. This academic fragmentation not only perpetuated disciplinary isolation but also tended to be reflected in the 'social isolation' of the different groups.

Part of the internal tensions in this regard were associated with a clash of internal mini-cultures between the majority of the faculty who were 'old school' employees from the state bureaucracy, and a relatively small cadre of PhD-qualified scientist/educators who had recently been appointed to leadership positions in the School.

Recent history had not been kind to Hawkesbury: over the years from the 1950s onwards in particular, the College had suffered significantly from strong competition for both resources and students, from the Faculties of Agriculture (or equivalent) of three universities within the State, as well as from one other technical institute and a couple of 'sub-tertiary' farmer training institutes. By 1978, such competitive pressures had left the research profile at Hawkesbury markedly diminished, the extension function virtually extinguished, the quality of the enrolling students less than optimal, the level of funding support in significant decline, and faculty morale at a disturbingly low ebb.

In secular Australia – and especially within the academy – the Hawkesbury community was unusual in the high proportion of its faculty who were overtly religious, in the sense that the often openly expressed their faith. It would be accurate to state that a strong Christian ethic prevailed across the School, with such values as duty, care, empathy, trust and integrity highly regarded as virtues. Personal relationships were also valued highly and the Australian commitment to democracy, equity and a 'fair go for all', clearly also prevailed within the institution.

Theory/practice and praxis

As mentioned at the start of the chapter, one of the very first things that we did at the launch of our endeavours was to agree to commit ourselves to becoming a 'whole community' of collaborators, learning with and from each other, as we investigated our seminal questions. In the process we would come to learn a lot about each other too (as well as about self), and about the way each of us learned, as well as about 'community'. We literally *learned to learn* our collective way forward, developing new modes of communicating with each other (and later beyond, into other communities) and sharing ideas and experiences as we went – even in the pre-email days.

We would invest considerable effort and resources into our own personal development – not only as individuals but also as a coherent community.

While initially maintaining the discipline-based departments, we also organized ourselves into a number of cross-disciplinary action researching teams to work on specific tasks, both fixed and emergent. In this manner we evolved a matrix organization in which every academic was concurrently a member of an academic department and a member of one or more task groups. These were much more than conventional 'committees'. Every group was established within the spirit of an action research team, concerned not just with the particular 'brief' – the matter to hand as it were – but also with the matter of action research praxis itself. A strong emphasis was placed on the need to integrate 'finding out' with 'taking action', and to incorporate critical reflections on both.

The theme for all of these groups was development through learning/researching. Teams were thus formed to 'develop' various aspects of curriculum, of resource accessibility, of the farm enterprises, of communication, of external 'networks of collaborators, and even our own community itself (with a programme of 'in-reaching'!), through learning.

It is important to emphasize that we saw our endeavours as responses to challenges we were identifying as matters 'real' to the everyday experiences of rural people in general and to farmers in particular.

And these challenges were clearly far more complex and messy than the technocentric view that prevailed as the conventional approach to development could entertain. We decided that we would do what we could to transcend these limitations from a position built, as emphasized earlier, on the twin foundations of (a) experiential learning and (b) systems principles – although any connections between these two that we made initially were purely intuitive. The powerful connections that we would establish between the two would come later.

Our decision to privilege experiential learning came essentially from the experiences of a number of the faculty with extension education, over in the School of Management, and familiarity of a number of us with the writings of Kolb (1976), Knowles (1975), Rogers (1969), and most especially, Freire (1972). Inspired also by the work of Reason and Rowan (1981), we were to accept experiential learning as a way of knowing that was complementary to the other two key ways of knowing – propositional learning and practical learning – with which we felt it needed to be integrated.

Meanwhile our choice of investigating the potential usefulness of systems approaches to agriculture and rural development also owed something to the personal experiences of some of us with various forms of systems analysis. We also learned from the writings of Spedding (1979), Ruthenburg (1971), and Dent and Anderson (1971), with reference to the nature and analysis of agricultural systems, and from Dillon and his colleagues (1978), and later Shaner et al (1982), Norman and Collinson (1985) and Fresco (1984) on the emerging field of farming systems research and development. The work of Conway (1985; 1987) on agroecosystem analysis and development, and especially with his recognition of sustainability as an 'emergent' property, would later prove to be of special significance.

Looking for a way forward from what we had identified through our researches as some of the inadequacies of the reductionist approaches to development that were prevailing in the late 1970s/early 1980s, we were inspired by Dahlberg's (1970) submission that 'most intellectual maps of agriculture fail to recognize in it the basic interface between human societies and their environment'. We would transcribe that into systems terms: agroecosystems lie at the interface between natural and social systems which not only act as the source of inputs and the sink for outputs, but also of 'perturbations' and 'forces', with the power to exert very significant influence on the performance and stability of the agroecosystems that they frame.

With this 'construct', we were departing in a number of significant ways from the views that were being promoted at that time by those writing about agroecology from

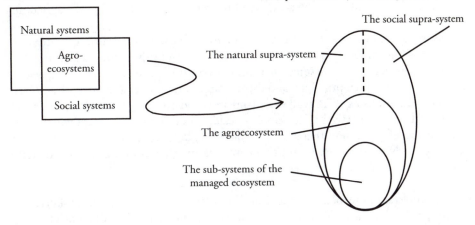

Figure 14.4 *Agroecosystems as the interface between natural and social systems*

an ecological perspective (rather than a system perspective) such as Cox and Aitken (1979) and Hart (1982), and later from the work of Lowrance et al (1984), and Altieri (1987). While the work of these and other authors was extremely insightful and very useful to us in framing our thoughts and practices, the focus of our concerns was somewhat different. Thus:

- we were as much concerned with the systems nature of the social and natural supra-systems as we were with agroecosystems per se;
- we viewed the managing sub-system essentially from a learning perspective – in other words, our central key concern was with the processes that farmers used to first understand the 'systems' nature of the world about them, and then to take action that was informed by those 'systems' understandings;
- the role of those who would graduate from our programmes would be on facilitating an awareness by farmers (and others in rural Australia) of the 'systems' nature of their farm enterprises and of the environments in which they managed those farms;
- we were therefore committed to providing a pedagogy that focused on the twin goals of 'learning' and of 'systems approaches' to agriculture.

Rather than 'motivating farmers to accept change' (Hildebrand, 1982) in their farming systems, we were most interested in having our graduates develop the competencies to help them learn how to approach their farms from a systems perspective.

After a number of pilot projects with the two-year diploma programme that we ran, we introduced the first undergraduate degree that Hawkesbury had ever offered in agriculture in 1981. The seven-semester programme would lead to Bachelor of Systems Agriculture, and it would differ radically from the six-semester diploma in applied science in virtually every way conceivable. It would be structured across three essential stages: three semesters on campus, followed by one semester living and working on a commercial farm somewhere within the state of New South Wales, and then three semesters spent working both on campus and in a variety of off-campus projects within the agricultural sector of the state.

We were somewhat ambivalent about the role of experiential learning strategies for this programme choosing to rely more heavily on them in the second and third stages while introducing them in a somewhat limited way, during the first stage. We believed that we needed to teach principles and theories appropriate at least to 'natural' and 'managed' ecosystems, using conserved woodland areas and our own college farms as 'experiential' field sites, respectively. The most radical innovation in this context would be the introduction of the notion that students could best learn about social systems by experientially studying themselves and their own learning processes, as they operated in 'learning groups'. In essence:

- We would shift the focus from teaching knowledge about, and technical skills related to, production agriculture, to facilitating the competency development of learners through guided experiences in the study of 'systems' – including themselves as social systems. The context would be 'persistence' rather straight productivity.
- All students would spend a dull semester living with, working for, and learning with and from the members of a 'host' farm family in rural New South Wales.
- As faculty, we would facilitate systems studies, meet with an assigned group of learners once a week, visit each individual student while they were out on their host farms, and prepare resource packages in our disciplines that students could access on demand.
- We would also continually monitor events while continuing with our collective learning as a faculty group.

While this was indeed the pedagogical pattern that was adopted in the School at the outset it was soon to take a significant turn for there were many among the faculty who were dissatisfied at the lack of connection between the experiential process that was serving the faculty community so well in their learning, and the School's 'new' pedagogical practices. With the focus on learning through the study of systems, students had no real practical output of their work in development terms, and the 'teachers' were still 'teaching' much more than 'facilitating' learning. There was wide acceptance within the Hawkesbury community that while the new programme was certainly 'different' it could still be significantly improved.

Furthermore, events happening in and to rural Australia were gaining a new momentum of urgency, as economic conditions continued to worsen, voices in opposition to agricultural practices continued to get ever louder, and rural people continued to become increasingly anxious to the point, often, of sheer desperation and despair. Conflict was becoming rife within the communities themselves, and the vulnerable and disenfranchised were clearly suffering most. The issue was clearly much more complicated than merely changing farming practices alone, even if such a shift was to embrace those organic, biodynamic or permacultural methods that some farmers and academics were beginning to promote.

Most significantly, the work that we were reading in the literature on systems approaches to agriculture at that time (1) remained systematic rather than systemic, (2) were either descriptive or highly mathematical, and/or (3) failed to capture the systemicity of the environmental 'supra-systems' in which the 'systems' operated. While, as mentioned earlier, Conway (1987) would later add some very valuable dimensions to the issue, as would Marten (1986), the work of the early writers in farming systems and agroecology still focused essentially on productivity as the focus of their work, with their research efforts restricted essentially to studies of relatively, simple sets of relationships.

Rather than being a matter of learning *about* agricultural, natural and social systems, perhaps the more fundamental issue was learning how to use systems methodologies.

Phase 2: Researching systems

Facilitating the learning of UGS graduate students (GS) and project collaborators (PCS) through the application of systems methodologies in 'real' world projects.

The first of the 'phase transitions' following the introduction of the new endeavours in 1978 started to manifest itself around 1985 – with seeds sown earlier. As it happened, it was soon after we had hosted an international conference on agricultural systems research for developing countries, hosted by the Australian Centre for International Agricultural Research (ACIAR). (Note: Within a further decade we would host two more international conferences – the Second World Congress for Experiential Learning in 1989 and, in 1994, the first meeting of the International Society for the Systems Sciences held outside the US.)

By 1985/1986 it had become apparent to most of us, that the 'systems' competencies that our students were gaining, were not that much more relevant to farmers than the technical knowledge and skills that we had replaced. It was also becoming obvious that in espousing the need to embrace 'social' and 'natural' system dimensions as well as agroecological ones in our curricula and the conceptual 'maps' that informed them, we needed to broaden our focus beyond agriculture *qua* agriculture – and certainly to broaden out methodological emphasis and indeed our praxis.

Context (*External environment*)

Through the mid- to late 1980s rural Australia came to be even more turbulent than before – with government policies based on market liberalization and structural adjustment being associated with even greater levels of anxiety among rural people.

At the same time however, a grass-roots movement of landcare was beginning to emerge. Traditionally 'ruggedly individualistic' farmers had begun to work together with other members of rural communities, and adopt cooperative group practices to reduce soil degradation across whole catchment areas, while also coming together to discuss other matters of environmental restitution and conservation.

Our students were being drawn into these activities during their host farm experiences, and of course the faculty were also 'seeing it first hand' as they travelled the state visiting their students. The latter were also beginning to access the emerging calls for 'sustainability' in the literature, with a seminal work on agricultural sustainability by Douglass (1984) being of particular importance with its emphasis on the multi-definitional nature of the concept.

The most significant aspect of our transition, however, was the explicit acknowledgement that our focus would shift from the 'systems' to 'people' – our emphasis on development would be changed from the development of things to the development of people, and especially of their learning capabilities.

Meanwhile, by the mid-1980s the activities of the 'Hawkesbury School' were beginning to attract considerable public attention: (1) through the extra-campus 'field' activities of students and faculty, who together had started to also conduct forums in rural districts on the 'hot issues of the day'; (2) through the publication of a number of articles in journals of education, agriculture and rural development; (3) through the presentation of seminars and conference papers by faculty, both within their professional societies and beyond, in industry-backed association meetings and other public forums; and (4) through exposure in the national press, including radio and television.

In the 1980s too, a number of international invitations were received that allowed faculty to engage with academic and other development initiatives in a number of different parts of the world including South East Asia, the Indian sub-continent, Europe, and the US.

It would be a serious omission not to mention the fact that not all of the responses from this increased public profile were positive, indeed one of the major challenges that we would now have to incorporate into our praxial developments as faculty, was the capability of handling criticism and conflict from often surprising sources. We were to discover that reason alone would not prevail! It was becoming very apparent that many others did not share the same view of the world that we were developing, and differences in worldviews, we were finding out, were rarely resolved by rational debate alone.

One of the first actions we took to 'signal' our intentions to change further was to argue within the institution for both our status and name to be changed. On this occasion reason did prevail and we were officially transformed from the School of Agriculture to the Faculty of Agriculture and Rural Development. The challenge now was to live up to the name, and embrace theories and practices that would be relevant to the more comprehensive compass.

Other changes within our internal milieu were soon to have an even greater impact. In 1987, the Faculty was greatly enhanced by the transfer to it from another Faculty, of the Centre of Social Ecology. This brought to the internal discourse a wonderfully vibrant and expanded 'set' of worldview perspectives from a faculty who were almost entirely social scientists, and, of whom, a significant proportion were women. In the following year, considerably reinforced by our new colleagues, we were able to accommodate the further momentous change of our incorporation, with two other regional CAEs, as the constituent members of the new University of Western Sydney as our response to the national fiat for all CAEs to become incorporated into universities – extant or new. At Hawkesbury this would now bring opportunities to introduce graduate degree programmes for coursework and research Masters as well as PhD programmes and allow access to federal funding for research, as well as to colleagues from a host of new academic disciplines.

The shift to university status would also introduce very significant turbulence with respect to critique of what we were doing within our faculty from what was now our own constituency, coincident with a time at which we were transformed from a relatively large fish in a small pond to a small fish in a very big pond. A dramatic shift in power and indeed in ethos had occurred through the amalgamation, and it would not be too long before the forces of conservatism would start to have an influence on the way that we did things in our group.

Theory/practice and praxis

The key intellectual contribution to our first phase transition, was initiated by two colleagues following their return from sabbaticals in England in 1983, where they had encountered the work of Peter Checkland and his colleagues at the University of Lancaster. The thrust of the Lancaster work was based on what was claimed to be a 'shift in systemicity from the world to the process of inquiring into that world' – a theoretical perspective that had been practically translated into what was being referred to as Soft Systems Methodology (SSM).

The essence of the approach was extremely appropriate to our developing circumstances. The Soft System approach to development is based, first and foremost, on 'learning' or 'researching' in contrast to the 'optimizing' focus of the so-called hard systems approaches. The emphasis is on situation-improving rather than on problem solving, and the process involves both the identification and participation of various stakeholders in the quest for desirable and feasible changes to 'situations of unease' to themselves. A key aspect of the methodology is the identification by various stakeholders that differences in opinions about what actually does constitute improvements in shared circumstances is essentially a function of differences in worldviews – in the language of the SSM practitioners, all transformations (T) are functions of *Weltanschaungen* (W). This would prove to be a critical insight for our work. So too would be our realization that not only was the SSM essentially a variation of the experiential learning process as we had come to envisage it, but so too were most forms of 'rational' inquiry.

Back at Hawkesbury, these two academics organized a small 'task force' to apply this methodology to search for improvements to our own activities as a Faculty that would constitute that which was both 'socially desirable' and 'culturally feasible'. A number of senior students were invited to join the task force from which a number of key recommendations for change eventually emerged. Again following the SSM format, these were then debated within the entire faculty group with the following outcomes emerging:

- A switch in the emphasis of the undergraduate programme away from the study of agricultural, natural and social systems 'out there', to the use of SSM to seek improvements in problematic situations identified by a host of stakeholders 'out there' – the shift from a focus on *researched agroecosystems* to *researching systems for improving rural development* (Figure 14.5). In this manner, students and faculty alike could develop praxes as 'participatory systems methodologists'.
- The development of a 'system of methodologies' that embraced a range of different 'researching systems' appropriate to address different types of issues. So in addition to learning how to use systems methodologies (especially SSM) our students (and we) would learn how to work with a whole system of different experiential methodologies with foci that ranged across a spectrum that included:
  - relatively simple puzzles to be resolved (reductionist science) and problems to be solved (reductionist technology);
  - system's performance to be optimized (the methods of 'hard systems');

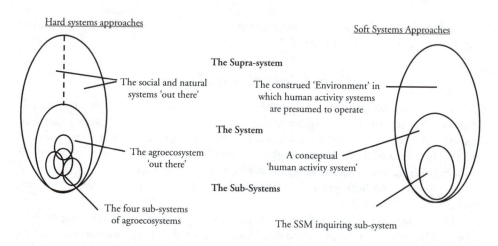

Figure 14.5 *The essence of the shift in systemicity from the world 'out there' to the process of seeking improvements to situations recognized 'out there' by those affected by them*

- complex situations to be improved (the methods of 'soft systems');
- complex conflict-laden situations to be managed appreciatively (the methods of 'critical systems' prompted by the work of Ulrich (1983) and Jackson (1985)).

- The greatly increased level of 'engagement' of faculty with other 'stakeholders' in the affairs of rural Australia, initially at least, through their involvement with undergraduate students as collaborators in the projects that the students initiated. Over time, this level of 'engagement' would be greatly enhanced through collaborations with graduate students in their participatory projects, as well as through self-initiated participatory research and consulting activities as individuals or as teams that often also involved researchers from other institutions. Through all of this work the focus on learning, systems and development would be maintained.
- The increased concern for philosophical dimensions of the work: both ethical and epistemological. From the establishment of ourselves as an action researching team, we had been concerned both with the nature of our own relationships and the normative aspects of our work together, as well as with nature of learning as well as knowledge itself. These concerns were never really far from the surface in influencing the activities that we designed and conducted. Working now with a focus on the idea of 'improvement' as a function of worldviews, it became apparent that we needed to be even more explicit in the way we dealt with the philosophical dimensions of the work that we did. In 1987 the 'in-reaching' task force organized a faculty workshop on 'Epistemology: An exploration of our beliefs about the nature of knowledge'.

The approach to development that we developed during this second phase of the endeavour was much more enduring than the first. Each year the students (often with faculty as colleagues) would conduct more than 100 action research projects in total, as they learned to apply various methodologies from the 'system of methodologies' that we had developed, with a special emphasis on SSM. A number of honours and graduate students, as well as faculty, also started to work with the more critical systems approaches being developed at the time by workers such as Jackson (1985) and Ulrich (1983). Yet for all of these efforts and the obvious impact that we were beginning to have in rural Australia, we were becoming increasingly sensitive to the fact that: while students and faculty were becoming more comfortable with the use of new systems methodologies, these 'process' benefits were not often being extended to our 'project collaborators'. They all too often continued to perceive themselves as the objects of research rather than as participant systems methodologists. They were certainly benefiting from their involvement in terms of the 'content outcomes' of the work, but there was not a general appreciation of the systemic nature of the methodological process. And indeed that was also disappointingly so for many of the students and faculty too.

While many were learning to become proficient in the use of systems methodologies, they were not necessarily *learning to be systemic* – learning to adopt a systemic/holistic view of the world. Indeed it was increasingly obvious to us as faculty that while the processes that we had developed were often very effective at revealing differences in worldviews between different stakeholders facing the same problematic issue, we were far less effective at facilitating changes in those worldviews. While both our students and we were becoming quite good at 'diagnosis of difference' and at recognizing their association with conflict, we were being far less effective at therapy (either curative or prophylactic)!

Perhaps of most concern to us, however, was that we were still failing to find ways of integrating methods for addressing the ethical and aesthetic dimensions of development into the methodologies that we were practicing and promulgating. The strange thing was that for a long time we failed to make the obvious connection between the intensity and dynamics of our own discourse as interrelating faculty-cum-students, and the need for similar 'inter-relational discourse' in, as well as about, the affairs of rural Australia.

Rather than being a methodological matter, perhaps the more fundamental target for change needed to be paradigmatic.

Phase 3: Critical systemic discourse

This phase involves facilitating the learning of UGS, PGS, PCS and a wide spectrum of other stakeholders in agriculture and rural affairs, through systemic discourse across an extensive 'actor' network.

In retrospect, the second phase transition at Hawkesbury probably began to become apparent in the early 1990s particularly as the levels of 'engagement' by faculty and graduate students with 'stakeholders' in the development of rural Australia started to reach significance – although, as before, many of the ideas and experiences that were seminal to the change could be traced to earlier years. The emergence of the third phase would be accompanied by an unexpected regression in some activities of the faculty, especially in the curriculum, back to Phase 1 – and in some ways even earlier, to pre-innovation days! To many observers, this was a considerable surprise, especially given the strong convergence between what was now beginning to happen in rural Australia and the continuing developments at Hawkesbury. And indeed given the role that many of the young graduates from our programme were beginning to have, as well as faculty and current students – both graduate and undergraduate.

Context (*External environment*)

In 1989, a very powerful (and unanticipated) coalition of the National Farmers' Federation (NFF) and the Australian Conservation Federation (ACF) submitted a proposal for a National Land Management Program to the federal government that resulted in the adoption of the Decade of Landcare and a $320 million allocation over ten years to support rural environmental restitution. As part of that initiative a new category of professional practitioner was created – the Landcare facilitator – to work with self-forming land care groups across the nation. Not surprisingly, Hawkesbury graduates would dominate the ranks of this new profession.

The Landcare movement would herald a range of new initiatives in rural Australia for which the momentum continues. Few services would remain unaffected as a new ethos of 'holistic development' began to take root across rural Australia. Major changes started to occur with the way governments started to allocate grants for conservation practices directly to local landcare groups in contrast to the previous route of state agencies or universities. Extension agencies, long focused on technical matters related to production and productivity, started to recognize the need to work 'more holistically', and even specialist agricultural scientists within the academy, and the state and federal research institutions, began to accept the need for multidisciplinary approaches to their work.

The publication, in 1987, of the Report of the World Commission on Environment and Development (Brundtland, 1987) with its emphasis on sustainability would provide a compelling new focus for calls for fundamental reforms in the way things were done in rural Australia. It would lead, many years later, to the federal government accepting the notion of Ecologically Sustainable Development (ESD) as the framework for national policy and strategy.

In the late 1980s, however, the view had started to emerge that farmers were stewards of the Australian heartland and managers of the landscape, as well as the producers of food and fibre to secure domestic markets and earn substantial foreign exchange on international markets. The awareness was growing that they should be rewarded for these multi-functions by the Australian society at large. Landcare was emerging as an issue that involved all Australians. As a movement it was also beginning to illustrate the significance of moral, ethical and aesthetic dimensions to the concept of 'development' and, in the process, challenging both the prevailing utilitarian ethos of production agriculture, and the prevailing reductionist paradigm of agricultural science.

Rural women started to find a collective voice as they started to establish groups of their own and connect up with one another through ever-extending networks in which a number of graduate students in the coursework Masters program were particularly active. And in 1991, the High Court of Australia passed the landmark judgment recognizing, and arguing for, the rights of aboriginals to 'native title' to land following years of appeals from aboriginal claimants. In both instances, these initiatives were also accompanied (and clearly amplified) by growing activism within Australian society at large.

Both sets of initiatives further amplified the calls from other activist groups, including those concerned with animal welfare, with food safety, and with environmental integrity, for ethical considerations that went beyond consequentialism to also address matters from a 'rights-based' (deontological) moral perspective.

A new *discourse* that was both *critical* and *practical* was beginning to characterize rural affairs in Australia around a theme of sustainability – a discourse that shared many aspects in common with the discourse that the faculty at Hawkesbury had established around 'persistence' back in the late 1970s. Stories were being told, listened to, and heard.

Ironically, while things were 'unfreezing' with respect to worldviews of agriculture and rural Australia, they were 're-freezing' within the academy, and a new 'old' discourse, that combined corporate style structural adjustment with a 'back–to-basics' educational ethos, was beginning to find voice. The innovativeness that had marked the CAE sector, in particular, started to be quenched as the 'culture' of the old traditional conservative universities started to pervade across what was now one academic sector. (It should be emphasized that all but a very small proportion of Australian universities were, as they remain, publicly funded state institutions.)

The University of Western Sydney was certainly not immune from the general trends toward both corporatization and conservatism, although it did take several years after incorporation for these effects to really manifest themselves as a cultural 'regression'. What did happen almost immediately, however, was an enormously demanding 'distraction' into matters concerned with the development of the new institution itself. Time previously spent in intellectual discussion with colleagues, or with students in projects, or, most importantly out in rural areas, was now spent in the plethora of committees that it takes to transform three 'technical institutions' into one comprehensive university. New rules and regulations for degree programs and courses were among the first issues to be placed on the agenda, along with faculty reward and recruitment procedures. Within a very short space of time indeed it was obvious that the practices and procedures that had supported our culture of innovation at Hawkesbury were very vulnerable indeed.

Two other matters were soon to further exacerbate the worsening circumstances of our 'terms of practice': severe fiscal restraint and new administrations (appointed with increasing frequency from the 'old' university sector) with power to flaunt. And flaunt they did. The undergraduate programme would become a major casualty of a new administrative culture that was both ill-inclined to support expensive visits by faculty to the countryside, and deeply suspicious and in the end non-supportive of experiential pedagogy. In the name of introducing 'rigour' to the educational practices of the old colleges, innovations were throttled, and in the name of fiscal responsibility, 'engagement' was virtually terminated.

For all that, we continued with our innovations, with a shift in focus towards to the graduate programs and emerging research and consulting programs.

Theory/practice and praxis

As with the first phase transition, our second 'turn' was informed as much by new (to us) theories and philosophies as it was by reflections on our practical experiences – which by now had extended beyond curriculum matters to also include many new initiatives of direct engagement with 'stakeholders' in both agriculture and rural affairs.

A significant cadre of graduate students, the largest percentage being international, added greatly to the continuing growth of both our scholarship and experiences – and thus praxis. And indeed the focus of the faculty would shift increasingly to graduate studies as coursework masters and research degrees, as it became increasingly difficult to continue the undergraduate curriculum in the face of the increasingly restrictive university rules and regulations. In many ways it was our helplessness in the face of these impositions that really sensitized us to the matter of institutional power and to the strong experiential parallels that we could draw between our own situation and that of rural people.

I might refer to the intellectual foundations of this second phase transition as an amalgam of a number of conceptual 'turns':

- *A critical* turn in both our understandings of systems (Ulrich, 1983; Jackson, 1985) and of emancipatory action research (Carr and Kemmis, 1983). The writings of Habermas (1972, 1984) would add crucially further to these matters.
- An *epistemological* turn in terms of our understanding of cognitive and moral development (Perry, 1970; Bateson, 1972; Kitchener, 1983; Salner, 1986; Fuller, 1988) with the work of Bernstein (1983), Belenky and her colleagues (1986) and Maturana and Varela (1988) all contributing very different but vital further perspectives.
- An *ethical* turn (MacIntyre, 1981; Thompson, 1988) where we would concern ourselves more explicitly with normative aspects of development and their role in public judgement concerning the constitution of 'improvements', with a particular emphasis on moral and aesthetic dimensions, and especially on 'ethics beyond utilitarianism'.
- An *institutional/political* turn (Giddens, 1984; Beck, 1992), and most fundamentally,
- A *paradigmatic* turn (Kuhn, 1962; Burrell and Morgan, 1979; Miller, 1983, 1985; Douglass, 1984; Brundtland, 1987; Milbrath, 1989). Where, in the previous phase the quest had been for methodological pluralism, the focus would now shift to the much more daunting task of paradigmatic pluralism.

The conceptual picture that a synthesis of the above matters was complex. The insights of many dozen other writers were also very influential in shaping our responses to the new challenges, but the ones cited above were of special significance to the emergence of the need, that we began to recognize, for us to learn to appreciate *a plurality of paradigms for responsible (sustainable?) development.* And once the features of a number of different paradigms did indeed become apparent to us, the challenge would come in trying to translate our understandings into practices appropriate to the participatory systemic approaches to development for which our graduates and we ourselves were becoming known.

It was this logic that drove the phase transition from the central concern with *systems methodologies* to *systemic development through critical discourse.*

As we confronted our own sense of growing unease with the inadequacies of our approach, we returned, in the true systemic spirit of recursiveness, to some of the earlier writers who had inspired us in different ways, and with different concepts. We then sought new connections between their ideas – adopted new worldviews perspectives if you will, to analyse and synthesize their insights. It was in this way that we came to inter-connect ideas on paradigms with key insights drawn from work on systemics, worldviews, sustainability and the cognitive process that Salner (1986) had referred to as *epistemic development.*

The complete story is too comprehensive to be told here in any detail. I will instead merely give a quick somewhat fragmented overview, complemented, naturally enough, with a last systems image. While it might appear now as a nice neat argument and smooth flow of logic, it was anything but that in practice – emerging only in 'fits and starts' over several years. Further details can be found in Bawden (1999).

- Douglass (1984) had presented the argument that three different 'schools' could be identified with respect to different interpretations of the concept of sustainable agriculture: (1) primary concern with capacity to continue to feed the world (technological sustainability), (2) primary concern to 'reduce non-harmonious practices' and minimize disruptions to 'biophysical ecological balances' (ecological sustainability) and (3) primary concern 'promoting vital, coherent rural cultures, and encouraging the values of stewardship, self-reliance, humility and holism' (holistic sustainability).
- From the work of Cotgrove (1982) and Miller (1983) on the psychology of 'environmental problem solving, the three 'schools' of Douglass could be re-conceptualized as the outcomes of three different cognitive styles. Their hypotheses allowed the identification of fourth 'mystic' style which would see the source of sustainability lying in 'individual consciences and morality' (Miller, 1983).
- The insights of Churchman (1971) and Checkland (1981) would allow a further interpretation of these 'schools' 'psychological styles' as *Weltanschauungen* or worldviews, which Kuhn (1962) had earlier recognized as being the essential features of paradigms. The work of Burrell and Morgan (1979) on sociological paradigms, introduced the notion that different paradigms could be distinguished from each other by differences in their epistemological, ontological, axiological and/or methodological foundations.
- Using ontological distinctions between reductionism and holism, and epistemological distinctions between objectivism and contextual relativism, our Hawkesbury team posited that the four 'schools' psychological styles' that they identified through the synthesis of previous work, could be re-interpreted as four different paradigms: (1) technocentrism (reductionism/objectivism), (2) ecocentricism (holism/objectivism), (3) egocentricism (reductionism/contextualism) and (4) holocentrism (holism/contextualism).
- This idea was then further extended to embrace the idea of 'paradigmatic development' reflecting epistemological and moral development, as first suggested by Perry, 1970. Over time, and as a function of epistemic challenges, individuals make changes in the epistemological/ontological beliefs that frame their worldviews together with changes in the value assumptions that they hold – 'progressing' from dualism (= objectivist reductionism), passing through multiplicity, and reaching contextualism. The sum of these changes, the Hawkesbury team posited, could be sufficient, under certain circumstances, to result in shifts in the paradigms to which each of us subscribe – and which might also result in collective paradigmatic shifts in situations where collaboration between individuals was close.
- In contrast to Kuhn (1972) and to Burrell and Morgan (1979), we also suggested that it was not necessary to reject one paradigm to shift to another, but that it was perfectly possible to hold on to different paradigms at the same time, using different ones contingently. Such paradigmatic pluralism would also endorse the methodo-

logical pluralism for which we had argued in an earlier phase of our work, while providing it with sounder philosophical foundations.

- The key to such epistemic development, we came to believe, was an enhanced appreciation of the nature of our worldviews and paradigms, and increased courage and capability to actually 'shift them'. The work of Kitchener (1983) and of Salner (1986) would prove very important in terms of the conceptualization of how increased awareness might be facilitated. Kitchener suggested that we are all capable of three levels of cognitive processing resulting in (1) knowledge about the matter to hand (cognition), (2) knowledge about the process of knowing (meta-cognition) and (3) knowledge about the nature of knowledge (epistemic cognition). Salner linked this concept with Perry's observations and suggested the crucial proposition that 'systems competencies only came through epistemic development itself achieved through experiential challenge'. Cognitive processing, we believed, could be equated to learning processes.
- This can now be captured with a systems image that reflects the experiential process at its core, working at all three levels of cognition and able to assume different paradigmatic states. In the model below (Figure 14.6), the four paradigms identified above are 'imagined' as the outcomes of four different 'windows on the world' through which the observers within a learning system 'see' the world that they are 'finding out about' as a prelude to 'taking action it'. The nature of this 'window' can be explored through epistemic processing while being influential at each of the other two levels as well.
- People could learn to develop systemic competencies, we were then to argue, through their participation in experiential collective learning situations where all three levels of cognitive processing were deliberately engaged, and where critical challenge and self-confrontation were pervasive. Such 'collectives' we would term 'critical learning systems' which could be conceptualized as an experiential process, in the familiar 'three dimensions', now referring to the three levels of cognitive processing.

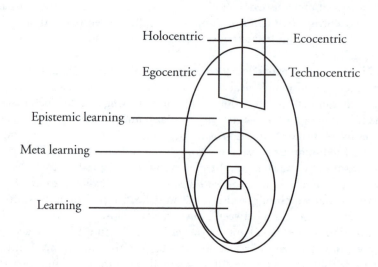

Figure 14.6 *A critical learning system*

'Critical learning systems' engage in critical discourse at all three levels, and, when 'developed', can utilize 'all four' paradigms (actually many more through the explication of further foundations). They concern themselves not just with 'matters to hand' and with all of the environments in which those matters are embedded, but also to their own development. They learn to address matters from ethical, aesthetic, and spiritual perspectives in addition to attending to the more conventional technical, social, economic and political dimensions.

Having said that, as the whole matter of politics and economics and technoscience, and all of the institutions that support them, are central to the need for fundamental reform in the context of sustainabilism (Bawden, they too must be embraced from a critical perspective and radical intent (Giddens, 1984), as indeed must the whole enterprise of modernity (Beck, 1992).

The aim of facilitation now is to assist in the development of 'critical learning systems' – strongly inter-connected and coherent collectives of people, aware of their own integrity and inherent systemicity, and active in systemic discourse appropriate to the issues in which they are embedded, including those of self-confrontation.

Postscript

And there it is – a brief tale from a road less travelled in which a group of us at Hawkesbury, over a period of 25 or so, changed our praxis in response to both the changing circumstances that we continued to face over that time, and the literature that we read as we went. Our twin foundations lay with experiential learning and systemics and, most significantly, we tried throughout to 'walk our talk'. We did the best that we could to develop as effective experiential learners ourselves – learning to interrelate theory with practice, reflection with action – even as we were learning to facilitate the experiential learning of others.

There have been some very significant, often dramatic, changes in rural Australia over that quarter century, and our students, graduates, project collaborators and we have been in thick of them. It is of course quite impossible to evaluate our actual impact but there is little doubt that we have made a difference. We have witnessed very significant shifts in the public discourse, in rural and urban society alike, with respect to 'sustainable developments in the countryside' and we have seen profound shifts in government policy and bureaucratic strategy. Most importantly, rural development in Australia now has a human face; it's about people trying to make sense of, and then take responsible actions in, complex, dynamic environments.

There is, however, a very long way yet to go. The changes that have occurred in development practices and policies, though very significant, have not yet really got to the heart of the matter that we really do not know what it is that we have got to do to achieve what Prozesky (1998) refers to as *inclusive wellbeing* – a focus for development that accords with the four basic ethical principles of *autonomy*, *dignity*, *integrity* and *vulnerability* that are being promoted in Europe as a normative framework for related discourse in bioethics and biolaw (Rendtorff and Kemp, 2000).

Furthermore, and tragically, we continue to see far too few attempts within the academy anywhere across the globe, to really develop alternative and complementary paradigms for sustainable development that are at once both scholarly and practical – and oh so urgently needed.

Let me leave the last words to Ulrich Beck who has done more than anyone to influence my own thinking, in the context of inclusive wellbeing, on the need to 'really' face up to the challenges of modernity and the re-design of relationships between 'people and their environments' in systemic, rather than plainly ecological, manner.

> The natural world, sapped by society and industrially endangered, has become the battle ground for its own survival, yet the ecological movement remains trapped in a naturalistic misunderstanding. It reacts to and acts upon a blend of nature and society that remains uncomprehended, in the name of a nature no longer extant, which is at the same time supposed to serve as model for the reorganization of an ecological society.
>
> The confusion of nature and society obscures from view another central political insight: the independence of destruction and protest. Protests against the despoliation of nature are culturally and symbolically mediated. They cannot be deciphered according to the calculus of hazards, but must be interpreted through the inner and personal experience of social ways of life.
>
> Beck, 1995

References

Altieri, M. A. (1987) *Agroecology: The Scientific Basis of Alternative Agriculture*, Westview Press, Boulder

Bateson, G. (1972) *Steps to an Ecology of Mind*, Jason Aronson, New Jersey

Bawden, R. J. (1999) 'A cautionary tale: The Hawkesbury experience', in van der Bor, W., Holen, P, Wals, A. E. J. and Leal Filho, W. (eds) *Integrating Concepts of Sustainability in Education for Agriculture and Rural Development*, Peter Lang, Frankfurt am Rhein

Beck, U. (1992) *Risk Society: Towards a New Modernity*. Sage Publications, London

Beck, U. (1995) *Ecological Politics in an Age of Risk*, Polity Press, Cambridge

Belenky, M. F., Clinchy, B. V., Goklberger, N. R. and Tarule, J. M. (1986) *Women's Ways of Knowing: The Development of Self Voice and Mind*, Basic Books, New York

Bernstein, R. J. (1983) *Beyond Objectivism and Relativism: Science, Hermeneutics and Praxis*, University of Pennsylvania Press, Philadelphia

Brundtland, G. (1987) *Our Common Future: Report of the World Commission on Environment and Development*, Oxford University Press, Oxford

Burrell, W. G. and Morgan, G. (1979) *Sociological Paradigms and Organisational Analysis*, Heinemann, London

Carr, W. and Kemmis, S. (1983) *Becoming Critical: Knowing through Action Research*, Deakin University Press

Checkland, P. B. (1981) *Systems Thinking, Systems Practice*, John Wiley, Chichester

Churchman, C. W. (1971) *The Design of Inquiring Systems*. Basic Books, New York

Conway, G. R. (1985) 'Agroecosystem analysis'. *Agricultural Systems*, vol 20, pp31–55

Conway, G. R. (1987) 'The properties of agroecosystems'. *Agricultural Systems*, vol 20, p31

Cox, G. W. and Aitken, M. D. (1979) *Agricultural Ecology: An Analysis of World Food Production Systems*, W. H. Freeman and Co, Oxford

Dahlberg, K. A. (1970) *After the Green Revolution*, Plenum Press, New York

Dent, J. B. and Anderson, J. R. (1971) *Systems Analysis in Agricultural Management*, John Wiley and Sons, Sydney

Dillon, J. L., Plucknett, D. L. and Vallaeys, G. J. (1978) *Farming Systems Research in the International Agricultural Research Centres*. Rome, Italy. Technical Advisory Committee to the Consultative Group on International Agricultural Research

Douglass, G. K. (1984) 'The meanings of agricultural sustainability', in Douglass, G. K. (ed) *Agricultural Sustainability in a Changing World Order*, Westview Press, Boulder, Colorado

Freire, P. (1972) *Pedagogy of the Oppressed*, Sheed and Ward, London

Fresco, L. (1984) *Comparing Anglophone and Francophone Approaches to Farming Systems Research and Extension*, 4th Annual Farming Systems Research Conference, Kansas State University.

Fuller, S. (1988) *Social Epistemology*, Indiana University Press, Bloomington, IN

Giddens, A. (1984) *The Constitution of Society*, Macmillan, London

Habermas, J. (1972) *Knowledge and Human Interests*, Heineman, London

Habermas, J. (1984) *The Theory of Communicative Action*, vol 1 'Reason and the rationalisation of society', Beacon Press, Boston

Hart, R. D. (1982) 'The ecological conceptual framework for agricultural research and development', in Shaner, W., Philipp, P. and Schmehl, W. (eds) *Readings in Farming Systems Research and Development*, Westview Press, Boulder

Hildebrand, P. (1982) 'Motivating farmers to accept change', in Shaner, W., Philipp, P. and Schmehl, W. (eds) *Readings in Farming Systems Research and Development*, Westview Press, Boulder

Jackson, M. C. (1985) 'Social systems theory on practice: The need for a critical approach', *International Journal of General Systems*, vol 10, p135

Kitchener, K. (1983) 'Cognition, meta-cognition and epsitemic cognition: A three level model of cognitive processing'. *Human Development*, vol 26, p222

Kolb, D. A. (1976) *The Learning Style Inventory: Technical Manual*, McBer and Company, Boston

Knowles, M. (1975) *Self-directed Learning: A Guide for Learners and Teachers*, Associated Press, New York

Kuhn, T. S. (1962) *The Structure of Scientific Revolutions*, The University Press, Chicago

Lowrance, R., Stinner, B. R. and House, G. J. (1984) (eds) *Agricultural Ecosystems: Unifying Concepts*, John Wiley, New York

MacIntyre, A. (1981) *After Virtue*, University of Notre Dame Press, Paris

Marten, G. (1986) 'Productivity, stability, sustainability, equitability and autonomy as properties of agroecosystem analysis', *Agricultural Systems*, vol 26, pp291–316

Maturana, H. R and Varela, F. J. (1988) *The Tree of Knowledge – The Biological Roots of Human Understanding*, New Science Library, Boston

Milbrath, L. W. (1989) Envisioning a Sustainable Society: Learning Our Way Out, SUNY Press, New York

Miller, A. (1983) 'The Influences of Personal Biases on Environmental Problem Solving, *Journal of Environmental Management*, vol 17, pp133

Miller, A, (1985) 'Psychological origins of conflict over pest control strategies', *Agricultural Ecosystems and Environment*, vol 12, p231

Norman, D. and Collinson, M. (1985) 'Farming systems research in theory and practice', in Remenyi, J. V. (ed) *Agricultural Systems Research for Developing Countries*. ACIAR, Canberra.

Perry, W. (1970) *Forms of Intellectual and Ethical Development in the College Years*, Holt, Rinehart and Winston, New York

Prozeski, M. (1998) *The Quest for Inclusive Well-being: Ground Work for an Ethical Renaissance*, Inaugural lecture, University of Natal: Professor of Comparative and Applied Ethics

Reason, P. and Rowan, J. (1981) *Human Inquiry; A Sourcebook of New Paradigm Research*. Wiley, Chichester

Remenyi, J. V. (1985) (ed) *Agricultural Systems Research for Developing Countries*, Australian Centre for International Agricultural Research, Canberra

Rendtorff, I. D. and Kemp, P. (2000) *Basic Ethical Principles in European Bioethics and Law*, Report to the European Commission of the BIOMED-II Project

Rogers, C. R (1969) *Freedom to Learn*, Charles Merrill, Columbus, Ohio

Ruthenburg, H. (1971) *Farming Systems in the Tropics*, Oxford University Press, Oxford

Salner, M. (1986) 'Adult-cognitive and epistemological development', Systems Research, vol 3, pp225–232

Schon, D. A. (1987) *Educating the Reflective Practitioner: Toward a New Design for Teaching and Learning in the Professions*, Jossey Bass, San Francisco

Shaner, W., Philipp, P. and Schmehl, W. (eds) (1982) *Readings in Farming Systems Research and Development*, Westview Press, Boulder

Spedding, C. R.W. (1979) *An Introduction to Agricultural Systems*, Applied Science Publishers, London

Thompson, P. B. (1988) 'Ethical issues in agriculture. The need for recognition and reconciliation', *Agriculture and Human Values*, vol 54, p415

Ulrich, W. (1983) *Critical Heuristics of Social Planning*, Haupt, Bern

15

Social Capital and the Collective Management of Resources

Jules Pretty

Introduction

The proposition that natural resources need protection from the destructive actions of people is widely accepted. Yet communities have shown in the past and increasingly today that they can collaborate for long-term resource management. The term social capital captures the idea that social bonds and norms are critical for sustainability. Where social capital is high in formalized groups, people have the confidence to invest in collective activities, knowing that others will do so too. Some 0.4–0.5 million groups have been established since the early 1990s for watershed, forest, irrigation, pest, wildlife, fishery and micro-finance management. These offer a route to sustainable management and governance of common resources.

From Malthus to Hardin and beyond, analysts and policy makers have widely come to accept that natural resources need to be protected from the destructive, yet apparently rational, actions of people. The compelling logic is that people inevitably harm natural resources as they use them, and more people therefore do more harm. The likelihood of this damage being greater where natural resources are commonly owned is further increased by suspicions that people tend to free-ride, both by overusing and underinvesting in maintaining resources. As our global numbers have increased, and as incontrovertible evidence of harm to water, land and atmospheric resources has emerged, so the choices seem to be starker. Either we regulate to prevent further harm, in Hardin's words (Hardin, 1968), to engage in mutual coercion mutually agreed upon, or we press ahead with enclosure and privatization to increase the likelihood that resources will be more carefully managed.

These concepts have influenced many policy makers and practitioners. They have led, for example, to the popular wilderness myth (Nash, 1973) – that many ecosystems are pristine and have emerged independent of the actions of local people, whether positive or negative. Empty, idle and 'natural' environments need protection – both from

Note: Reprinted from *Science*, vol 302, Pretty, J., 'Social Capital and the Collective Management of Resources', pp1912–1915, copyright © (2003) with permission from AAAS

harmful large-scale developers, loggers and ranchers, and from farmers, hunters and gatherers (Callicott and Nelson, 1998). Since the first national park was set up at Yellowstone in 1872, some 12,750 protected areas of greater than 1000 hectares have been established worldwide. Of the 7322 protected areas in developing countries, where many people rely on wild resources for food, fuel, medicine and feed, 30 per cent covering 6 million km² are strictly protected, permitting no use of resources (Pretty, 2002).

The removal of people, often the poorest and the indigenous (Posey, 1999), from the very resources on which they most rely has a long and troubling history, and has framed much natural resource policy in both developing and industrialized countries (Gadgil and Guha, 1992). Yet common property resources remain immensely valuable for many people, and exclusion can be costly for them. In India, for example, common resources have been estimated to contribute some US$5 billion per year to the income of the rural poor (Beck and Naismith, 2001).

An important question is: could local people play a positive role in conservation and management of resources? And if so, how best can unfettered private actions be mediated in favour of the common good? Though some communities have long been known to manage common resources such as forests and grazing lands effectively over long periods without external help (Ostrom, 1990), recent years have seen the emergence of local groups as an effective option instead of strict regulation or enclosure. This 'third way' has been shaped by theoretical developments both on governance of the commons and on social capital (Singleton and Taylor, 1992; Ostrom et al, 2002). These groups are indicating that, given good knowledge about local resources, appropriate institutional, social and economic conditions (O'Riordan and Stoll-Kleeman, 2002), and processes that encourage careful deliberation (Dryzek, 2000), then communities can work together collectively to use natural resources sustainably over the long term (Uphoff, 2002).

Social capital and local resource management groups

The term social capital captures the idea that social bonds and norms are important for people and communities (Coleman, 1988). It emerged as a term following detailed analyses of the effects of social cohesion on regional incomes, civil society and life expectancy (Putnam, 1993, 2000; Wilkinson, 1999). As social capital lowers the transaction costs of working together, it facilitates cooperation. People have the confidence to invest in collective activities, knowing that others will also do so. They are also less likely to engage in unfettered private actions with negative outcomes, such as resource degradation (Pretty and Ward, 2001; Agrawal, 2002). Four features are important: relations of trust; reciprocity and exchanges; common rules, norms and sanctions; connectedness in networks and groups.

Relations of trust lubricate cooperation, and so reduce transaction costs between people. Instead of having to invest in monitoring others, individuals are able to trust them to act as expected, thus saving money and time. But trust takes time to build, and is easily broken. When a society is pervaded by distrust or conflict, cooperative arrangements are unlikely to emerge (Wade, 1994). Reciprocity increases trust, and refers to simultaneous exchanges of goods and knowledge of roughly equal value, or continuing relationships over time (Coleman, 1988). Reciprocity contributes to the development of long-term obligations between people, which helps in achieving positive environmental outcomes.

Common rules, norms, and sanctions are the mutually agreed upon or handed-down drivers of behaviour that ensure group interests are complementary with those of individuals. These are sometimes called the rules of the game (Taylor, 1982), and give individuals the confidence to invest in the collective good. Sanctions ensure that those who break the rules know they will be punished. Three types of connectedness (bonding, bridging and linking) have been identified as important for the networks within, between and beyond communities (Woolcock, 2001). Bonding social capital describes the links between people with similar objectives and is manifested in local groups, such as guilds, mutual-aid societies, sports clubs and mothers' groups. Bridging describes the capacity of such groups to make links with others that may have different views, and linking describes the ability of groups to engage with external agencies, either to influence their policies or to draw on useful resources.

But do these ideas work in practice? First, there is evidence that high social capital is associated with improved economic and social wellbeing. Households with greater connectedness tend to have higher incomes, better health, higher educational achievements, and more constructive links with government (Putnam, 1993; Wilkinson, 1999; Krishna, 2002; Ostrom et al, 2002; Pretty, 2002). What, then, can be done to develop appropriate forms of social organization that structurally suit natural resource management?

Collective resource management programmes that seek to build trust, develop new norms, and help form groups have become increasingly common, and are variously described by the terms community-, participatory-, joint-, decentralized- and co-management. They have been effective in several sectors, including watershed, forest, irrigation, pest, wildlife, fishery, farmers' research and micro-finance management (Table 15.1). Since the early 1990s, some 400,000–500,000 new local groups were established in varying environmental and social contexts (Pretty and Ward, 2001), mostly evolving to be of similar small size, typically with 20–30 active members, putting total involvement at some 8–15 million households. The majority continue to be successful, and show the inclusive characteristics identified as vital for improving community wellbeing (Flora and Flora, 1993), and evaluations have confirmed that there are positive ecological and economic outcomes, including for watersheds (Krishna, 2002), forests (Murali et al, 2002) and pest management (Pontius et al, 2001).[1]

Further challenges

The formation, persistence and effects of new groups suggests that new configurations of social and human relationships could be prerequisites for long-term improvements in natural resources. Regulations and economic incentives play an important role in encouraging changes in behaviour, but although these may change practices, there is no guaranteed positive effect on personal attitudes (Gardner and Stern, 1996). Without changes in social norms, people often revert to old ways when incentives end or regulations are no longer enforced, and so long-term protection may be compromised.

Table 15.1 *Social capital formation in selected agricultural and rural resource management sectors (since early 1990s)*

Countries and programmes	Numbers of local groups (thousand)
Watershed and Catchment Groups	
Australia (4500 Landcare groups containing about one third of all farmers), Brazil (15,000–17,000 microbacias groups), Guatemala and Honduras (700–1100 groups), India (30,000 groups in both state government and non-governmental organization (NGO) programmes), Kenya (3000–4500 Ministry of Agriculture catchment committees), US (1000 farmer-led watershed initiatives)	54–58
Irrigation water users' groups	
Sri Lanka, Nepal, India, the Philippines, Pakistan (water users groups as part of government irrigation programmes)	58
Microfinance Institutions	
Bangladesh (Grameen Bank and Proshika), Nepal, India, Sri Lanka, Vietnam, China, the Philippines, Fiji, Tonga, Solomon Islands, Papua New Guinea, Indonesia and Malaysia	252–295
Joint and Participatory Forest Management	
India and Nepal (joint forest management and forest protection committees)	35
Integrated Pest Management	
Indonesia, Vietnam, Bangladesh, Sri Lanka, China, the Philippines, India (farmers trained in Farmer Field Schools)	18–36

Note: This table suggests that 417,000–482,000 groups have been formed. Additional groups have been formed in farmers' research, fishery and wildlife programmes in a wide variety of countries.
Source: see Pretty and Ward (2001)

However, there remains a danger of appearing too optimistic about local groups and their capacity to deliver economic and environmental benefits, as divisions within and between communities can result in environmental damage. Moreover not all forms of social relations are necessarily good for everyone. A society may have strong institutions and embedded reciprocal mechanisms yet be based on fear and power, such as feudal, hierarchical and unjust societies. Formal rules and norms can also trap people within harmful social arrangements, and the role of men may be enhanced at the expense of women. Some associations may act as obstacles to the emergence of sustainability, encouraging conformity, perpetuating inequity, and allowing certain individuals to shape their institutions to suit only themselves, and so social capital can also have its 'dark side' (Portes and Landolt, 1996).

Social capital can help to ensure compliance with rules and keep down monitoring costs, provided networks are dense, there is frequent communication and reciprocal arrangements, small group size and lack of easy exit options for members. However, factors relating to the natural resources themselves, particularly whether they are stationary, have high storage capacity (potential for biological growth), and clear boundaries, will also play a critical role in affecting whether social groups can succeed can keep the costs of enforcement down and ensure positive resource outcomes (Stern et al, 2002).

Communities also do not always have the knowledge to appreciate that what they are doing may be harmful. For instance, it is common for fishing communities to believe that fish stocks are not being eroded, even though the scientific evidence indicates otherwise. Local groups may the support of higher-level authorities, for example with legal structures that give communities clear entitlement to land and other resources, and insulation from the pressures of global markets (Ostrom, 1990; Ostrom et al, 2002). For global environmental problems, such as climate change, governments may need to regulate, partly because no community feels it can have a perceptible impact on a global problem. Thus effective international institutions are needed to complement local ones (Haas et al, 1993).

Nonetheless, the ideas of social capital and governance of the commons, combined with the recent successes of local groups, offer routes for constructive and sustainable outcomes for natural resources in many of the world's ecosystems. To date, however, the triumphs of the commons have been largely at local to regional level, where resources can be closed access, and where institutional conditions and market pressures are supportive. The greater challenge will centre on applying some of these principles to open access commons and worldwide environmental threats, and creating the conditions by which social capital can work under growing economic globalization.

Note

1   See the following websites for more data and evaluations on the ecological and economic impact of local groups: (1) Sustainable agriculture projects – analysis of 208 projects in developing countries in which social capital formation was a critical prerequisite of success, see www2.essex.ac.uk/ces/ResearchProgrammes/subheads4food-prodinc.htm. (2) Joint forest management (JFM) projects in India. For impacts in Andhra Pradesh, including satellite photographs, see www.ap.nic.in/apforest/jfm. htm. For case studies of JFM, see www.teriin.org/jfm/cs.htm and www.iifm.org/databank/jfm/jfm.html. See also Murali et al (2002, 2003). (3) For community IPM see www. communityipm.org/ and Pontuis et al (2001). (4) For impacts on economic success in rural communities, see Narayan and Pritchett (1997) and Donnelly-Roark and Ye (2002). (5) For Landcare programme in Australia, where 4500 groups have formed since 1989, see www.landcareaustralia.com.au/projectlist.asp and www.landcareaustralia.com.au/FarmingCaseStudies.asp.

References

Agrawal A. (2002) 'Common resources and institutional sustainability', in Ostrom, E. et al (eds) *The Drama of the Commons*, National Academy Press, Washington, DC

Beck, T. and Naismith, C. (2001) *World Development*, vol 29 (1), pp119–133

Callicott, J. B. and Nelson, M. P. (eds) (1998) *The Great New Wilderness Debate*, University of Georgia Press, Athens

Coleman, J. (1988) *American Journal of Sociology*, vol 94, suppl. S95

Donnelly-Roark, P. and Ye, X. (2002) *Growth, Equity and Social Capital: How Local Level Institutions Reduce Poverty*, World Bank, Washington, DC. Available at http://poverty.worldbank.org/library/view/13137

Dryzek, J. (2000) *Deliberative Democracy and Beyond*, Oxford University Press, Oxford

Flora, C. B. and Flora, J. L. (1993) *American Academy of Political and Social Sciences*, vol 529, pp48–55

Gadgil, M. and Guha, R. (1992) *This Fissured Land: An Ecological History of India*, Oxford University Press, New Delhi

Gardner, G. T. and Stern, P. C. (1996) *Environmental Problems and Human Behavior*, Allyn and Bacon, Needham Heights

Haas, P. M., Keohane, R. O. and Levy, M. A. (1993) 'The effectiveness of international environmental institutions', in Haas, P. M. et al (eds) *Institutions for the Earth*, MIT Press, Cambridge

Hardin, G. (1968) *Science*, vol 162, p1243

Krishna, A. (2002) *Active Social Capital*, Columbia University Press, New York

Murali, K. S., Murthy, I. K. and Ravindranath, N. H. (2002) *Environmental Management and Health*, vol 13, pp512–528

Murali, K. S. et al (2003) *International Journal of the Environment and Sustainable Development*, vol 2, p19

Narayan, D. and Pritchett, L. (1997) *Cents and Sociability: Household Income and Social Capital in Rural Tanzania*, World Bank Policy Research Working Paper 1796, World Bank, Washington, DC. Available at http://poverty.worldbank.org/library/view/6097/

Nash, R. (1973) *Wilderness and the American Mind*, Yale University Press, New Haven

O'Riordan, T. and Stoll-Kleeman, S. (2002) *Biodiversity, Sustainability and Human Communities*, Earthscan, London

Ostrom, E. (1990) *Governing the Commons*, Cambridge University Press, New York

Ostrom, E., Dietz, T., Dolak, N., Stern, P. C., Stonich, S. and Weber, E. U. (eds) (2002) *The Drama of the Commons*, National Academy Press, Washington, DC

Pontius, J., Dilts, R. and Bartlett, A. (2001) *From Farmer Field Schools to Community IPM*, FAO, Bangkok

Portes, A. and Landolt, P. (1996) *The American Prospect*, vol 26, pp18–21

Posey, D. (ed) (1999) *Cultural and Spiritual Values of Biodiversity*, IT Publications, London

Pretty, J. (2002) *Agri-Culture: Reconnecting People, Land and Nature*, Earthscan, London

Pretty, J. and Ward, H. (2001) *World Development*, vol 29, pp209–227

Putnam, R. (2000) *Bowling Alone*, Simon and Schuster, New York

Putnam, R. D. (1993) *Making Democracy Work*, Princeton University Press, Princeton, NJ

Singleton, S. and Taylor, M. (1992) *Journal of Theoretical Politics*, vol 4, pp309–324

Stern, P. C., Dietz, T., Dolak, N., Ostrom, E. and Stonich, S. (2002) 'Knowledge and questions after 15 years of research', in Ostrom, E. et al (eds) *The Drama of the Commons*, National Academy Press, Washington, DC

Taylor, M. (1982) *Community, Anarchy and Liberty*, Cambridge University Press, Cambridge

Uphoff, N. (ed) (2002) *Agroecological Innovations*, Earthscan, London

Wade, R. (1994) *Village Republics*, ICS Press, San Francisco

Wilkinson, R. G. (1999) *Annals New York Academy of Sciences*, vol 896, pp48–63

Woolcock, M. (2001) *Canadian Journal of Policy Research*, vol 2, pp11–17

16

Gateway to the Global Garden: Beta/Gamma Science for Dealing with Ecological Rationality[1]

Niels Röling

You cannot opt out of science.

James Lovelock

You cannot use the same methods that got us into the problem to get us out of it.

Albert Einstein

Introduction

The 'eco-challenge' was invented by Jane Lubchenco (1998) who, as President of the American Association for the Advancement of Science, was confronted with the prospect that now the US had won the Cold War, science had run out of things to do. The social contract of science and society's willingness to pay were at stake. However, the mounting evidence of the rapid decline of the Earth's eco-systems convinced Lubchenco that we are entering 'the age of the environment' and that the ensuing 'eco-challenge' will provide a new social contract for science. This lecture is an effort to explore the role of science and social science, and especially of their interdisciplinary encounter that we now call beta/gamma science, in dealing with the anthropogenic eco-challenge.

It is not the place here to dwell at length on the ecological predicament that confronts us. Suffice it to refer to '*World Resources 2000–2001: People and Ecosystems, the Fraying Web of Life*', a report on world ecosystems released jointly by the United Nations Development Programme (UNDP), the United Nations Environment Programme (UNEP), The World Bank and World Resources Institute in September 2000. One hundred and seventy-five scientists contributed to the report which took two years to produce.

The report reveals a widespread decline in the conditions of the world's ecosystems due to increasing resource demands, and warns that if the decline continues it could

---

Note: Reprinted from Eighth Annual Hopper Lecture by Röling, N. copyright © (2000) with permission from the University of Guelph, Canada

have devastating implications for human development and the welfare of species. The report examines coastal, forest, grass land, freshwater and agricultural ecosystems, and analyses their health on the basis of their ability to produce the goods and ecological services that the world currently relies on. These include the production of food, the provision of pure and sufficient water, the storage of atmospheric carbon, the maintenance of biodiversity and the provision of recreation and tourism opportunities. Most ecosystems are described as being in fair but declining condition. This conclusion is underpinned by statistics such as the following:

- Half the world's wetlands were lost during the last century.
- Logging and conversion have shrunk the world's forests by as much as half. Some 9 per cent of the world's tree species are at risk of extinction; tropical deforestation may exceed 130,000km²/annum.
- Fishing fleets are 40 per cent larger than the oceans can sustain. Nearly 70 per cent of the world's major marine fish stocks are over-fished or are being fished at their biological limit.
- Twenty per cent of the world's fresh water fish are extinct, threatened or endangered.
- Soil degradation has affected two-thirds of the world's agricultural lands in the last 50 years.
- Since 1980, the global economy has tripled in size and population has grown by 30 per cent to 6 billion people.

Without further analysis, I will assume that the ecological basis for human society is under threat of rapid anthropogenic deterioration. Ecological services, including access to water, food, clean air, productive resources and energy, as well as health, genetic integrity, effective carbon recycling, protection against cosmic rays, climate stability and biodiversity, will feature increasingly prominently on the political agenda at local, national and international levels. In fact, it is likely that sustaining the ecological basis for human life will become the highest priority human project in the foreseeable future.

The 'Global Garden' in the title of the lecture reflects my conviction that the Earth must be looked upon as a garden tended by human collective action. The metaphor of the garden is instigated by fact that no ecosystem, be it a wetland, forest, mountain range, ocean or watershed will continue to exist or be regenerated unless people deliberately set out to create the conditions for it and agree to act collectively to that end. The world's ecosystems require increasing interactive design and management. We can no longer afford to ignore opportunities for actively enhancing ecological services, for example by using roofs and walls of buildings for carbon sequestration and oxygen production by green plants. Not only conservation and regeneration, but active design and construction will be required.

Against this background, a number of observations can be made:

- Humans have become a major force of nature (Lubchenco, 1998). The eco-challenge is driven by human activity. Improving the situation is a reflexive exercise and requires dealing with human behaviour. Hence people can be considered a 'reflexive major force of nature'.

- The current human project is largely driven by economic concerns. We measure the effectiveness of our politicians against economic criteria such as employment, incomes and inflation. The feedback we get about the state of the world via the media largely reports on economic indicators. Ecological issues provide worrisome noise but are not part of current political and governance systems. For example, a recent unexpected financial windfall for the Dutch Government of several billions of dollars did not lead to any increased expenditure on reducing carbon dioxide emissions, although the Netherlands is very far from honouring its Kyoto commitments. There is as yet no political advantage to be gained as yet with ecological issues.

- The societal means of communication and of shaping the collective agenda are dominated by efforts to make people consume more and increase their use of natural resources and ecological services. For example, people and especially children are exhorted to consume processed foods and drinks which add value for manufacturers but have already caused an increasing incidence of obesity and are likely to be responsible for the rapid increases in asthma, allergies and intolerances recorded in industrial societies. The US diet is now seen as a major threat to American public health.

- We have elaborate and widespread *scientific knowledge* about the biophysical world and about causal relationships. This knowledge has served us to develop a sophisticated technology for instrumentally manipulating and controlling these causal relationships. This is the basis for widespread welfare and opulence. We also have developed elaborate and widespread *economic knowledge* and practices for managing the economy and for optimizing human outcomes in situations of scarcity on the basis of the operation of market forces. What we have not developed is widespread *reflexive knowledge* about ourselves and our collective action which could be the basis for effectively dealing with the eco-challenge. Although humans have become a major force of nature, we lack the intellectual instruments to learn to deal with that force. Yet, increasingly, success on that score is a condition for our survival. Neither purely scientific nor economic knowledge, alone or in combination, can be expected to provide the basis for getting us out of our ecological predicament. There is no technical fix and the market fails when it comes to the eco-challenge. In fact, our predilection to technical solutions and reliance on market forces increasingly seem part of the problem (Funtowicz and Ravetz, 1993; Beck, 1994). This is not to say that technology and economics can not be part of the solution. What this lecture will emphasize is that technology and economics can only be applied towards a sustainable society within a framework of collective action that overrides instrumental and economic rationality.

We have been changing our domain of existence in a direction where it no longer can support human life. The signals to this effect are becoming increasingly loud and clear. Yet our knowledge and organization, that is, our ability to act effectively in this new domain of existence, is compromised by the fact that our current theories are largely developed to deal with yesteryear's challenges: lack of control over nature (science) and scarcity (economics). We now face a different challenge. The context has changed. And human survival depends on our ability to change our paradigm so as to effectively deal with that change (after Kuhn, 1970). That is the challenge this lecture sets itself: the search for a more appropriate widely shared paradigm that provides the 'reflexive major force of nature' with the intellectual tools to pull itself out of the predicament it has created for itself.

The paper explores the hypothesis that cognition is such an intellectual tool. Biologists have recognized it as 'the basic process of life' (Maturana and Varela, 1992). This biological perspective allows a sharp definition of ecological rationality. The basis of all social science, including economics, is the explanation of human action on the basis of the operation of a cognitive system (Rosenberg, 1988). In computer simulation, Multi Agent Systems (MAS) are rapidly replacing static linear models (Gilbert and Troitzsch, 1999). The Agents are autonomous and cognitive. Ecologists increasingly recognize the importance of the humans in determining ecosystem outcomes, and have identified adaptive management based on social learning as the key ingredient of a sustainable society (Holling, 1995). Learning is the property of cognitive systems. Sustainable natural resource management is increasingly seen as the emergent property of 'soft systems' (Checkland, 1981; Röling and Jiggins, 1998) and of overcoming social dilemmas (Ostrom, 1992; Steins, 1999; Maarleveld, 2000). Social learning, soft systems and overcoming social dilemmas all refer to processes by which individual cognitive agents realize their common fate and agree to engage in collective action. Conversely, pathologies resulting from failed development and resource degradation, such as frustration and marginalization, seem to reflect defective (collective) cognitive functioning (Merton, 1957; Van Haaften et al, 1998). Finally, an analysis of religious peak experiences (Maslow, 1964) leads to a description in terms of cognition.

In all, a unifying and powerful paradigm seems to be emerging for looking at the eco-challenge. That paradigm focuses on the behaviour of perceiving, intentional and reasoning beings engaged in collective action, that is, on (*collective*) *cognitive agents*, as a, if not the, key ingredient in sustainable society. Widespread understanding of (collective) cognition seems to be a likely condition for managing change and hence the gateway to the global garden. One objective of this lecture is to assemble bits of this paradigm into a more coherent whole as the contribution to a story that could ultimately be widely shared. Another objective is to point to some of the research issues suggested by this paradigm that could propel agricultural science as a champion in dealing with the eco-challenge

The Santiago Theory of Cognition

Though non-living, the cybernetic system or thermostat clearly has a rudimentary cognitive structure and provides a recognizable starting point for a discussion of human cognition. The thermometer *perceives* what is happening in its *domain of existence*; the set temperature represents *emotion* (*what is desired*); and the furnace allows *action* that affects the domain and is again perceived by the thermometer (*feedback*). The simple thermostat is helpful because it highlights the main elements and draws attention to the role of a governor which allows communication and comparison among the elements of the system, and which can trigger action. Further requisites are an apparatus for taking action and a throughput of energy. The simple thermostat makes us aware of what it *cannot* do, compared to even the simplest cellular organism. It cannot adapt or learn.

There is limited dynamic interplay with the domain of existence. If a fire broke out next to the thermostat, the mechanism would stop working because it would be 'warm enough'.

Typically, neo-classical economics operates at the level of the thermostat with its assumption of a given utility function or preferences (emotion), perfect information (perception) and rational choice (action). This economic thermostat is also blind and dumb, in the sense that destruction of the human habitat counts as desirable economic growth, for example, driving an automobile contributes more to desirable growth of the gross national product than riding a bicycle. Models of higher system levels, cells and more complex living systems, presume the elements of the simple thermostat, but go considerably further. We capture these higher levels with the cognitive system according to the Santiago Theory of Cognition. This theory was developed by two Chilean biologists and summarized succinctly by Capra (1996, p257):

> In the emerging theory of living systems, mind is not a thing, but a process. It is cognition, the process of knowing, and it is identified with the process of life itself. This is the essence of the Santiago theory of cognition, proposed by Humberto Maturana and Francisco Varela (1992).

Their starting point was the question: how do organisms perceive? Take a frog looking at a fly. Their research showed that the image of the fly can not be projected on the central nervous system of the frog. In fact, the physical processes that govern the image of the fly (light waves) are totally different from the neurological processes that determine the image created in the central nervous system of the frog. One could say that the central nervous system is informationally closed. There is no way that the fly can be 'objectively' projected. But the presence of a fly can trigger change in the central nervous system of the frog. The frog 'does not bring forth *the* fly, but *a* fly'. The active construction of reality is not a human prerogative but a quality of all living organisms.

But, say Maturana and Varela, the frog does not bring forth *any* fly (as pure relativists would have us believe). It brings forth a fly the frog can catch and eat. Organisms and their environment are *structurally coupled*. They maintain this coupling through mutual perturbation. The process by which organisms bring forth *a* world allows them to maintain structural coupling with their environment. This leads Maturana and Varela to their startling and powerful definition of *knowledge as effective action in the domain of existence*.

This definition is startling, not only because so many people think of knowledge as the prerogative of *Homo sapiens*, but also because we are taught to believe that as scientists we are building a store of ever expanding objective knowledge. Maturana and Varela change that metaphor. A store of knowledge developed in an old context can become a blinding insight, and a downright barrier to taking effective action in a changing context or a new domain. We indeed have developed an enormous body of knowledge. But we appear to have very little knowledge, in the sense of effective action in our new domain of existence – a domain marked by anthropogenic destruction of the conditions for life.

According to Maturana's and Varela's definition, cognition is the very process of life. Mind is immanent in matter at all levels of life. There is no organism that is not capable of cognitive action, that is, of assessing experience according to some emotion and taking action accordingly. 'The new concept of cognition, according to the Santiago theory, is much broader than that of thinking. It involves perception, emotion, and action – the entire process of life' (Capra, 1996, p170). We observe that the system includes (1) an organism or *agent* that can perceive the environment and take action in it; and (2) the *domain of existence* with which the agent is structurally coupled. We could further distinguish (3) an *ecosystem*, that is, a space in which multiple agents interact and mutually determine each other's domain of existence.

It is important to note here that Maturana in particular does not accept the identification of 'emotion' with 'intentionality'. Nor, further, does he accept a definition of 'intentionality' as implying some a priori setting of an objective to be attained, that motivates action (as, for example, Searle (1984) would argue in his discussion of consciousness and the perception of freely willed action). The biological basis of cognitive system implies on the contrary, that the triggered response of a perceiving organism's cognitive processes to its environment is necessarily something that occurs 'in the moment'. Learning, that is, occurs in the continuous present and is necessarily adaptive (Jiggins, et al, 1999).[2]

Ecological rationality

The rationality implied by the Santiago Theory of Cognition, we shall call it *ecological rationality*, is very different from the economic rationality implied in rational choice theory. The ultimate rationality of the cognitive agent is to maintain structural coupling with its domain of existence. For the agent this implies that it has the capacity to take effective action in its domain by reducing discrepancy between perception, emotion and action, and by being perturbed by, or by perturbing the domain of existence. Taking effective action therefore implies the following capacities:

- *Control:* act to make outcomes satisfy emotion-based purpose.
- *Adapt:* adapt emotion based purpose to the opportunities (perceived to be) offered by the environment, or to the outcomes that can be elicited (feedback).
- *Learn:* develop perception to fit the opportunities or threats in the environment and adapt action and purposeful behaviour to changed perception.
- *Evolve or innovate:* adapt the ability to take effective action to the perceived and/or experienced threats and opportunities in the domain of existence.
- *Mutate or reflexively manage the cognitive system itself:* when structural coupling cannot be achieved through the above four capacities, deliberately manage the configuration of the elements of the cognitive system, for example by changing blinding insights, or by making the context visible.

Rationality is the effective pursuit of structural coupling. This implies that the three elements, perception, emotion and action, tend towards (cognitive) consistency, a term coined by Leon Festinger (1957) many years ago. But it also implies that the consistency achieved can be broken open. Hence rationality requires a remarkable mix of consolidation and self-renewal. I shall come back to this point later when I discuss blinding insights.

The concept of rationality implied by the Santiago theory not only seems much richer than the simple goal-seeking rational choice theory used by economics, it also seems a sounder guide for people as a 'major force of nature'. It goes beyond a focus on the selfish optimization of outcomes (as presumed by 'economic man') or on the strategic rationality of the players in the marketplace who seek to win in competition (Platteau, 1996, 1998). Ecological rationality seeks first of all to maintain structural coupling and is based on bringing forth a world that allows structural coupling to be maintained. Hence rationality can not be based on invariable objectives or preferences (such as optimizing target variables and suppressing all others by using the best technical means, satisfying utility functions, or winning from competitors). Rationality is much more self-interested, it is adaptive if that is what is required.

This paper departs from the assumption that, in the final analysis, people are eco-logically rational. This not only implies that ecological rationality overrides instrumen-tal and competitive or strategic rationality, in that it is logically more comprehensive and superior, but also that humans, as all other organisms, are 'wired' to be ecologically rational. But to realise that potential and remain structurally coupled, they must bring forth an appropriate world. Given that collectively humans have become a destructive major force of nature, the world they bring forth must allow them to deal *reflexively* with collective cognition.

The change of paradigm

In his studies of strategies to deal with the Spruce Bud Worm in the New Brunswick forests, the Canadian Allan Miller (1983, 1985) has suggested a graph, which was fur-ther developed by Bawden (1997) and slightly adapted by me, which allows us to visual-ize the change of paradigm that seems to be required for a reflexive major force of nature (Figure 16.1).

The first quadrant (I) represents the blinding insight based on reductionism and positivism. Miller called it the 'technocentric' quadrant. His colleagues operating in it were likely to counsel chemical spraying (or providing economic incentives for it) as the best strategy for dealing with the Spruce Bud Worm.

The second quadrant (II) is still based on positivism, but has moved on to a holistic perspective. Ecosystem approaches fall into this quadrant. But people as cognitive and intentional beings do not feature in this type of thinking. This is not to deny the fact that ecologists, such as Holling (1995) and his colleagues, conclude that ecosystems in which humans play dominant roles can be maintained only if those humans engage in adaptive management and in the kind of social learning that allows human institutions to engage in adaptive management (Jiggins and Röling, 2000). Miller observed that colleagues operating on the premises of quadrant II were likely to advocate Integrated Pest Management approaches to combat the Spruce Bud Worm.

Quadrant III is less straightforward. It represents a holistic approach coupled to a constructivist epistemology. Thus it gives space to soft system thinking and methodol-ogy (Checkland, 1981; Checkland and Scholes, 1990), to social learning in the sense of humans' collective learning to manage themselves, to futures that emerge from human interaction among multiple stakeholders, and to communicative rationality (Habermas, 1984, 1987). In this quadrant one can effectively look at humans as intentional and

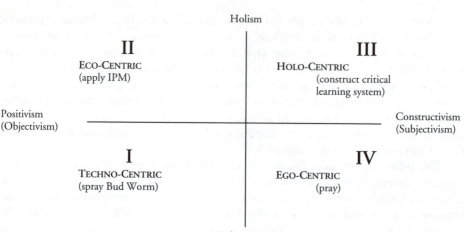

Source: Based on Miller (1983, 1985) and Bawden (1997)

Figure 16.1 *A typology showing the paradigm shift between blinding insights (the case of the Spruce Bud worm)*

learning cognitive agents, and at human organizations as collective or inter-subjective cognitive agents. Miller observed few colleagues whose Spruce Bud Worm management was based on the assumptions of this quadrant. But collective action negotiated among the multiple stakeholders in the Spruce forest is probably what he would have seen. After all, the outbreaks of Spruce Bud Worm were the result of humans planting enormous tracts with one species of tree, so the real solution was not a question of dealing with the worm but with human systems.

Quadrant IV is a difficult one. It could represent the condition in which people still subscribe to constructivism but have learned or accepted that ecologically rational collective action is not within the grasp of human beings. Miller typifies the action in this quadrant as 'pray'. But perhaps this is also the quadrant which provides a home for those who consider spirituality as a key ingredient in ecological rationality and as the step beyond soft systems (Van Eijk, 1998; Auerbach, 1999; Wielinga, pers comm). In that view, a direct link between manifest and explicit discursive consciousness, on the one hand, and deeper layers of implicit knowledge (Broekstra, 1998) and transcendental consciousness, on the other, is a feature without which it is impossible to break through the current blinding insights of science and neo-classical economics to achieve ecological rationality.

Maslow (1964, p61), who has analysed religious peak experiences of a great number of people whom he met during his career as a (humanist) psychologist, names the following key aspects reported by most of his respondents:

- the clear perception that the universe is one whole of which one is an integral part;
- non-judgemental, or comparative, total acceptance of everything;

- we are lifted to greater than normal human heights and able to see things beyond normal human concerns;
- the experience justifies life as such, the world is experienced as good, beautiful, and never as bad or undesirable. Existence becomes sufficient in itself.

'The complete human person at certain moments is able to perceive the unity of the cosmos, to fuse with it and rest in it, for the moment totally satisfied in his desire for being at one' (p92). In Japanese Zen Buddhism, such a religious peak experience is called Satori, a 'transformation of paradigm' that is the object of the quest of those who practice Zen (Suzuki, 1994, p46). It seems that such spirituality is the well-spring of ecological rationality. Ecological rationality is the practical implication of religious experience.

The Miller/Bawden scheme allows me to make a bold statement. The eco-challenge requires that we move from Quadrant I, through Quadrant II, to at least Quadrant III, and some would argue, to Quadrant IV. My experience suggests that those who are still in Quadrants I or II find it impossible to understand Quadrant III. Hence any university or other institution concerned with training professionals for managing the human use of land, natural resources or ecological services should make the move to Quadrant III a priority. But this is running ahead of my story.

General systems theory (Boulding, 1968) distinguishes seven levels of theoretical discourse: (1) static structures (frameworks); (2) simple dynamic systems (clockworks); (3) self-regulating, cybernetic systems (thermostats); (4) self-maintaining living structures (cells) and (5, 6 and 7) more complex living and self-organizing adaptive systems. Each higher level presumes the lower one. Adequate theoretical models extend to the fourth level and not much beyond. The level of the clockwork is the level of classical natural science. I assert that the movement from Quadrant I through Quadrant II to Quadrant III represents reflexive moving up system levels from, respectively, the clockwork and the self-regulating thermostat, to self-organizing adaptive reflexive systems.

Paradigm shift in economics?

Before I conclude the discussion of paradigm shifts, I would like to draw attention to the fact that Figure 16.1 largely refers to science and not necessarily to economics. During my life as a faculty member in an agricultural university, I have become convinced that the impact of classical economic thinking (with its assumption of egoism and strategic calculation) on the reflexive understanding of students and faculty is perhaps greater than positivism and reductionism and much less easy to discuss. Students believe in strategic narratives and find it naive to believe that agreement or institutions can override selfishness (Röling and Maarleveld, 1999). Ridley (1995) has also struggled with this question. I have encountered many economists who hold the positivist notion that the market is a force of nature, much like gravity, which is not amenable to human deliberate decision making and who are totally oblivious of the debate going on in economics itself.

The knowledge of human knowledge claims a place of its own in economics. Beyond the walls of our discipline, spectacular progress is taking place in the field of empirical research into human knowledge – the so-called cognitive sciences. In the light of such advances, the old and classical axiom that nothing scientific can be said beyond the axioms of substantive rationality now looks very much like the protective belt of a degenerating programme (Tamborini, 1997).

The formulation of 'bounded rationality' by Simon (1969) and the identification of institutions as mediators of information by North (1990) have led recently to numerous new perspectives with respect to the cognitive basis of economics. In an attack on methodological individualism, Arrow (1994) concluded that 'social variables, not attached to particular individuals, are essential for studying the economy or any social system, and that, in particular, knowledge and technical information have an irremovably social component, of increasing importance over time'.

In a distinguished lecture on economics in government, Aaron (1994) chastises his discipline for failure to take the formation of preferences seriously ('the recalcitrant refusal of economists to venture beyond a model of human behaviour others see as seriously incomplete'). He also laments the reliance on a model of utility that has no relation to current psychology, and draws attention to the following claims as directly relevant to economics:

- people never know the full consequences of their actions;
- the human brain does not contain a central processing unit, a giant server supervising many workstations. A more useful metaphor is of the brain as a massive parallel processing unit (Clark, 1997);
- people have a capacity for self-reference through which they can judge their own lives and relationships;
- humans derive satisfaction from helping each other;
- people care about others as ends, not only as means;
- people derive enormous satisfaction from interpersonal relationships;
- the satisfaction people take from setting goals and achieving them has erroneously been singled out as the most important;
- the most palpable reality of all our lives is internal conflict.

Aaron argues that each person operates more than one utility function, including self-respect, profit, others' regard and social capital. The trade-offs that are made among them, and the utility function that is determinant at any time, is an empirical, contextual outcome of contingent history, opening to the economist 'a vast scope for theoretical imagination'.

Satz and Ferejohn (1994) conclude their social theoretical analysis of rational choice theory by pointing out that that theory is most credible under conditions of scarcity where human choice is constrained. Without constraints, agents will not behave as the theory predicts. 'We need a background theory to identify in just which contexts a psychological interpretation of rational choice theory makes sense'.

More recently, Amartya Sen (1999), the 1998 Nobel Laureate for economics, has criticized the assumptions about human cognition on which neo-classical economics has been founded: the homo economicus, the rational egoist driven by selfish motives, and the Paretan criterion that general wellbeing increases if the wellbeing of at least one individual increases while the wellbeing of others does not decrease. According to Sen, such assumptions about human behaviour are too narrow. Behaviour is not only motivated by selfish, but also by other interests, such as the group interest, one's social position, ethics, etc. Instead of the deliberate fiction of 'homo economicus', Sen pleads for a more pluriform approach in which other motives are also taken into account.

In all, modern economics is struggling to move away from the axioms about cognition on which much of its theory is based. However, the very fact that economics finds this struggle so difficult illustrates the strangle-hold of its assumptions. The older 'strategic narratives' of economics still guide ordinary people's expectations about human nature and are embedded in most of our institutions, such as political parties (Röling and Maarleveld, 1999). We have created environments that encourage people to act selfishly and strategically and have now reified this self-referential world to an extent where we cannot imagine to venture beyond it. Effectively dealing with the eco-challenge requires a deliberate programme of economic research that allows us to embed paradigms and institutions based on 19th century economic rationality into a wider conception of ecological rationality. That indeed is the mission of ecological economics (O'Connor, 1998), not to be confused with environmental economics, which tries to place environmental concerns within the confines of the neo-classical assumptions. For the time being, neo-classical economics provides a formidable blinding insight, not only with respect to individual, but also collective behaviour.

At the collective level, the market is an arena of selfish individuals competing to gain advantage in the access to scarce goods and services. The important assumption of neo-classical economics ever since Adam Smith and Jeremy Bentham is that the emergent property of this interaction is 'the greatest good for the greatest number', hence an optimal allocation in conditions of scarcity. This leads to a blinding insight that has provided the basis for an arrogant programme of global societal design dominated by the International Monetary Fund (IMF), the World Bank and business corporations. David Korten (1995, p71) has formulated this blinding insight as follows:

- people are by nature motivated primarily by greed;
- the drive to acquire is the highest expression of what it means to be human;
- the relentless pursuit of greed and acquisition leads to socially optimal outcomes;
- it is in the best interest of human societies to encourage, honour and reward the above values.

These assumptions underpin the following 'beliefs espoused by free-market ideologues' (p70):

- Sustained *economic growth*, as measured by gross national product, is the path to human progress.
- *Free markets*, unrestrained by government, generally result in the most efficient and socially optimal allocation of resources.
- *Economic globalization*, achieved by removing barriers to the free flow of goods and money anywhere in the world, spurs competition, increases economic efficiency,

creates jobs, lowers consumer prices, increases consumer choice, increases economic growth, and is generally beneficial to almost everyone.
- *Privatization*, which moves functions and assets from governments to the private sector, improves efficiency.
- The primary responsibility of government is to provide the infrastructure necessary to advance commerce and enforce the rule of law with respect to *property rights and contracts.*

Korten (1995) convincingly shows how these assumptions have legitimated the destructive role of business corporations in destroying livelihoods and agroecosystems, in channelling wealth from the poor to the rich, and in undermining the vulnerable biosphere on which we all depend. Clearly neo-classical economics and the free market ideology which it underpins provide for an untenable design of global society. Inventing a better design is the challenge ecological economics has set itself. But inventing such a design is obviously not just the task of economists, it is the next major global human project.

Praxis for analysing blinding insights

Bawden (2000) has provided us with a tool to consider the reflexive cognitive agent. Praxis is practice informed by theory, it is the practice which emerges from deliberate, interactive and mindful iteration through major anchor points of cognition and decision making: context, values, theory and action.

One easily recognizes the elements of the cognitive system, except that the elements are now coined in terms of inter-subjective discourse. Consistency-seeking iteration through the elements gradually gels into a blinding insight. But changes in context, and to a lesser extent, values, theory or action allow people to break through the bondage of self-referentiality. Hence there is a pulse of consolidation and self-renewal. This pulse is, to my opinion, of key importance for research by a reflexive major force of nature (also Eshuis, in preparation). Bawden's notion of praxis is helpful for analysing blinding insights and for thinking about how we get out of them.

By way of illustration I will describe a typical agricultural blinding insight and the way we are getting out of it. Wageningen University's most famous professor, Cees de Wit, defined agriculture with breath-taking and destructive clarity (1974):

Agriculture is harnessing the sun's energy through plants for human purposes.

As a result of the work of de Wit's students, Wageningen has become famous/notorious for its crop growth simulation models that focus on abiotic variables and natural laws. These models integrate the hard agricultural sciences into interdisciplinary efforts that allow prediction, for example, of global maximally attainable yields, and hence of the human population that the earth can support. This approach has had a tremendous impact, been heuristic in generating a great deal of research, and refreshing in toppling established truths. An example is the thesis that it is not lack of water but lack of nutrients that limits biomass production in the Sahel.

If we look at de Wit's perspective on agriculture, we can make the following observations:

- The *context* is determined by abiotic factors and the dynamic forces that govern it are natural laws. Hence the system level is the clockwork, or at most the thermostat (Boulding, 1968). The context is assumed to exist irrespectively of the human observer and to be objectively knowable by scientific research (positivism).
- The *theory* concerns harnessing the sun's energy through plants and only addresses the abiotic factors and natural laws, and not, for example, other cognitive agents in the domain. The action is technical or instrumental.
- This action serves human purposes (*values*). In Wageningen practice, these purposes are assumed and translated into achieving maximally attainable production, and more recently, under the impact of the sustainability debate, into resource efficiency. The scientist's purpose is to develop the best technical means to achieve these purposes.
- In its systems orientation, De Wit's perspective is an achievement in terms of overcoming reductionist science and moving to Quadrant II (Figure 16.1) and, through its systems approach, has provided a basis for interdisciplinarity among the hard sciences.

As can be imagined, de Wit's perspective has been very influential and still is. It has, however, a number of distinct disadvantages and blind spots. Although De Wit was an active socialist, his perspective does not include people. Hence people do not feature in the models, but utility functions have been built in. There is a general recognition that, once scientists have developed the best technical means, communication specialists like myself come in handy to deliver them to ultimate users.

Another problem with de Wit's perspective is that the limited definition of the context does not take into account its ecosystemic nature. Hence, according to De Wit-type models, the Earth can support as many people as can be fed by producing the maximally attainable grain equivalent yield of 10 tonnes per ha on the available arable land. This calculation does not take into account the likelihood that such production would probably destroy the complex web of life, and the biosphere as a complex adaptive system on which we depend (Capra, 1996).

Totally ignoring the existence of people in the domain of existence also leads to remarkable distortions. Thus, given its exposure to sunlight, the water available and the nature of its soils, West Africa can be predicted to be able to produce sufficient food for its rapidly rising population by 2040. However, this expectation does not take into account the intractability of the 'soft side of land' (Röling, 1997). Agriculture is embedded in rights to land, in knowledge and technology and the societal capacity to change them, in the cultural traits of local people, and in the institutions and infrastructure (including markets) of the societies concerned. It has so far apparently not been rational for West African farmers to embark on the kind of science based agriculture that would be required to reach 10 tonnes per ha, even though they have been exceptionally innovative in coping with change (Mazzucato and Niemeijer, 2000).

It does not require great leaps of fantasy to see that De Wit's blinding insight and its domination of the Wageningen intellectual scene are a serious handicap for the University in providing intellectual leadership for dealing with the eco-challenge and in attracting the exciting young people that want to make a difference.

It is interesting to observe how this blinding insight is now slowly being undermined. The main factor, in my experience, is that the models are not being used. Thus scientists who have spent years of their lives believing that their 'systems approach' could be the basis for policy based on scientific truth are slowly coming to the sobering realization that policy makers in both industrial and developing countries are not using their models and that their work has had little impact. In the ensuing uncertainty one observes a remarkable receptivity for other ideas, for example an interest in Figure 16.1. This does not mean, of course, that the overwhelming thrust of the university has shifted from technical positivist science. The 'life sciences' (narrowly defined as biotechnology) are the latest enthusiasm and are seen as the future basis for the social contract of agricultural science. I agree with Lubchenco (1998) that the eco-challenge seems a much more lucrative mission.

But even when the eco-challenge is accepted as *the* problem, this does not necessarily mean that the core of the problem, human collective behaviour itself, is recognized. An illustrative example is the conclusion of the report *World Resources 2000–2001: People and Eco-system, the Fraying Web of Life*, that I referred to earlier. The advance internet billing of the report says:

> One of the most important conclusions is that there is a lack of much of the baseline knowledge that is needed to properly determine ecosystem conditions on a global, regional or even local scale... The dimensions of this information gap are large and growing, rather than shrinking as we would expect in this age of satellite imaging and the Internet.

In all, the report provides 'impetus for the Millennium Ecosystem Assessment, a plan put forward by governments, UN agencies and leading scientific organizations, to allow an on-going monitoring and evaluation of the health of the world's ecosystems'.

In other words, recognition of the eco-challenge as a key threat to human survival leads to calls for more scientific research on the state of ecosystems, to use Lubchenco's phrase, 'to tell people what is out there'. It is the same as trying to reduce the incidence of lung cancer by more research into the relationship between smoking and mortality. We now know that what is needed is a focus on smoking itself, and on the corporate behaviours that stimulate it. We now accept that cigarette manufacturers are liable for the consequences of smoking. We forbid advertisement for smoking and label packages of cigarettes with a health warning. We are still far from drawing similar conclusions with respect to the anthropogenic eco-challenge. For example, a recent overview of the environmental fate and toxicology of organophosphate pesticides (Ragnarsdottir, 2000) cannot but leave one with the expectation that the time is not far off that pesticide companies will be in the same position as cigarette manufacturers now.

Lest we become depressed, it is perhaps good, at this point, to report another major departure from De Wit's blinding insight that is taking place in Wageningen (and many other agricultural universities that I know of). It is the spontaneous development of 'beta-gamma' approaches in important agricultural science fields, where beta stands for the natural sciences and gamma for the social sciences. Together, beta and gamma sciences are becoming increasingly involved in the 'interactive design' of technology, farming systems, knowledge systems, natural resource use and other forms of 'land use negotiation' (Leeuwis, 1999). Note that we do not talk of land use *planning* any longer (Brinkman, 1994).

The beta/gamma focus was not developed by social scientists. It is more accurate to say that in departments such as irrigation, entomology, forestry, soil and water conservation, ecology, and spatial planning, maverick scientists and doctoral students began to take seriously developments which were taking place in the field (ie the context) and which emerged from the recognition that the 'delivery' of science based technological means to 'ultimate users' such as farmers, simply does not work (Chambers and Ghildyal, 1985, Chambers and Jiggins, 1987).

We are still in the middle of these developments, and in the middle of trying to develop adequate curricula to prepare our graduates for beta/gamma tasks. Some professors still are reluctant to give up achievements hard won in positivist pursuits. Lest I am misunderstood, I do not reject their positivist pursuits, but their efforts to block significant development. I strongly believe in the usefulness of maintaining a capacity to do pure, laboratory, on-station, hard science, under positivist/realist assumptions.

The exciting thing is that a new *additional* area of science professionalism is emerging. When it comes to designing effective action in the domain of existence, pure science has an important role, but *in addition*, we have to deal with people's objectives and opinions as 'extended facts', with 'self-appointed activists' as extended peers (Funtowicz and Ravetz, 1993), and with shared cognitions, intentionality, and institutions as essential design ingredients. In such design, people are not objects that can be instrumentally or even strategically manipulated. They must participate. Agricultural science is, to a large degree, interactive (Röling, 1996). It is not natural laws that determine the direction in which natural systems evolve, but human intentionality, agreement, conflict, and, hopefully, forward looking collaborative adaptive management (Buck et al, 2001).

Pathology of the collective cognitive system: The eco-challenge and mental health

In our analysis of collective cognitive systems and ecological rationality, it is of interest to examine pathology. In terms of cognitive agents, we can describe the responses as follows (Röling, 1970): *innovation* is the adaptation of action or of the ability to act so as to satisfy changed goals. Earlier, we called this control. Innovation might express itself in the adoption of new technologies, emigration, legal redistribution of access to assets and power, or even magic, developing supernatural means to achieve new goals. *Ritualism* is the rejection of new goals to comply with existing ability to act. Old forms of action are fixated. This fixation is considered a substitute goal response. Development can be marked by extreme 'traditionalism' or fundamentalism, often an expression of frustration. *Retreatism* is the rejection of the new goal *and* of the existing ability to act. It represents withdrawal resulting in apathy. A typical response is fatalism, the belief that external forces determine one's outcomes, but other responses include voluntary isolation, and escapism into alcohol, cults or religious extremism. Such responses are often seen as the best adaptation people can make to a hopeless situation. Seligman and Hager (1972) have called this learned helplessness. *Rebellion* is the rejection of the institutional arrangements within which the ability to act is defined (eg access to resources). In a way, rebellion is an innovative response. *Conformity* is adherence to both the new goal and the existing ability to act. This seeming paradox can be explained by the fact that people do not allow themselves to be motivated (and frustrated) by all possible goals (eg standards of living to which they are exposed). Goals are limited to those perceived as attainable in one's own situation. One feels poor only in relation to the outcomes 'relevant' others experience (the so-called reference group). Merton's typology highlights some interesting aspects of collective cognition:

- 'Cultural goals' are seen as more prone to change than the opportunities to satisfy them. The tension raises the salience of the human mechanisms that regulate motivation in view of realisable opportunity.
- If we consider current global interconnectedness to be increasing, yet inequality in access to resources, capital and other opportunities, and enjoyment of benefits, also to be increasing, then we may anticipate widespread 'deviant' adaptations that threaten the achievement of a global system based on ecological rationality. The eco-challenge can not be tackled without alleviating global inequalities and poverty.
- To the extent that ecological rationality means adapting cultural goals to limitations in outcomes, that is, taking less and/or giving more, it requires (self-) management of cognitive inconsistency to prevent pathological adaptations. This in turn implies reshaping criteria for status and achievement, new enthusiasms and new ideas about what is worthwhile, and perhaps new social institutional devices that replace the religion of old.

In all, the pathologies of the cognitive system suggest that environmental degradation and frustration can elicit adaptive responses that reduce the resilience of collective cognitive agents to deal with ecological surprises.

Taking stock: The future is a human artefact

Humans have become a major force of nature (Lubchenco, 1998). The future is a human artefact. There is no God, science, or miraculous emergent property that is going to get us out of the eco-challenge. Unless we take it upon ourselves purposefully to grapple with the future, there won't be one. In that sense, 'we cannot opt out of science'. But 'we cannot get out of a problem by the same methods that got us into it'. Hence it is the nature of science that needs to change. The focus shifts from manipulating things to reflectively learning to deal with our own behaviour. Take forests. Everyone agrees there should be forests. But forests can only exist if people take purposeful concerted action to create and maintain them. The default is no forest (Keiter and Boyce, 1991; Röling, 2000).

The one element of praxis that can blow up our current blinding insights is change in *context*. As Kuhn (1970) established in his *Scientific Revolutions*, a blinding insight ('normal science' as he calls it) starts off by ignoring contextual signals that are inconsistent with it. Gradually, however, as the brunt of evidence becomes overwhelming, theory, values and action must adapt to context. Of course, many societies, such as the Medieval Nordic communities on Greenland (Pain, 1993), have collapsed because they maintained their blinding insights in the face of evidence of their inadequacy. The ability of elites to maintain their lifestyles long beyond the time when it is prudent to do so is an important factor in explaining such irrational behaviour. In that respect, the moneyed elites and corporations in the global economy can be expected to be formidable sources of resistance to change.

Perceiving the change in context and making it visible by scientists is definitely one important necessary, but not sufficient, condition for change. That monitoring must reverberate throughout society and become a basis for social learning, if it is to be effective (Guijt, in preparation). If humans are a major force of nature, and if the conditions for human life are threatened by human activity, then what we must change is human activity, not ecosystems. Or better, we must adaptively manage the interface, duality, or structural coupling between human collective cognitive agents and the ecosystems on which they depend. This conclusion has major implications.

The *first implication* is that we cannot continue to consider the earth as a substrate for human activity or use it as a resource for only human ends, and then rely on the resilience of the web of life to deliver the ecological services on which humans as biological agents depend. People have *de facto* taken responsibility for the direction in which the earth evolves. In this sense we can say that ecosystems, including the earth, have become soft systems (Checkland, 1981): whether by default or intention, their future states emerge from human interaction. It this interaction that requires deliberate management at the level of the ecosystems under threat.

Such collective action makes special demands on the nature of human collective cognition. It requires shared sense making, conflict resolution, negotiated agreement and accommodation, and deliberate concerted action among the stakeholders in the system. Our survival as a species is no longer only a question of learning about our environment, but increasingly one of being able to collectively learn about and control our own collective behaviour. Beck (1994) calls this 'reflexive modernization'. We shall speak of 'social learning'. While social understanding has not been more than a marginal influence over science and public policy during the era during which people imposed increasing technical control over the biophysical world, understanding ourselves as a unique, reflexive, cognitive system is now becoming vitally important to our own survival. This is an unprecedented change in context that science nor economics is able to address, and a very important challenge for agricultural universities.

The *second implication* is that people's activities must increasingly be guided by ecological rationality. Though by no means an unknown consideration in human experience, it will require a major transformation of the current rather single-minded pursuits of instrumental control and economic growth, and of the institutions dedicated to fostering those enthusiasms. In terms of the elements of praxis, we are dealing with a fundamental change in *values*.

The *third implication* is that humans must deliberately develop a 'soft side' to the sustainable management of the biosphere (Röling, 1997; Jiggins and Röling, 2000). The soft side of land refers to the human knowledge, technology, institutions, resource allocation and so forth from which land use emerges. Ifugao in the northern Philippines illustrates this concept (Box 16.1).

Box 16.1 *The Ifugao Rice Terraces as a monument to the 'soft side' of land*

The rice terraces developed by the Ifugaos during the course of 2000 years have been declared a Heritage Site by the United Nations Educational, Scientific and Cultural Organization (UNESCO). The sight of entire mountainsides covered by terraces awes the visitor, not only because of their beauty or the enormous effort that must have been involved in their construction, but also because of the ingenuity, organization and collective management that such a structure requires. Unlike the pyramids and other world wonders built by tyrants who used slaves for their own glory, the Ifugao terraces are due to voluntary collaboration and organization. Careful study reveals that the 'hard' terrace system of irrigation channels, walls, protective forests and so on, has its counterpart in complex social institutions and human cognition involving spirits and gods, rituals, work organization, discipline, leadership, shared experiential knowledge and values. The fact that the hard system now is collapsing can be traced to the erosion of the *social* system that ensured its upkeep.

People have always created their own world according to their shared enthusiasms. During the age of religion, they built cathedrals, mosques and temples and organized according to commands construed as given by god. In the industrial age, science had impact through the emergence of actor networks that replicate laboratory findings on a societal scale (Callon and Law, 1989). Our current enthusiasm is to transform the whole world into a global, competitive marketplace dedicated to satisfying consumptive needs. But we have now entered a new context. We shall have to design a soft side of the Earth, new institutions, knowledge, learning processes, language and so on that allow humanity to tend the global garden. It is a very hard thing to do and the transformation required makes ours an exciting but also depressing future.

An agenda for agricultural and environmental research

The fact that we have international institutions that focus on establishing a global market, and that democratically elected governments assume that the free market is the best design for society at the regional, national and global level, is itself an indication that people are part of collective cognitive systems. Even when mobility was determined by the speed of the horse, collective enthusiasms with enormous consequences for how society was designed, such as Christianity and Enlightenment, swept huge tracts of the surface of the earth.

We must now learn to deliberately manage our collective cognitive systems themselves. Fundamentalism, be it Mohammedan, Christian, or neo-liberal Republican, is a major risk for human survival because it detracts from responsiveness. The same can be said of corporations and other agents that have vested interests in the status quo. We have designed a highly interdependent soft side of the globe. Yet it is increasingly clear that the present set-up is very vulnerable. An increase in the price of fossil fuel by a few pennies already leads to major upheavals as we have seen recently. A piquant detail that emerged is that the engines of Dutch fishing boats are not only too powerful for sustainable fishing, but, at high fuel prices, too expensive to run, given the value of the fish that can be caught with them. From the reaction of farmers, the same appears to be true for the tractors that produce our cheap food. Our infrastructure, capital investment, organizations, marketing chains, insurance systems and so forth increasingly seem to be built on quicksand and not resilient enough to withstand climate change, let alone other major ecological surprises.

The resilience of the global human system depends on our ability to deliberately manage, that is, quickly adapt, the collective cognitive system and the institutions we have built to underpin it. This means widely shared relativism with respect to the substantive content of our current enthusiasms and blinding insights. The difficult question is how such readiness to change collective cognitive systems can be brought about. For example, most democratic societies accept that millions of dollars are being spent on elections or on advertising to increase consumption, but very few would accept deliberate use of public and commercial media for purposes of promoting ecological rationality. In fact, the democratic process would prevent such use. An example is a public advertising campaign for promoting ecologically produced food for which the Dutch Ministry of Agriculture had been provided funds. Protests by conventional farmers ensured that the advertisements could not make comparisons between ecologically and conventionally produced food. Hence the whole thing became a nonsense.

The resilience of collective human cognitive systems is a depressing political minefield. Luckily the scene is not without points of hope. One mechanism that seems to be effective in fostering transformation is the emergence of self-appointed activists, informal protest groups, NGOs, voluntary environmental organizations, and nature conservation associations that are supported by a generous public that feels that something is amiss even though it does not know what to do about it. It is such groups that were able to cause the Seattle and Prague hiccups in the myopic transformation of the world into a global market place. It seems that continued generous funding of such dynamics by governments, foundations and the public is one condition for the emergence of a new order.

But there is more. The fact that people like myself are beginning to buy into the ecological agenda, that corporate leaders are beginning to strategically take into account the uncertainties caused by the eco-challenge, and that the public is beginning to wake up to the quality of food, means a very gradual change in the mainstream which must eventually have political repercussions. It also means that research money is becoming increasingly available for forward looking studies that deal with the eco-challenge. I will give four examples of studies that I am involved in myself:

1 The European Commission has provided 1.2 million Euros for a study in five European countries of watershed management as the emergent property of multistakeholder interaction, of the most effective facilitation of that interaction, and of the policy implications. A similar proposal failed three times before, but now the funding mechanism was considered 'ripe for such ideas'. The Dutch partners include not only my university but also a commercial consultancy company that is already actively

involved in the participatory management of water quantity and the transition from water retention to 'space for water'. This new perspective has emerged as a result of the impossibility to manage the consequences of recent freak weather events, not to mention the fact that the Netherlands are slowly sinking and the sea level is slowly rising.

2   Wageningen University has provided about 250,000 Euros for an interdisciplinary project called 'Convergence of Sciences: Inclusive Innovation Processes for Integrated Crop and Soil Management'. The focus of the study is on interdisciplinarity, especially among technical and social scientists; on including farmers and other stakeholders in technology development (where 'technology is very broadly defined to include marketing, organization and collective action); and a shift in focus from a prescribed input base to complex agroecological development. Hence the focus is not on developing 'the best technical means', but on creating insight in how to generate learning systems to deal with intractable problems such as Striga parasitism of sorghum in West Africa.

3   The Dutch Ministry of Agriculture and a voluntary association called the KNHM, an organization that emerged in the years of the great land reclamations and clearings at the end of the last century, have provided seed money (Euro 25,000) for exploratory field experiments to develop methodologies for effective discourse among rural and urban people about the use of green space. The Ministry, the KNHM and a provincial government that is involved all emphasized that they have received funds for innovation in managing the land, and that they are looking for ideas on how to spend them. In fact, moves are afoot to spend up to 10 per cent of the public funds allocated to agricultural research to 'demand-driven research' formulated in regional 'knowledge centres' set up to formulate that demand.

4   The Farming Systems Research Group of the French national research organization INRA will soon publish a book (LEARN@Paris, in press) that has been edited by a group of international scientists and that focuses on how the agricultures in industrial countries learn to deal with their new context. It was no problem to identify high quality contributions from all over the world.

As a result of such experiences, I am convinced that agricultural universities have increasing opportunities to use their expertise in managing the land to buy into the increasing public concern about the eco-challenge and not only to attract 'green' research funding, but also the top students who are driven by issues instead of self-interest. If Lubchenco's (1998) ideas about the eco-challenge as a new basis for a social contract for science are right, they seem to be especially applicable to agricultural universities. If we miss the boat, it is because of our blinding insights of yesteryear.

Let me end this lecture by providing some examples of research issues that focus on the transformation of collective cognitive systems. I invite you to improve upon my effort.

•   *Eco-indicator development.* We currently use economic variables, such as inflation rate, gross national product, Dow-Jones Index, and so on as indicators of societal well-being. We have very few indicators that reflect the eco-challenge, except perhaps the level of traffic pollution on inversion days and the quality of water for swimming. It is high time to develop indicators for the quality of the eco-systems and ecological services on which we depend.

- *Mechanisms involved in the governance of collective cognitive consistency.* The eco-challenge as a new context should lead to a process of adaptation in the direction of a new cognitive consistency. Processes that seem to be involved in governance mechanisms leading to such new consistency can be expected to include credibility, legitimation, collective simple heuristics (Gigerenzer et al, 1999) that allow dismissal or acceptance of signals about the new context, and the perception of realistic alternatives for action. So far we seem to have no overall shared theory that allows us to be self-aware of the mechanisms by which we tolerate inconsistencies and build new consistencies.

- *The development of methods for facilitating collective action.* There is an increasing acceptance of the notion that innovation (eg sustainability) emerges from interaction among stakeholders in the theatre of innovation (Engel and Salomon, 1997). Rapid Appraisal of Agricultural Knowledge Systems (RAAKS) has been developed as a soft systems methodology for the participatory analysis of their interaction by complementary stakeholders involved in improving innovative performance (Engel and Salomon, 1997). Considerable recent advances have been made in methods for large group interventions, such as open space technology (eg Bunker and Alban, 1997; Holman and Devane, 1999). But the facilitation of change processes (or 'change management') is only beginning to be a subject of (action) research and field experimentation, especially where collective action with respect to natural resource management is concerned (Groot, in preparation, Groot and Maarleveld, 2000; Buck, 2000).

- *Participatory change management.* Much research has implicitly been directed at, or carried out in collaboration with, public agencies. However, given the predominance of the private sector, it seems necessary to focus more on the private sector. Of course, many agricultural universities are already collaborating with the private sector in that they are obtaining private funds for research that is of direct benefit to the private sector. However, it seems necessary to explore, with the private sector, the opportunities for continuity in the new context. My assumption is that most corporations are more interested in continuity than short-term profit.

- *The interface between ecosystems and human institutions.* Earlier, I have mentioned the 'soft side of land'. And some interesting research shows the importance of that concern. For example, Mazzucato and Niemeijer (2000) have shown that Burkinabe farmers' extended families and friendship networks allow them to expand their access to land and fallowing as a key device for maintaining soil fertility, provide them with a flexible source of labour, the key input into their agriculture, and give them reliable access to food when their crops fail. But there is much of the soft side of land that we do not understand. For example, there is a tendency of agricultural research to focus on the farm level and to assume that farmers are the primary decision makers in agriculture. This obscures the larger social networks and institutions in which they are embedded and that to a large extent determine their choices.

- *Collective cognition and human institutions.* Mary Douglas (1986, p128) has expressed her distaste for the prevailing rational choice theory:

> Only by deliberate bias and by extraordinarily disciplined effort has it been possible
> to erect a theory of human behaviour whose formal account of reasoning only con-
> siders the self-regarding motives, and a theory that has no possible way of including
> community-mindedness or altruism, still less heroism, except as an aberration...
> For better or for worse, individuals really do share their thoughts and they do to
> some extent harmonize their preferences, and they have no other way to make the
> big decisions except within the scope of the institutions they build.

She sets out to amend such un-sociological views of human cognition and traces
resistance to 'the idea of a supra-personal cognitive system' to the enthusiasm of our
society for individualism. The commitment that subordinates individual interests
to a larger social whole must be explained. Douglas therefore considers 'the role of
cognition in forming the social bond'. The whole system of knowledge is a collective good that the community is jointly constructing. It is this process that is centre
of Douglas' book. 'Half of our task is to demonstrate this cognitive process at the
foundation of social order. The other half of our task is to demonstrate that the
individual's most elementary cognitive process depends on social institutions' (p45).
Douglas (1996) has been influential through her 'cultural theory' (reviewed by
Oversloot, 1998). In brief, Douglas argues that our preferences are largely the product of our social relations. Social relations are embedded in what Douglas calls 'forms
of social life that recur: the individualist, sectarian (or egalitarian) and hierarchical'
(p7). These 'forms of social life' can be seen as a typology formed by two dimensions: group and grid. The former describes the extent to which individuals form
part of bounded units, and the latter the extent to which the life of the individual
is determined by rules. The resulting typology is shown in Figure 16.2. For Douglas, the typology has predictive value. For example, 'the competitive (individualist)
society celebrates its heroes, the hierarchy celebrates its patriarchs and the sect its
martyrs' (p80).

The possibilities of linking Figures 16.2 and 16.1 are intriguing. Does a technocentric blinding insight build on individualism? Does the holocentric view, and
its focus on cooperative choices implied in ecological rationality, assume some subjugation of the individual to an agreed collective order, or hierarchy? An interesting group to study such issues are Dutch conventional farmers. They fiercely espouse
individualism, consider themselves free entrepreneurs and abhor any submission to
hierarchy. They are, however, totally immersed in a context dominated by marketing mechanisms that systematically destroy their livelihoods and freedom of choice.
Their conditions are becoming more miserable every year, while society blames them
for destroying age-old landscapes, reducing biodiversity, poisoning and polluting
the environment and producing unsafe foods.

- *Institutions and the perception of nature*. Thompson et al (1990), quoted by Oversloot (1998), link the typology in Figure 16.2 to 'myths of nature'. Thus a perspective on nature as benign, robust and tolerant fits with individualism. Because it is
  robust, nature does not need protection. A view of nature as fairly tolerant, but
  perverse (if you treat it badly, the damage cannot be repaired) fits with hierarchy.
  Because nature must be handled with care, control by the group over individuals is
  required. Nature is extremely vulnerable (ephemeral) for egalitarians. It must be

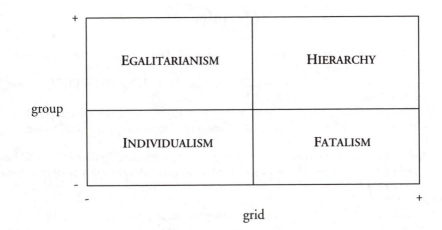

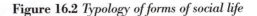

Source: Based on Oversloot (1998) and Douglas (1996)

Figure 16.2 *Typology of forms of social life*

treated with extreme care. But egalitarians do not have the means to prescribe others how they should behave. The only thing they can do is to proselytize and lead an exemplary life. Finally, the idea that nature is capricious is consistent with fatalism, but also with spiritualism. In short, research on the possible link between 'myths of nature' and cultural predilections seems relevant for our ability to deal with the eco-challenge.

Conclusion

My conclusion can be brief. The eco-challenge presents a unique opportunity for agricultural universities to provide leadership. But such leadership cannot be based on a narrow definition of life sciences as the search for the best biotechnological means. That is only more of the same and a sure road to marginalization. Biologists Maturana and Varela (1992) have opened an exciting perspective on cognition as the very process of life. An equally exciting new mission for agricultural science emerges if we define 'life sciences' in that perspective.

Notes

1   I gratefully acknowledge helpful suggestions of Russ Dilts, Rhodora Gonzalez, Janice Jiggins, Marleen Maarleveld and Ricardo Ramirez and Jim Shute.
2   We are grateful to Professor Dr Ray Ison of the Open University, Milton Keynes, UK, for pointing this out.

References

Aaron, H. (1994) 'Public policy, values and consciousness. Distinguished lecture on economics and government', *Journal of Economic Perspectives*, vol 8, pp3–21

Arrow, K. (1994) 'Methodological individualism and social knowledge (Richard T. Ely Lecture)'. *AEA Papers and Proceedings*, vol 84, pp1–9

Auerbach, R. (1999) 'Design for participation in ecologically sound management of South Africa's Mlazi River Catchment', Wageningen University, Wageningen, Published Doctoral Dissertation

Bawden, R. (1997) 'The community challenge: The learning response', Keynote Plenary Address to the Annual International Meeting of the Community Development Society, Athens, Georgia, July 1997

Bawden, R. (2000) 'The importance of praxis in changing forestry practice (preliminary title)', Invited Keynote Address for 'Changing Learning and Education in Forestry: A Workshop in Educational Reform', held at Sa Pa, Vietnam, 16–19 April, 2000

Beck, U. (1994) 'The reinvention of politics: Towards a theory of reflexive modernisation', in Beck, U., Giddens, A. and Lash, S. (eds) *Reflexive Modernisation: Politics, Tradition and Aesthetics in the Modern Social Order*, Stanford University Press, Stanford, pp1–55

Boulding, K. (1968) 'General systems theory: The skeleton of science', in Buckley, W. (ed) *Modern Systems Research for the Behavioural Scientist*, Aldine Publishing Co, Chicago, pp3–11

Brinkman, R. (1994) 'Recent developments in land use planning', Keynote address at the 75-year anniversary conference of Wageningen Agricultural University, in Fresco, L. O. et al (eds) *Future of the Land: Mobilising and Integrating Knowledge for Land Use Options*, John Wiley and Sons, Chichester

Broekstra, G. (1998) 'An organisation is a conversation', in Grant, D., Keeney, T. and Oswick, C. (eds) *Discourse and Organisation*, Sage Publications, London

Buck, L. E. (2000) 'Facilitating knowledge systems. The case of an agro-forestry network in the North-western USA', Preliminary title, Cornell University, Department of Natural Resources, unpublished PhD Dissertation

Buck, L. E., Geisler, C.G., Schelhas, J.W. and Wollenberg, E. (eds) (2001) *Biological Diversity: Balancing Interests though Adaptive Collaborative Management*, CRC Press, Boca Raton

Bunker, B. B. and Alban, B.T. (1997) *Large Group Interventions. Engaging the Whole System for Rapid Change*, Jossey-Bass Publishers, San Francisco.

Callon, M. and Law, J. (1989) 'On the construction of socio-technical networks: Content and context revisited', *Knowledge in Society: Studies in the Sociology of Science Past and Present*, vol 8, pp57–83

Capra, F. (1996) *The Web of Life. A New Synthesis of Mind and Matter*, HarperCollins Publishers (Flamingo), London

Chambers, R. and Ghildyal, R. (1985) 'Agricultural research for resource-poor farmers: A parsimonious paradigm', IDS, Discussion paper 220, Brighton, Sussex

Chambers, R. and Jiggins, J. (1987) 'Agricultural research for resource-poor farmers. Part I: Transfer-of-technology and farming systems research. Part II: A parsimonious paradigm', *Agricultural Administration and Extension*, vol 27, pp35–52 (Part I) and vol 27, pp109–128 (Part II)

Checkland, P. (1981) *Systems Thinking, Systems Practice*, John Wiley, Chichester

Checkland, P. and Scholes, J. (1990) *Soft Systems Methodology in Action*, John Wiley and Sons, Chichester

Clark, A. (1997) *Being There: Putting Brain, Body and World Together Again*, The MIT Press, Cambridge, Massachusetts

De Wit, C. T. (1974) 'Public address at the occasion of the Dies Natalis of the "Landbouwhoge-school"', Wageningen Agricultural University, Wageningen

Douglas, M. (1986) *How Institutions Think*, University of Syracuse Press, Syracuse , New York

Douglas, M. (1996) *Thought Styles*, Thousand Oaks, London

Engel, P. G. H. and Salomon, M. (1997) *Facilitating Innovation for Development. A RAAKS Resource Box*, KIT, Amsterdam

Eshuis, J. (in preparation) 'New institutional arrangements in the relations between nature and culture', Published Doctoral Dissertation in preparation. Wageningen University, Wageningen

Festinger, L. (1957) *A theory of Cognitive Dissonance*, Row and Peterson, Evanston, Illinois

Funtowicz, S. O. and Ravetz, J. R. (1993) 'Science for the post-normal age', *Futures*, vol 25, pp739–755

Gigerenzer, G., Todd, P. M. and the ABC Research Group (1999) *Simple Heuristics that Make us Smart*, Oxford University Press, New York and Oxford

Gilbert, N. and Troitzsch, K. (1999) *Simulation for the Social Scientist*, Open University Press, Buckingham

Groot, A. E. (in preparation). 'Participatory action research: Case studies from Africa (prelimi-nary title)'. Doctoral dissertation in preparation. Agricultural University, Wageningen

Groot, A. and Maarleveld, M. (2000) *Demystifying Facilitation in Participatory Development, Gatekeeper Series no 89*, IIED, London

Guijt, I. (in preparation) 'Participatory monitoring: From darling of the donors to essence of social learning (preliminary title)', Published doctoral dissertation in preparation. Wagenin-gen University, Wageningen

Habermas, J. (1984) *The Theory of Communicative Action. Volume 1: Reason and the Rationalisa-tion of Society*, Beacon Press, Boston

Habermas, J. (1987) *The Theory of Communicative Action. Volume 2: Life World and System. A Critique of Functionalist Reason*, Beacon Press, Boston

Holling, C. S. (1995) 'What barriers? What bridges?' in Gunderson, L. H., Holling, C.S. and Light, S. S. (eds) *Barriers and Bridges to the Renewal of Ecosystems and Institutions*, Colombia Press, New York, pp3–37

Holman, P. and Devane, T. (eds) (1999) *The Change Handbook. Group Methods for Shaping the Future*, Berrett Koehler Publishers, San Francisco

Jiggins, J., Hubert, B. and Collins, M. (1999) 'Globalisation and technology: The implications for learning processes in developed agriculture', in Learning Research Network (LEARN@ Paris) (ed) *Cow up a Tree: Knowing and Learning for Change in Agriculture. Case Studies from Industrial Countries*, INRA, Paris

Jiggins, J. and Röling, N. (2000) 'Adaptive management: Potential and limitations for ecological governance. *International Journal of Agricultural Resources, Governance and Ecology (IJARGE)*, vol 1

Keiter, R. B. and Boyce, M.S. (1991*) The greater Yellowstone Ecosystem: Redefining America's Wil-derness Heritage*, Yale University Press, Boston

Koestler, A. (1967) *The Ghost in the Machine*, Arkana, London, pp45–59

Korten, D. (1995) *When Corporations Rule the World*, Kumarian Press, West Hartford, Connecticut

Kuhn, T. S. (1970) *The Structure of Scientific Revolutions. 2nd Ed*, University of Chicago Press, Chicago

LEArning Research Network (LEARN@Paris (ed) *Cow up a Tree: Knowing and Learning for Change in Agriculture. Case Studies from Industrial Countries*, INRA, Paris

Leeuwis, C. (ed) (1999) *Integral Design: Innovation in Agriculture and Resource Management*, Mansholt Study 15. Wageningen University, Wageningen

Lubchenco, J. (1998) 'Entering the century of the environment: A new social contract for sci-ence', *Science*, vol 279, pp491–496

Maarleveld, M. (2000) 'Social learning in dilemmas in natural resource management. The case of subterranean drinking water resources in Gelderland, the Netherlands (preliminary title)', Published doctoral dissertation. WAU, Wageningen

Maslow, A. (1964) *Religion and Peak Experience*, Kappa, Delta, Phi, West Lafayette

Maturana, H. R. and Varela, F. J. (1987, and revised edition 1992) *The Tree of Knowledge, the Biological Roots of Human Understanding*, Shambala Publications, Boston, Massachusetts

Mazzucato, V. and Niemeijer, D. (2000) 'Rethinking soil and water conservation in a changing society. A case study of eastern Burkina Faso', Published doctoral dissertation. Wageningen University, Wageningen

Merton, R. (1957) *Social Theory and Social Structure*, Free Press, Glencoe

Miller, A. (1983) 'The influence of personal biases on environmental problem-solving', *Journal of Environmental Management*, vol 17, pp133–142

Miller, A. (1985) 'Technological thinking: Its impact on environmental management', *Environmental Management*, vol 9, pp179–190

North, D. C. (1990) *Institutions, Institutional Change and Economic Performance*, Cambridge University Press, New York

O'Connor, M. (1998) 'Pathways for environmental evaluation, a walk in the (hanging) gardens of Babylon', Paper at the 5th Biannual Meeting of the International Association for Ecological Economics, 'Beyond Growth: Policies and Institutions for Sustainability', Santiago, Chile, 15–19 November, 1998

Ostrom, E. (1990, 1991, 1992) *Governing the Commons. The Evolution of Institutions for Collective Action*, Cambridge University Press, New York

Oversloot, H. (1998) 'De culturele theorie wellwillend blicht door een agnosticus', *Tijdschrift voor Beleid, Politiek en Maatschappij*, vol 5, pp2–14

Pain, S. (1993) '"Rigid" cultures caught out by climate change', *New Scientist*, 5 March

Platteau, J. P. (1996) 'The evolutionary theory of land rights as applied to Sub-Saharan Africa: A critical assessment', *Development and Change*, vol 27, pp29–86

Platteau, J. P. (1998) 'Distributional contingencies of dividing the commons', Invited paper for Research School CERES Seminar 'Acts of Man and Nature? Different constructions of natural and resource dynamics', Bergen, the Netherlands, 22–24 October, 1998

Ragsnarsdottir, K. V. (2000) 'Environmental fate and toxicology of organo-phosphate pesticides', *Journal of the Geological Society*, vol 157, pp859–876

Ridley, M. (1995) *The Origins of Virtue*, Penguin Books, Harmondsworth

Röling, N. (1970) 'Adaptations in development: A conceptual guide for the study on non-innovative responses of peasant farmers', *Economic Development and Cultural Change*, vol 19, pp71–85

Röling, N. (1996) 'Towards an interactive agricultural science', *European Journal of Agricultural Education and Extension*, vol 2, pp35–48

Röling, N. (1997) 'The soft side of land. Socio-economic sustainability of land use systems', Invited Paper for the Conference on Geo-Information for Sustainable Land Management, held at the International Institute for Aerospace Survey and Earth Sciences (ITC), Enschede, the Netherlands, 17–21 August 1997. Published in *ITC Journal*, Special Congress Issue on Geo-Information for Sustainable Land Management, Volume 1997, no 3–4, pp248–262

Röling, N. (2000) 'Changing forestry education. Enhancing beta/gamma professionalism', Invited Keynote Address for 'Changing Learning and Education in Forestry', held at Sa Pa, Vietnam, 16–19 April 2000

Röling, N. and Jiggins, J. (1998) 'The ecological knowledge system', in Röling, N. and Wagemakers, A. (eds) *Facilitating Sustainable Agriculture. Participatory Learning and Adaptive Management in Times of Environmental Uncertainty*, Cambridge University Press, Cambridge, pp283–307

Röling, N. and Maarleveld, M. (1999) Facing strategic narratives: An argument for interactive effectiveness, *Agriculture and Human Values*, vol 16, pp295–308

Rosenberg, A. (1988) *Philosophy of Social Science*, Westview Press, Boulder

Satz, D. and Ferejohn, J. (1994). 'Rational choice and social theory', *The Journal of Philosophy*, vol 91, pp71–88.

Searle, J. (1984) *Minds, Brains and Science*, The 1984 Reith Lectures, BBC, London

Seligman, M. and Hager, J. (1972) *Biological Boundaries of Learning*, Appleton-Century-Crofts, New York.

Sen, A. (1999) *Development as Freedom*, Anchor Books, New York

Simon, H. (1969) *The Sciences of the Artificial*, MIT Press, Cambridge, Massachusetts

Steins, N. A. (1999) 'All hands on deck. An interactive perspective on complex common-pool resource management base on case studies in coastal waters of the Isle of Wight (UK), Connemara (Ireland) and the Dutch Wadden Sea. Wageningen (NL)', Published Doctoral Dissertation, Wageningen University, Wageningen.

Suzuki, D. T. (1972; 1994) *Living by Zen. A Synthesis of the Historical and Practical Aspects of Zen Buddhism*, Samuel Weiser, York Beach, Maine

Tamborini, R. (1997) 'Knowledge and economic behaviour. A constructivist approach', *Journal of Evolutionary Economics*, vol 7, pp49–72

Thompson, M., Ellis, R. and Wildavsky, A. (1990) *Cultural Theory*. Westview, Boulder, Colorado

Van Eijk, T. (1998) 'Farming systems research and spirituality: An analysis of the foundations of professionalism in developing sustainable farming systems', Published doctoral dissertation, Wageningen Agricultural University, Wageningen

Van Haaften, E. H., van de Vijver, F., Leenaars, J. and Driessen, P. (1998) 'Human and biophysical carrying capacity in a degrading environment: The case of the Fulani in the Sahel', *The Land*, vol 2, pp39–51

17

Ecological Basis for Low-toxicity Integrated Pest Management (IPM) in Rice

Kevin Gallagher, Peter Ooi, Tom Mew, Emer Borromeo, Peter Kenmore and Jan-Willem Ketelaar

Introduction

This chapter focuses on a case studies arising from an Asian based Integrated Pest Management (IPM) programme. It is an in-depth analysis of the well-researched and widespread rice-based IPM.

Rice production is a highly political national security interest that has often justified heavy handed methods in many countries to link high yielding varieties, fertilizers and pesticides to credit or mandatory production packages and led to high direct or indirect subsidies for these inputs. Research, including support for national and international rice research institutes, was well-funded to produce new varieties and basic agronomic and biological data. Vegetable production on the other hand has been led primarily by private sector interests and local markets. Little support for credit, training or research has been provided. High usage of pesticides on vegetables has been the norm due to lack of good knowledge about the crop, poorly adapted varieties and a private sector push for inputs at the local kiosks to tackle exotic pests on exotic varieties in the absence of well-developed management systems.

However, other pressures are now driving change to lower pesticide inputs on both crops. Farmers are more aware of the dangers of some pesticides to their own health and their production environment. The rise of Asian incomes has led to a rise in vegetable consumption that has made consumers more aware of food safety. Cost of inputs is another factor as rice prices fall and input prices climb. More farmers are producing vegetables for urban markets, so driving competition to lower input costs as well. Highly variable farm-gate prices for vegetables make farmers' economic decisions to invest in pesticide applications a highly risky business. Research on vegetables is beginning to catch up with rice allowing for better management of pests through prevention and biological con-

Note: Reprinted from *The Pesticide Detox* by Pretty, J. (ed), copyright © (2004) Earthscan, London

trols. IPM programmes in both crops aim to reduce the use of toxic pesticide inputs and the average toxicity of pest management products that are still needed whilst improving profitability of production.

Integrated pest management in rice

This chapter has been prepared to provide a conceptual guide to the recent developments in rice IPM within an ecological framework. It is not a 'how to' guide but rather a 'why to' guide for IPM programmes that are based on ecological processes and work towards environmentally-friendly and profitable production. We provide a broad overview of IPM practices in rice cultivation including its ecological basis, decision-making methods, means of dissemination to farmers and future needs to improve these practices. The breadth of pest problems, including interaction with soil fertility and varietal management are discussed in depth. Although the main focus is on Asian rice cultivation, we also provide examples of rice IPM being applied in other regions are given.

IPM in rice has been developing in many countries since the early 1960s. However, much of the development was based on older concepts of IPM including intensive scouting and economic thresholds that are not applicable under all conditions (Morse and Buhler, 1997) or for all pests (eg diseases, weeds), especially on smallholder farms where the bulk of the world's rice is grown and which are often under a weak or non-existing market economy. During the 1980s and 1990s, important ecological information became available on insect populations that allowed the development of a more comprehensive ecological approach to pest management, as well as greater integration of management practices that went beyond simple scouting and economic threshold levels (Kenmore et al, 1984; Gallagher, 1988; Ooi, 1988; Graf et al, 1992; Barrion and Litsinger, 1994; Rubia et al, 1996; Settle et al, 1996).

Since then, an ecological and economic analytical approach has been taken for management considers crop development, weather, various pests and their natural enemies. These principles were first articulated in the Indonesian National IPM Programme, but have expanded as IPM programmes have evolved and improved. Currently programmes in Africa and Latin America now use the term Integrated Production and Pest Management (IPPM), and follow these principles: grow a healthy soil and crop; conserve natural enemies; observe fields regularly (soil, water, plant, pests, natural enemies); and farmers should strive to become experts. Within these principles, economic decision making is still the core of rice IPM but incorporates good farming practices as well as active pest problem solving within a production context.

IPM in rice seeks to optimize production and to maximize profits through its various practices. To accomplish this, however, decision making must always consider both the costs of inputs and the ecological ramifications of these inputs. A particular characteristic of Asian rice ecosystems is the presence of a potentially damaging secondary pest, the rice brown planthopper (BPH), *Nilaparvata lugens*. This small but mighty insect has in the past occurred in large scale outbreaks and caused disastrous losses (IRRI, 1979). These outbreaks were pesticide-induced and triggered by pesticide subsidies and policy mismanagement (Kenmore, 1996). BPH is still a localized problem, especially where pesticide overuse and abuse is common, and therefore can be considered as an ecological focal point around which both ecological understanding and management are required for profitable and stable rice cultivation. BPH also becomes the major entry point for all IPM educational programmes since it is always necessary to prevent its outbreak during crop management. Other pests which interact strongly with management of inputs are rice stemborers and the various diseases discussed below.

A major issue when considering IPM decision making is one of paths to rice production intensification. In most cases, intensification means use of improved high yielding varieties, irrigation, fertilizers and pesticides – as was common in the Green Revolution. However, two approaches to intensification should be considered. The first is input intensification in which it is important to balance optimal production level against maximizing profits and for which higher inputs can destabilize the production ecosystem. The second route to intensification is one of optimizing all outputs from the rice ecosystem to maximize profits. In many lowland flooded conditions, this may mean systems such as rice–fish or rice–duck that may be more profitable and less risky, yet require lower inputs (and often resulting in lower rice yields). In areas where inputs are expensive, where the ecosystem is too unstable (because of drought, flood) to ensure recovery of input investments, or where rice is not marketed, then such a path to intensification may be more beneficial over time. However, such a system has a different ecology due to the presence of fish or duck, and therefore will involve a different type of IPM decision making.

Ecological basis of rice IPM

IPM in much of Asian rice is now firmly based on an ecological understanding of the crop and its interaction with soil nutrients and crop varieties. We present below an ecological overview of our current understanding of how the rice ecosystem operates during the development of the crop.

The rice ecosystem in Asia is indigenous to the region and its origins of domestication date back 8000 years to the Yangtze Valley in southern China (Smith, 1995), and more widely some 6000 years ago (Ponting, 1991). Cultivation practices similar to those of today were reached by the 16th century (Hill, 1977). This period of time means that rice plants, pests and natural enemies existed and coevolved together for thousands of generations. Rice ecosystems typically include both a terrestrial and an aquatic environment during the season with regular flooding from irrigation or rainfall. These two dimensions of the rice crop may account for the extremely high biodiversity found in the rice ecosystem and its stability even under intensive continuous cropping – and contrasts with the relative instability of rice production under dryland conditions (Cohen et al, 1994). The irrigated rice systems in Africa, Americas and Europe also include this aquatic and terrestrial element within which high levels of biodiversity are also found.

Insects

Studies by Settle and farmer research groups in Indonesia (Settle et al, 1996) show that flooding of fields triggers a process of decomposition and development of an aquatic food-web, which results in large populations of detritus-feeding insects (especially Chirono-mid and ephydrid flies). These insects emerge onto the water surface and into the rice canopy in large numbers, very early in the growing season, providing critical resources to generalist predator populations long before 'pest' populations have developed.

This is quite different from the usual predator-prey models taught in most basic IPM courses and provides a mechanism to suggest that natural levels of pest control in tropical irrigated rice ecosystems are far more stable and robust than purely terrestrial agroecosystems. This stability, however, was found to be lower in rice landscapes that are subject to long (more than three month) dry seasons and where rice is planted in large-scale synchronous monocultures, as well as in areas where farmers use pesticides intensively. Increased amounts of organic matter in the soil of irrigated rice fields, by itself a highly valuable practice for sustainable nutrient management, has the additional advantage of boosting both populations of detritus-feeding insects and insect predators, and thereby improving natural levels of pest control (Settle et al, 1996).

A second consideration for rice IPM is the ability of most rice varieties to compensate for damage. The rice plant rapidly develops new leaves and tillers early in the season replacing damaged leaves quickly. The number of tillers produced is always greater than the number of reproductive tillers allowing for some damage of vegetative tillers without effecting reproductive tiller number. The flag leaf contributes to grain filling but the second leaf provides photosynthates as well, while lower leaves are actually a sink that compete with the panicle. Finally, photosynthates appear to move from damaged reproductive tillers to neighbouring tillers so that total hill yield is not as severely impacted as expected when a panicle is damaged by stemborers.

Thus, early season defoliators (such as whorl maggot, case worms and armyworms) cause no yield loss up to approximately 50 per cent defoliation during the first weeks after transplanting (Shepard et al, 1990; Way and Heong, 1994) although higher damage occurs when water control is difficult. As early tillering is also higher than what the plant can ultimately support reproductively, up to 25 per cent vegetative tiller damage by stemborers ('deadhearts') (caused by *Scirpophaga* spp, *Chilo* spp, and *Sesamia* spp) can be tolerated without significant yield loss (Rubia et al, 1996). Significant damage (above 50 per cent) to the flag leaf by leaffolders (*Cnaphalocrocis mdeinalis* and *Marasmia* spp) during panicle development and grain filling can cause significant yield loss, although this level of damage is uncommon where natural enemies have been conserved (Graf et al, 1992). Late season stemborer damage (white heads) also causes less damage than previously expected such that up to 5 per cent white heads in most varieties does not cause significant yield loss (Way and Heong, 1994; Rubia et al, 1996).

The conspicuous rice bug (*Leptocorisa oratorius*) is another major target for insecticide applications. However, in a recent study involving farmers and field trainers at 167 locations, van den Berg and Soehardi (2000) have demonstrated that the actual yield loss in the field is much lower than previously assumed. The rice panicle normally leaves part of its grain unfilled as if to anticipate some level of loss (Morrill, 1997). Numerous parasitoids, predators and pathogens present in most rice ecosystems tend to keep these potential pests at low densities (Shepard and Ooi, 1991; Barrion and Litsinger, 1994; Loevinsohn, 1994; Ooi and Shepard, 1994; Matteson, 2000).

Thus, under most situations where natural enemies are conserved, little yield loss is expected from typical levels of insect pests. Up until recently, insecticide applications for early defoliators, dead-hearts and white-heads often led to lower natural enemy populations allowing the secondary pest, rice brown planthopper (*Nilaparvata lugens*), to flare up in massive outbreaks (Rombach and Gallagher, 1994). Work by Kenmore et al (1984) and Ooi (1988) clearly showed the secondary pest status of brown planthoppers. Although resistant varieties continue to be released for brown planthopper, the highly migratory sexual populations were found to have high levels of phenotypic variation and highly adaptable to new varieties. Although wrongly proposed to be 'biotypes', it was found that any population held significant numbers of individuals able to develop on any gene for resistance (Claridge et al, 1982; Sogawa et al, 1984; Gallagher et al, 1994). Huge outbreaks have not reoccurred in areas where pesticide use has dropped due either to changes in policy regulating pesticides in rice or due to educational activities.

A few minor pests are predictable problems and therefore should be considered for preventive action with natural enemies, resistant varieties, or specific sampling and control. These include black bug (*Scotinophara* spp), gall midge (*Orseolia oryzae*), and rice hispa (*Dicladispa* spp) which are consistently found in certain regions; thrips (*Stenchaetothrips biformis*) whereas drought causes leaf-curling that provides them a habitat; armyworms (*Mythimna* spp and *Spodotera* spp) in post-drought areas that are attracted by high levels of mobilized nitrogen in the rice plant and panicle cutting armyworms cause extreme damage.

Green leafhoppers (*Nephotettix* spp) are important vectors of tungro (see below) but by themselves rarely cause yield loss. White-backed planthoppers (*Sogatella* spp) are closely related to brown planthoppers in terms of population dynamics and are not usually a major yield reducing pest. Rice water weevil (*Lissorhoptrus oryzophilus*) introduced from the Caribbean area in the US and north-east Asia is a problem pest requiring intensive sampling (Way et al, 1991) that deserves greater research on its natural enemies. In upland ecosystems, white grub species and population dynamics are not well studied and are difficult to manage. Way et al (1991) provide an overview of insect pest damage dynamics, while Dale (1994) gives an overview of rice insect pest biology.

Diseases

The need to grow more rice under increasingly intensive situations leads to conditions that favour diseases. High planting density, heavy inputs of nitrogen and soil fertility imbalance result in luxuriant crop growth conducive to pathogen invasion and reproduction. This is made worse by genetic uniformity of crop stand that allows unrestricted spread of the disease from one plant to another, together with continuous year-round cropping that allows carry over of the pathogen to succeeding seasons. Reverting to the less intense, low yield agriculture of the past may be out of the question, but a thorough understanding of the ecological conditions associated with the outbreak of specific diseases may lead to sustainable forms of intensification. We briefly describe the specifics for three major diseases of rice, namely, rice blast, sheath blight and rice tungro disease.

Blast (*Pyricularia grisea, Magnaporthe grisea*) occurs throughout the rice world but is usually a problem in areas with a cool, wet climate. It is a recognized problem in upland ecosystems with low input use and low yield potential, as well as in irrigated ecosystem with high input use and high yield potential (Teng, 1994). Fertilizers and high planting density are known to exacerbate the severity of infection. Plant resistance is widely used to control the disease, but varieties often need to be replaced after a few seasons because pathogens quickly adapt and overcome the varietal resistance. Recent work by IRRI and the Yunnan Agricultural University demonstrated that the disease can be managed effectively through varietal mixtures (Zhu et al, 2000).

Sheath blight (*Rhizoctonia solani*) is a problem during warm and humid periods and is also aggravated by dense planting and nitrogen inputs above 100kg ha^{-1}. No crop plant resistance is known for sheath blight. A number of bacteria (*Pseudomonas* and *Bacillus*) isolated from the rice ecosystem are known to be antagonistic to the pathogen. Foliar application of antagonistic bacteria at maximum tillering stage appeared to effect a progressive reduction of disease in the field over several seasons (Du et al, 2001). Incorporation of straw and other organic matter, with its effect on soil fertility, pH, and possibly on beneficial micro-organisms may reduce sheath blight incidence in the long term.

Rice tungro disease caused by a complex of two viruses transmitted by the green leafhopper (*Nepthettix virescen*) is a destructive disease in some intensively cultivated areas in Asia where planting dates are asynchronous (Chancellor et al, 1999). Overlapping crop season provide a continuous availability of host that enables year round survival of the virus and the vector. Controlling the vector population with insecticide does not always result in tungro control. Synchronous planting effectively puts the disease at manageable levels. When and where planting synchrony is not possible, resistant varieties are recommended. In addition to varieties with certain degree of resistance to the vector, varieties highly resistant to the virus itself became available recently. Farmers should also employ crop or varietal rotation, and rogue intensively.

Fungicidal control of blast and sheath blight is increasing in many intensified rice areas. It is extremely important that these fungicides be carefully screened not only for efficacy as fungicides but also for impact on natural enemies in the rice ecosystem. One example is the release of iprobenfos as a fungicide for blast control. Iprobenfos is an organophosphate that was originally developed for brown planthopper control and is highly toxic to natural enemies. Its use in the rice ecosystem is likely to cause ecological destabilisation and consequent outbreaks of brown planthopper. Fungicides should also be carefully screened for impact on fish, both to avoid environmental damage in aquatic systems and to avoid damage to rice–fish production.

In general, clean and high quality seed with resistance to locally known diseases is the first step in rice IPM for diseases. An appropriate diversification strategy (varietal mixture, varietal rotation, varietal deployment, crop rotation) should counter the capacity of pathogens to adapt quickly to the resistance of the host. Management of organic matter has to be geared not only towards achieving balanced fertility but also in enhancing the population of beneficial microorganisms.

Farmers in Korea who face heavy disease pressure can learn to predict potential outbreaks using educational activities that combine various weather and agronomic input parameters with disease outcomes. Computer based models are also being commercially sold to predict disease potential based on meteorological monitoring. With increasing nitrogen applications, however, greater disease incidence can also be expected.

Weeds

The origin of puddling for lowland rice cultivation is thought to have been invented to create an anaerobic environment that effectively kills several weeds including weedy and red rice. In most IPM programmes for lowland rice, weed management has therefore been closely considered part of agronomic practices during puddling and later during aeration of the soil with cultivators. At least two hand weedings are necessary in most crops, and considered in many countries economically viable due to low labour cost or community obligations to land-less who are then allowed to participate in the harvest. With raising labour costs, decreasing labour availability and more effective herbicides this situation is rapidly changing to one of using one or two applications of pre- or post-emergence herbicides. As in the case of fungicides, it is critical that these herbicides do not upset natural enemies, fish or other beneficial/non-target organisms in the aquatic ecosystem including micro-organisms. In the case of upland rice, similar changes are rapidly occurring although better dry land cultivators are already developed for inter-row cultivation as an alternative to herbicides.

Non-herbicide but low labour weed management methods are also emerging from the organic agriculture sector. The International Association of Rice Duck Farming in Asia supports research and exchanges among mostly organic farmers. In rice–duck farming, a special breed of duck is allowed to walk through the field looking for food that is either broadcast or naturally occurring, and the action of walking up and down the rows is sufficient to control most weeds. In Thailand, mungbean and rice are broadcast together with some straw covering in rainfed rice fields. When the rains come, both crops germinate. If there is abundant rain, the mungbean will eventually die and become part of the mulch, but if the rain is insufficient for the rice then the mungbean will be harvested.

No-till, no-herbicide combined with ground cover from winter barley straw or Chinese milky-vetch is being used in South Korea in both conventional and organic systems. Organic farmers in California use a water management system in which there is a period of deep (30cm) flooding followed by complete drying – the rice can take the changes but young weeds cannot. A widely adopted method in Central Thailand involves growing rice from ratoons. After harvest the stubble is covered with straw and then irrigated which allows the rice plant to emerge. This method not only controls weeds effectively but also increases organic matter and requires no tilling.

However, for the majority of rice cultivation, labour saving often means moving towards direct seeded rice and thus more weed problems. Red rice (weedy off-type of rice) is already the key pest in most of the Latin American direct seeded rice production areas. It seems clear that more direct seeding will lead to more herbicide use in rice production. Yet herbicide resistance is also sure to eventually emerge and there are obvious health and environmental costs associated with some herbicides. Thus it is important that IPM for rice weeds be improved and considered in the broadest terms (eg promoting modern rice varieties that are red in colour among consumers may be part of the solution to red rice problems). Crop rotations are feasible in only some areas, while simple line sowers or tractor sowing in rows combined with manual or tractor cultivation may provide some solutions for lowland and upland rice.

Genetically modified herbicide-resistant rice will be eventually on the market, but Asian consumer preference may not favour these varieties. However, the resulting increase in herbicide use could have obvious adverse effects on the aquatic systems that are associated with most rice production. In addition, a major problem of herbicide resistant rice is the possibility of the transfer of gene resistance to weedy rice, though such transfers would not occur to wild grass species. Use of herbicide resistant rice in monocropping could also create in long-term serious problems of glyphosate resistance in weed species previously susceptible to the herbicide. The ecosystem level interactions of herbicide resistant rice will need careful assessment prior to their use.

Community pests

Insects, diseases (with the exception of tungro virus) and weeds in rice ecosystems are generally managed with decisions on individual farms or plots. However, some pests, particularly rats, snails and birds, require community-level planning and action. Management of these pests requires facilitation of community organisations not generally supported by extension services with the possible exception of some multi-purpose cooperatives and water-user associations.

Numerous species of rats occur in rice fields and can cause considerable damage. Rats migrate from permanent habitats to rice fields as food supply changes throughout a yearly cycle, with rice plants most preferred after the panicles have emerged. Some natural enemies of rats, particularly snakes, are harmed by pesticides and often killed by farmers, thus resulting in more rats. The most effective management strategies are to ensure baits are appropriate to species present, and then carry out continuous trapping along feeding routes, fumigation or digging of rat holes, and establishing early season bait stations using second generation anticoagulant baits (although more toxic zinc phosphide and repackaged and unlabelled aldicarb is still commonly seen but strongly discouraged in most countries due to deaths of children and small livestock). Community programmes can include educational activities on rat biology and behaviour (Buckle, 1988), and an emphasis on action during the early season vegetative stage is considered the key to rat management (Buckle and Smith, 1994; Leung et al, 1999). An innovative owl habitat programme in Malaysia has been successful in increasing owl populations to control rats in rice and plantation crops.

The golden apple snail, *Pomacea canaliculata*, was originally introduced to rice growing areas as an income generating activity for a caviar look-alike given its brightly pink coloured egg clusters. It has since become widespread from Japan to Indonesia and is now one of the most damaging pests of rice. It was introduced without appropriate tests in any country even though on quarantine lists of several countries. The snail feeds on vegetation in aquatic environments, including newly transplanted rice seedlings up to about 25 days when the stems become too hard. Without natural enemies and with highly mobile early stages that flow with irrigation water, the golden snail spreads rapidly.

Pesticides are often used before transplanting or direct seeding, mainly highly toxic products such as endosulfan, organo-tin products, and metaldehyde. These products have serious health implication and also cause death of potential fish predators and natural enemies early in the season (Halwart, 1994). The use of bamboo screens as inlets to fields to inhibit snail movement is reported as the first line of snail defence. Draining fields that have several shallow ditches where the snails will congregate allows for faster collection or ease of herding ducks in fields to eat the snails. In Vietnam, snails are reported to be collected, chopped, cooked and used as fish food to such an extent that they are now a declining problem.

Birds can be very damaging especially when occurring in large flocks. The Red-billed Quelea, *Quelea quelea*, in sub-Saharan Africa and various species in Asia are known as consistent problems in rice ecosystems. In most Asian countries and in Chad, netting is used to trap large numbers of birds for sale as food. Mass nest destruction is also possible for some species. In Asia, these methods have effectively reduced pest bird populations to very low numbers. In Africa, the capture method may bring benefits to local people in terms of income or a good protein addition to the diet, but the impact on pest bird populations has been small. During the ripening period in north-east Asia, some fields are protected by being covered with bird nets. Reflective ribbons or used video or cassette tape are widely used to scare birds in Asia. Sound cannons and owl or hawk look-alikes are also used in many countries, though some birds become quickly habituated to mechanical devices. Use of poisoned baits, and destruction of bird nesting habitat, are discouraged both because they are seldom effective and also because of the potential negative effect on non-target species in adjacent aquatic environments.

Does IPM work for rice farmers?

Although there is a large amount of grey literature (see www.communityipm.org) related to rice IPM impact among farmers, there are few peer-reviewed published data. This is in part a reflection of the financial and technical difficulty of conducting these studies. Longitudinal studies in agriculture are notoriously difficult due to seasonal changes. Latitudinal studies (comparisons across sites) are also difficult due to the fact that finding an identical IPM and non-IPM control is rarely possible given the diversity of ecological and social conditions. Nonetheless, such evidence as does exist indicates considerable benefits for rice IPM farmers.

The first, and perhaps strongest indicator, is the greatly reduced incidence of brown planthopper. Wide area outbreaks accompanied with massive losses have no longer been experienced during the past 15 years since IPM programmes have become widely implemented in both policy and field training. In most cases, changes in policy involved removal of pesticide subsidies, restrictions on outbreak-causing pesticides, and investment in biological research and educational programmes for decision makers, extension and farmers. These policy changes most often came about as a result of successful small-scale field trials. The FAO Inter-Country Programme for Rice IPM in south and south-east Asia headed by Peter Kenmore brought policy makers in contact with researchers and farmers who could explain from their own experience the ecological basis of farm

Table 17.1 *Financial analysis of ten IPM field school alumni and ten non-alumni farms from impact assessment in Lalabata, Soppeng, Ujung Pandang, South Sulawesi, Indonesia*

	IPM Alumni (Rp. 000 ha^{-1})	Non-alumni (Rp. 000 ha^{-1})
Ploughing	105	84
Planting	113	102
Weeding	49	47
Harvest	67	59
Seeds	18	21
Urea	80	96
SP36	30	12
KCl	25	12
ZA	41	0
Pesticides	7	28
Irrigation	25	25
Total costs	560	501
Yield (kg ha^{-1})	6633	5915
Returns	2786	2485
Income	2226	1983
Difference	+243	

Note: farm gate rice price Rp. 420/kg
Source: FAO, 1998

ing with IPM methods. The banning of 57 pesticides and removal of pesticide subsidies known to cause brown planthopper outbreaks in 1987 in Indonesia by the former President Suharto came about after cabinet officials were brought into a dialogue with both senior Indonesian and IRRI scientists and farmer groups who had shown the outbreak effects of the pesticides and their ability produce high rice yields without these pesticides (Eveleens, 2004).

The second indication comes from case study literature (FAO, 1998). Table 17.1 gives a typical result found across hundreds of communities surveyed in rice IPM programmes. This shows the key changes in practices, especially the common outcome of investing less in pesticides and more in fertilizers (including P and K). Other large-scale studies provide similar data, although a recent study in Vietnam notes an increase in the use of fungicides (FAO, 2000). The authors have noted that with higher levels of fertilizers (as would be found in Vietnam) such increases in fungicide are predictable. These data also reveal the multidisciplinary aspect of rice IPM in that it encourages farmers to look beyond the pest complex into the multiple parameters for achieving a profitable high yielding crop.

Getting IPM into the hands of farmers

'IPM is not for farmers but is by farmers' is often noted in IPM programmes. Getting IPM into the hands of farmers, however, is not always easy. Several methods have been developed with various levels of information and completeness. Most agricultural extension services now recognize the importance of natural enemies and are quick to point out the need to conserve them, even though their co-promotion of various insecticides, fungicides and herbicides is at odds with this apparent awareness of natural enemies. Work by Heong and others from the Rice IPM Network (Heong et al, 1998; Heong and Escalada, 1999; Huan et al, 1999) has developed interesting radio messages to get the word out on a large scale that early spraying of insecticides during the first 40 days of the crop is not only unnecessary but increases the risk of higher pest populations later in the crop. The radio messages are accompanied by field-based plant compensation participatory research groups in many cases (Heong and Escalada, 1998). This programme has been effective in increasing awareness of the adverse effects of insecticides on natural enemies and the role of plant compensation in recovering without yield loss from early season pest damage and has resulted in reduced early insecticide sprays.

Study groups of various types are now common in many rice systems. There are reported from organic agriculture, rice–duck groups, Australian rice farmer association and many others. The FAO Community IPM Programme in Asia (Matteson et al, 1994) has promoted study groups now called 'Farmer Field Schools' under which structured learning exercises in fields ('schools without walls') are used to study both ecosystem level dynamics transferable to other crops (predation, parasitism, plant compensation) as well as specific rice IPM methods. Already, more than 1.5 million farmers have graduated from one or more season long Field Schools in Asia over the past decade with good cost effectiveness as an extension methodology (Ooi et al, 2001).

Community based study groups, study circles, field schools and other approaches are now being integrated with wider community based organizations, such as IPM clubs, water-user groups, women's organizations and local farmer unions (Pretty and Ward, 2001). With the large-scale training and visit style extension programmes generally being phased out in most countries, it will be necessary for local communities to become organized in ways in which they can increasingly cover their own costs for experts. Primary school programmes on IPM are also emerging in Thailand, Cambodia, the Philippines and other countries as part of environmental education curriculum related to Asian rice-culture. Such programmes as Farmer Field Schools in many countries or Landcare in Australia and the Philippines are providing innovative models in community based study and action.

The future of IPM in rice in Asia, if not globally, should see the phasing out of all Class Ia, Ib and II products, while phasing in production methods that allow for whole ecosystem approaches. Organic pest management (OPM) along side the rapid expansion of certified organic rice production is certainly an area fertile for research and training in addition to a modernized IPM approaches.

Future needs in rice IPM

There is still much room for improvement for IPM. Indeed, the ecological view of rice and vegetables presented must be given greater support by international and national scientists and policy makers to widen economic and ecosystems benefits already being realized by some farmers. A new CD-ROM produced by IRRI is beginning to bring together basic rice information in an accessible format, while the World Vegetable Centre in Taiwan has developed a web-based study programme. Both programmes could be helpful in training extension staff but still remain distant from farmers. Major other challenges remain. Post-harvest pests are still a problem and deserve greater research on non-toxic management methods, and environmentally friendly methods of controls for all types of pests, especially weeds and fungal pathogens, are required to reduce the pressure on the natural resources.

Some countries are calling for major changes. South Korea has banned pesticide use in Seoul's watersheds and is promoting organic agricultural investments to ensure both clean water and high levels of production. Other communities are moving away from grain maximisation to diversification such as rice–fish–vegetable culture as a response. This is expected to increase as demand for more profitable non-grain products increases and nitrogen use is reduced to lower environmental impacts and incidence of expensive-to-control fungal pathogens. However, IPM development is required in more countries. These programmes should ensure that educational systems (both formal and non-formal) are responding to the future needs of reducing the environmental impact of agriculture while improving yields. IPM is clearly a major aspect of this education.

There is a need to phase in new plant protection methods and products including subsidising commercialisation of locally produced products such as pheromones, attractants, natural enemies, pest-exclusion netting (for insects and birds), high-quality seed, improved disease resistance and balanced soil fertility products. High foreign exchange costs for imported pesticides and increasing consumer awareness of the social costs arising from pesticides and inorganic fertilizers can be expected to drive rice IPM system development. The trend will be towards lower impact and local production of environmentally friendly pest management. A significant redefinition of IPM to exclude Class I and most Class II products could be a most important step to revitalize private sector, research and extension IPM activities.

References

Barrion, A. T. and Litsinger, J. A. (1994) 'Taxonomy of rice insect pests and their arthropod parasites and predators', in Heinrichs, E. A. (ed) *Biology and Management of Rice Insects*, Wiley Eastern Limited, pp13–362

Buckle, A. P. (1988) 'Integrated management of rice rats in Indonesia', *FAO Plant Protection Bulletin*, vol 36, pp111–118

Buckle, A. P. and Smith, R. H. (1994) *Rodent Pests and their Control*, CAB International, Wallingford, UK

Chancellor, T. C. B., Tiongco, E. R., Holt, J., Villareal, S. and Teng, P. S. (1999) 'The influence of varietal resistance and synchrony on tungro incidence in irrigated rice ecosystem in the

Philippines', in Chancellor, T. C. B., Azzam, O. and Heong, K. L. (eds) *Rice Tungro Disease Management.* Proceedings of the International Workshop on Tungro Disease Management, 9–11 November 1998, IRRI, Los Banos, Laguna, the Philippines. Makati City, the Philippines, International Rice Research Institute, pp121–127

Claridge, M. F., Den Hollander, J., and Morgan, J. C. (1982) 'Variation within and between populations of the brown planthopper, *Nilaparvata lugens* (Stal)', in Knight, W. J., Pant, N. C., Robertson, T. S. and Wilson, M. R. (eds) *1st International Workshop on Leafhoppers and Planthoppers of Economic Importance,* Commonwealth Institute of Entomology, London, pp36–318

Cohen, J. E., Schoenly, K., Heong, K. L., Justo, H., Arida, G., Barrion, A. T. and Litsinger, J. A. (1994) 'A food web approach to evaluate the effect of insecticide spraying on insect pest population dynamics in a Philippine irrigated rice ecosystem', *Journal of Applied Ecology*, vol 31, pp747–763

Dale, D. (1994) 'Insect pests of the rice plant: Their biology and ecology', in Heisricks, E. A. (ed) *Biology and Management of Rice Insects*, Wiley Eastern Ltd

Du, P. V., Lan, N. T. P., Kim, P. V., Oanh, P. H., Chau, N. V. and Chien, H. V. (2001) 'Sheath blight management with antagonistic bacteria in the Mekong Delta', in Mew, T. W., Borromeo, E. and Hardy, B. (eds) (2000) *Exploiting Biodiversity For Sustainable Pest Management.* Proceedings of the Impact Symposium on Exploiting Biodiversity for Sustainable Pest Management, 21–23 August 2000, Kunming, China. International Rice Research Institute, Makati City, the Philippines

Eveleens, K. (2004) *The History of IPM in Asia*, FAO, Rome

FAO (1998) *Community IPM: Six Cases from Indonesia,* FAO Technical Assistance: Indonesian National IPM Programme, FAO, Rome

FAO (2000) *The Impact of IPM Farmer Field Schools on Farmers' Cultivation Practicies in their Own Fields.* A report submitted to the FAO Intercountry Programme for Community IPM in Asia by J. Pincus, Economics Department, University of London

Gallagher, K D. (1988) 'Effects of host plant resistance on the microevolution of the rice brown planthopper, *Nilaparvata lugens* (Stal) (Homoptera: Delphacidae)', PhD dissertation, University of California, Berkeley

Gallagher, K. D., Kenmore, P. E. and Sogawa, K. (1994) 'Judicious use of insecticides deter planthopper outbreaks and extend the life of resistant varieties in Southeast Asian rice', in Denno, R. F. and Perfect, T. J. (eds) *Planthoppers; Their Ecology and Management*, Chapman & Hall, New York, pp599–614

Graf, B., Lamb, R., Heong, K. L. and Fabellar, L. (1992) 'A simulation model for the populations dynamic of rice leaf folders (Lepidoptera) and their interactions with rice', *Journal of Applied Ecology*, vol 29, pp558–570

Halwart, M. (1994) 'The golden apple snail, *Pomacea canaliculata* in Asian rice farming systems: Present impact and future threat', *International Journal of Pest Management*, vol 40, pp199–206

Heong, K. L. and Escalada, M. M. (1998) 'Changing rice farmers' pest management practices through participation in a small-scale experiment', *International Journal of Pest Management*, vol 44, pp191–197

Heong, K. L., Escalada, M. M., Huan, N. H. and Mai, V. (1998) 'Use of communication media in changing rice farmers' pest management in the Mekong Delta, Vietnam', *Crop Protection*, vol 17, pp413–425

Heong, K. L. and Escalada, M. M. (1999) 'Quantifying rice farmers' pest management decisions: Beliefs and subjective norms in stem borer control', *Crop Protection*, vol 18, pp315–322

Hill, R. D. (1977) *Rice in Malaya: A Study in Historical Geography*, Oxford University Press, Kuala Lumpur

Huan, N. H., Mai, V., Escalada, M. M. and Heong, K. L. (1999) 'Changes in rice farmers' pest management in the Mekong Delta, Vietnam', *Crop Protection*, vol 18, pp557–563

International Rice Research Institute (IRRI) (1979) *Brown Planthopper: Threat to Rice Production in Asia*, Los Baños, the Philippines

Kenmore, P. E. (1996) 'Integrated Pest Management in rice', in Persley, G. J. (ed) *Biotechnology and Integrated Pest Management*, CAB International, Wallingford, pp76–97

Kenmore, P. E., Carino, F. O., Perez, C. A., Dyck, V. A. and Gutierrez, A. P. (1984) 'Population regulation of the brown planthopper within rice fields in the Philippines', *Journal of Plant Protection in the Tropics*, vol 1, pp19–37

Leung, L., Singleton, K.-P., Sudarmaji, G. R. (1999) 'Ecologically-based populations management of the rice-field rat in Indonesia', in Singleton, G. R., Hinds, L., Herwig, L. and Zhang, Z. (eds) *Ecologically-based Rodent Management*, ACIAR, Canberra, Australia, pp305–318

Loevinsohn, M. E. (1994) 'Rice pests and agricultural environments', in Heinrichs, E. A. (ed) *Biology and Management of Rice Insects*, Wiley Eastern Limited, pp487–515

Matteson, P. C. (2000) 'Insect pest management in tropical Asian irrigated rice', *Annual Review of Entomology*, vol 45, pp549–574

Matteson, P. C., Gallagher, K. D. and Kenmore, P. E. (1994) 'Extension of integrated pest management for planthoppers in Asian irrigated rice: Empowering the user', in Denno, R. F. and Perfect, T. J. (eds) *Ecology and Management of Planthoppers*, Chapman and Hall, London, pp656–668

Morrill, W. L. (1997) 'Feeding behavior of *Leptocorisa oratorius* (F.) in rice', *Recent Research Developments in Entomology*, vol 1, pp11–14

Morse, S. and Buhler, W. (1997) *Integrated Pest Management: Ideas and Realities in Developing Countries*, Lynee Rienner, Boulder, Colorado

Ooi, P. A. C. (1988) 'Ecology and Surveillance of *Nilaparvata lugens* (Stal) – Implications for its Management in Malaysia', PhD dissertation, University of Malaya

Ooi, P. A. C. and Shepard, B. M. (1994) 'Predators and parasitoids of rice insect pests', in Heinrichs, E. A. (ed) *Biology and Management of Rice Insects*, Wiley Eastern Limited, pp585–612

Ooi, P. A. C., Warsiyah, Nanang Budiyanto, and Nguyen, Van Son (2001) 'Farmer scientists in IPM: A case of technology diffusion', in Mew, T. W., Borromeo, E. and Hardy, B. (eds) *Exploiting Biodiversity for Sustainable Pest Management*. Proceedings of the Impact Symposium on Exploiting Biodiversity for Sustainable Pest Management, 21–23 August 2000, Kunming, China. Makati City, the Philippines, International Rice Research Institute, Los Banos, pp207–215

Ponting, C. (1991) *A Green History of the World: The Environment and the Collapse of Great Civilizations*, Penguin Books, London

Pretty, J. and Ward, H. (2001) 'Social capital and the environment', *World Development*, vol 29, pp209–227

Rombach, M. C. and Gallagher, K. D. (1994) 'The brown planthopper: Promises, problems and prospects', in Heinrichs, E. A. (ed) *Biology and Management of Rice Insects*, Wiley Eastern Limited, pp693–711

Rubia, E. G., Heong, K. L., Zalucki, M., Gonzales, B., and Norton, G. A. (1996) 'Mechanisms of compensation of rice plants to yellow stem borer *Scirpophaga incertulas* (Walker) injury', *Crop Protection*, vol 15, pp335–340

Settle, W. H., Ariawan, H., Tri Astuti, E., Cahyana, W., Hakim, A. L., Hindayana, D., Sri Lestari, A. and Pajarningsih (1996) 'Managing tropical rice pests through conservation of generalist natural enemies and alternative prey', *Ecology*, vol 77, pp1975–1988

Shepard, B. M., Justo, H. D., Rubia, E. G. and Estano, D. B. (1990) 'Response of the rice plant to damage by the rice whorl maggot, *Hydriella philippina* Ferino (Diptera: Ephydridae)', *Journal of Plant Protection in the Tropics*, vol 7, pp173–177

Shepard, B. M. and Ooi, P. A. C. (1991) 'Techniques for evaluating predators and parasitoids in rice', in Heinrichs, E. A. and Miller, T. A. (eds) *Rice Insects: Management Strategies*, Springer-Verlag, New York, pp197–214

Smith, B. D. (1995) *The Emergence of Agriculture,* Scientific American Library, New York

Sogawa, K., Kilin, D., and Bhagiawati, A. H. (1984) 'Characterization of the brown planthopper population on IR42 in North Sumatra, Indonesia', *International Rice Research Newsletter*, vol 9

Teng, P. S. (1994) 'The epidemiological basic for blast management', in Zeigler, R. S., Leong, S. A. and Teng, P. S. (eds) *Rice Blast Disease*, CAB International, Wallingford

van den Berg, H. and Soehardi (2000) 'The influence of the rice bug *Leptocorisa oratorius* on rice yield', *Journal of Applied Ecology*, vol 37, pp959–970

Way, M. J. and Heong, K. L. (1994) 'The role of biodiversity in the dynamics and management of insect pests of tropical irrigated rice – A review', *Bulletin of Entomology Research*, vol 84, pp567–587

Way, M. O., Grigarick, A. A., Litsinger, J. A., Palis, F. and Pingali, P. L. (1991) 'Economic thresholds and injury levels for insect pests of rice', in Heinrichs, E. A. and Miller, T. A. (eds) *Rice Insects: Management Strategies*, Springer-Verlag, New York, pp67–106

Zhu, Y., Chen, H., Fen, J., Wang, Y., Li, Y., Cxhen, J., Fan, J., Yang, S., Hu, L., Leaung, H., Meng, T. W., Teng, A. S., Wang, Z. and Mundt, C. C. (2000) 'Genetic diversity and disease control in rice', *Nature*, vol 406, pp718–722

Part 4

Perspectives from Industrialized Countries

Introduction to Part 4: Perspectives from Industrialized Countries

Jules Pretty

Part 4 of the *Reader in Sustainable Agriculture* contains four articles that address the particular challenge of industrialized countries. These are by Jules Pretty, Dana Jackson, Tim Lang and Michael Heasman, and Jack Kloppenburg and colleagues, and address critical issues from landscapes to diets, and from nature to foodsheds.

In the first article, which comprises an early part of the 2002 book, *Agri-Culture*, Jules Pretty explores a landscape perspective on agricultural systems, and indicates how much has been lost during the modern agricultural experiment. In recent years, connections to nature and places, both of which are regularly shaped by agricultural practices, have become neglected and eroded. Recent thinking and policy has separated food and farming from nature, and then accelerated the disconnectedness. At the same time as real commons have been appropriated, by enclosures or prairie expansion, the metaphorical food commons have also been stolen away. Food now largely comes from dysfunctional production systems that harm environments, economies and societies, and yet we seem not to know, or even to care overmuch. The environmental and health costs of losing touch are enormous. The consequences of food systems producing anonymous and homogenous food are obesity and diet-related diseases for about a tenth of the world's people, and persistent poverty and hunger for another seventh.

So does sustainability thinking and practice have anything to offer? Can it help to reverse the loss of trust so commonly felt about food systems, and prevent the disappearance of landscapes of importance and beauty? Can it help to put nature and culture back into farming? Can it help to produce safe and abundant food? Several themes are important in these modern disconnections. One is that accumulated and traditional knowledge of landscapes and nature is intimate, insightful and grounded in specific circumstances. Communities sharing such knowledge and working together are likely to engage in sustainable practices that build local renewable assets. Yet, industrialized agriculture, also often called modernist because it is single-coded, inflexible and monocultural, has destroyed much place-located knowledge. In treating food simply as a commodity, it threatens to extinguish associated communities and cultures altogether by conceiving of nature as being separate from humans. Natural landscapes and sustainable food production systems will only be recreated if we can create new knowledges and understanding, and develop better connections between people and nature.

In the second article, Dana Jackson of the Land Stewardship Project describes the way that farms can be developed as part of natural habitats. As she says, 'it's hard to imagine what it must have looked like when Europeans first settled the mid-west, when it was a wilderness with prairie, forest, clean streams and herds of buffalo. Too quickly it became dominated by agriculture.' The remaining wildlands are preserved and protected in parks and reserves, but that leaves the great majority of land directly shaped by the business of food production. Jackson introduces an alternative vision for this agriculture that is inspired by Aldo Leopold, and which indicates that farming and natural areas should be interspersed, not separated. There are many benefits of thinking differently – the benefits of biodiversity for farming itself, and the effects of more sustainable farming on biodiversity. We must teach, as Jackson says, that 'the land is one organism'.

The third article is drawn from Tim Lang and Michael Heasman's book *Food Wars*, and focuses on diet and health, and what has occurred during the recent experiment with modern agricultural systems that has focused primarily on increasing food production without concern for what has been lost. Modern agriculture has produced a commodity-based food system, and the diets of the majority of people in industrialized countries have shifted enormously. Food-related ill-health now exceeds the costs of smoking in Europe and North America, with obesity, type II diabetes, cardiovascular diseases, diet-related cancers and osteoporosis on the increase. There are now some 1.7 billion people worldwide who are overweight – in a world where 800 million people remain hungry on a daily basis. Lang and Heasman set out details of this nutrition transition, and indicate precisely how diet composition has changed. We are now in a world where there are large numbers of underfed, overfed and badly fed, and this of course raises significant questions for policy makers who are largely yet to grasp the need for fundamental change.

In the final article, Jack Kloppenburg and co-authors set out the compelling concept of the foodshed, and indicate just how connections to food and place can make a difference for both consumers and producers. As they say, 'if we are to become native to our places, the foodshed is one way of envisioning that beloved country'. They, too, though, show that fundamental changes are required if we are to evolve more ecologically and socially responsible agricultural and food systems. This raises questions about the nature of economies as a whole – are they simply geared in such a way so as to prevent sensible outcomes for people and nature? Or can they be changed too? The idea of the foodshed, a socio-geographic space, in which human activity is embedded in the natural integument of the particular place, is powerful and could help to provoke some serious rethinking about whole agricultural systems and their sustainability.

18

Landscapes Lost and Found

Jules Pretty

This common heritage

In a bend of the river stands an ancient, open meadow. This 30 hectares of The Fen, as it is known hereabouts, is a relic. For 600 years, the flint church tower has gazed through village trees upon an ever-changing agricultural landscape. This common, though, has survived intact. It is parcelled into 180 fennages, or rights to graze cattle, and so is in common ownership. When the harsh easterly winds drive down from Scandinavia, the grass crunches underfoot, and the pasture hollows are thick with ice. On a summer's day, you walk the same route past carpets of yellow buttercups, or divert past an enclosed hay meadow dotted with purple bee orchids. In autumn, after a few days of rain, the river floods and spills upon the pastures, lighting the landscape with the colour of the sky. In the long evenings, bats flit through clouds of insects, and owls hoot in search of scurrying prey. Splashes from the river remind us of the mysterious lives of otters. This Fen is different from the surrounding farmland, and it has been this way for centuries.

Other things are important about this common meadow. It links local people with nature, and as it is used and valued as a common, so it connects rights' owners and users with one another. In recent years, though, both of these types of connection have been widely neglected and consequently eroded – to our loss, and to the loss of nature at large. As food has become a commodity, most of us no longer feel a link to the place of production and its associated culture. Yet agricultural and food systems, with their associated nature and landscapes, are a common heritage and thus, also, a form of common property. They are shaped by us all, and so in some way are part of us all too. Landscapes across the world have been created through our interactions with nature. They have emerged through history, and have become deeply embedded in our cultures and consciousnesses. From the rural idylls of England to the diverse *satochi* of Japan, from the terraced rice fields and tree-vegetable gardens of Asia to the savannahs of Africa and forests of the Amazon, they have given collective meaning to whole societies, imparting a sense of permanence and stability. They are places that local people know, where they feel comfortable, where they belong.

Note: Reprinted from *Agri-culture: Reconnecting People, Land and Nature*, by Pretty, J., copyright © (2004) Earthscan, London

When we feel that we have ownership in something, even if technically and legally we do not, or that our livelihood depends upon it, then we care. If we care, we watch, we appreciate, we are vigilant against threats. But when we know less, or have forgotten, we do not care. Then it is easier for the powerful to appropriate these common goods and so destroy them in pursuit of their own economic gain. For more than 100 centuries, cultivators have tamed the wilderness – controlling and managing nature, mostly with a sensitive touch. But all has changed in the last half per cent of that time. The rapid modernization of landscapes in both developing and industrialized countries has broken many of our natural links with land and food, and so undermined a sense of ownership, an inclination to care, and a desire to take action for the collective good.

Sometimes the disconnection is intentional. The state has special terms for people who use resources without permission and for land not conforming to the dominant model. They are wild settlers, poachers or squatters, they are traditional or backward, and their lands are wastelands. Landscapes are cleaned up of their complexity, and of both their natural and social diversity. Hedgerows and ponds are removed, but so are troublesome tribes and the poorest groups. In these landscapes are real and metaphorical commons. Most of the 700,000 villages of India have, or had, commons – officially designated by name, but vital sources of food, fuel, fodder and medicines for many local people. In northern Europe, open-field or common farming sustained communities for millennia; in southern Europe, huge tracts of uplands are still commonly grazed. In England and Wales, there are still more than 8000 commons, covering 500,000 hectares, each embodying permanence in the landscape and continuity over generations. Most are archaic reminders of another age in an increasingly industrialized landscape.

Recent thinking and policy has separated food and farming from nature, and then accelerated the disconnectedness. At the same time as real commons have been appropriated, by enclosures or prairie expansion, the metaphorical food commons have also been stolen away. Food now largely comes from dysfunctional production systems that harm environments, economies and societies, and yet we seem not to know, or even to care overmuch. The environmental and health costs of losing touch are enormous. The consequences of food systems producing anonymous and homogenous food are obesity and diet-related diseases for about a tenth of the world's people, and persistent poverty and hunger for another seventh.

So does sustainability thinking and practice have anything to offer? Can it help to reverse the loss of trust so commonly felt about food systems, and prevent the disappearance of landscapes of importance and beauty? Can it help to put nature and culture back into farming? Can it help to produce safe and abundant food? These are some of the questions addressed in this chapter about what I believe to be agriculture's most significant revolution. Several themes will reoccur. One is that accumulated and traditional knowledge of landscapes and nature is intimate, insightful and grounded in specific circumstances. Communities sharing such knowledge and working together are likely to engage in sustainable practices that build local renewable assets. Yet, industrialized agriculture, also called modernist in this chapter because it is single-coded, inflexible and monocultural, has destroyed much place-located knowledge. In treating food simply as a commodity, it threatens to extinguish associated communities and cultures altogether by conceiving of nature as being separate from humans. Natural landscapes and sustainable food production systems will only be recreated if we can create new knowledges and understanding, and develop better connections between people and nature.

The world food problem

But why should this idea of putting nature and culture back into agriculture matter? Surely we already know how to increase food production? In developing countries, there have been startling increases in food production since the beginning of the 1960s, a short way into the most recent agricultural revolution in industrialized countries, and just prior to the Green Revolution in developing countries. Since then, total world food production grew by 145 per cent. In Africa, it is up by 140 per cent, in Latin America by almost 200 per cent, and in Asia by a remarkable 280 per cent. The greatest increases have been in China – an extraordinary five-fold increase, mostly occurring in the 1980s and 1990s. In the industrialized regions, production started from a higher base – yet in the US, it still doubled over 40 years, and in western Europe grew by 68 per cent.[1]

Over the same period, world population has grown from three to six billion.[2] Again, though, per capita agricultural production has outpaced population growth. For each person today, there is an extra 25 per cent of food compared with people in 1961. These aggregate figures, though, hide important differences between regions. In Asia and Latin America, per capita food production has stayed ahead, increasing by 76 and 28 per cent, respectively. Africa, though, has fared badly, with food production per person 10 per cent less today than in 1961. China, again, performs best, a trebling of food production per person over the same period. Industrialized countries as a whole show similar patterns, roughly a 40 per cent increase in food production per person.

Yet these advances in aggregate productivity have only brought limited reductions in incidence of hunger. At the turn of the 21st century, there were nearly 800 million people hungry and lacking adequate access to food, an astonishing 18 per cent of all people in developing countries. A third are in East and Southeast Asia, another third in South Asia, a quarter in sub-Saharan Africa, and a 20th each in Latin America and the Caribbean, and in North Africa and Near East. Nonetheless, there has been progress to celebrate, as incidence of undernourishment stood at 960 million in 1970, comprising a third of people in developing countries at the time. Since then, average per capita consumption of food has increased by 17 per cent to 2760 kilocalories (kcal) per day – good as an average, but still hiding a great many people surviving on less: 33 countries, mostly in sub-Saharan Africa still have per capita food consumption under 2200kcal per day. The challenge remains huge.[3]

There is also significant food poverty in industrialized countries. In the US, the largest producer and exporter of food in the world, 11 million people are food insecure and hungry, and a further 23 million are hovering close to the edge of hunger – their food supply is uncertain but they are not permanently hungry. Of these, 4 million children are hungry, and another 10 million are hungry for at least one month each year. A further sign that something is wrong is that one in seven people in industrialized countries are now clinically obese, and that five of the ten leading causes of death are diet-related – coronary heart disease, some cancers, stroke, diabetes mellitus and arteriosclerosis. Alarmingly, the obese are increasingly outnumbering the thin in some developing countries, particularly in Brazil, Chile, Colombia, Costa Rica, Cuba, Mexico, Peru and Tunisia.[4]

So, despite great progress, things will probably get worse for many people before they get better. As total population continues to increase, until at least the latter part of the 21st century, so the absolute demand for food will also increase. Increasing incomes will mean people will have more purchasing power, and this will increase demand for food. But as our diets change, so demand for the types of food will also shift radically. In particular, increasing urbanization means people are more likely to adopt new diets, particularly consuming more meat and fewer traditional cereals and other foods – what Barry Popkin calls the nutrition transition.[5]

One of the most important changes in the world food system will come from an increase in consumption of livestock products. Meat demand is expected to double by 2020, and this will change farming systems.[6] Livestock are important in mixed production systems, using foods and byproducts that would not have been consumed by humans. But increasingly farmers are finding it easier to raise animals intensively, and feed them with cheap cereals. Yet this is very inefficient: it takes 7kg of cereal to produce one kilogram of feedlot beef, 4kg to produce one of pork, and 2kg to produce one of poultry. This is clearly inefficient, particularly as alternative and effective grass-feeding rearing regimes do exist.[7]

These dietary changes will help to drive a total and per capita increase in demand for cereals. The bad news is that food consumption disparities between people in industrialized and developing countries are expected to persist. Currently, annual food demand in industrialized countries is 550kg of cereal and 78kg of meat per person. By contrast, in developing countries it is only 260kg of cereal and 30kg of meat per person. These gaps in consumption ought to be deeply worrying to us all.

Commons and connections

For most of our history, the daily lives of humans have been played out close to the land. Since our divergence from apes, humans have been hunter-gatherers for 350,000 generations, then mostly agriculturalists for 600, industrialized in some parts of the world for eight to ten, and lately dependent on industrialized agriculture for just two generations.[8] We still have close connections to nature. Yet many of us in industrialized countries do not have the time to realize it. In developing countries, many are still closely connected, yet are tragically locked into poverty and hunger. A connectedness to place is no kind of desirable life if it brings only a single meal a day, or children unable to attend school for lack of food and books, or options for wage earning that are degrading and soul-destroying.

For as long as people have managed natural resources, we have engaged in forms of collective action. Farming households have collaborated on water management, labour sharing and marketing; pastoralists have co-managed grasslands; fishing families and their communities have jointly managed aquatic resources. Such collaboration has been institutionalized in many local associations, through clan or kin groups, water users' groups, grazing management societies, women's self-help groups, youth clubs, farmer

experimentation groups, church groups, tree associations and labour-exchange societies.

Through such groups, constructive resource management rules and norms have been
embedded in many cultures – from collective water management in Egypt, Mesopota-
mia and Indonesia to herders of the Andes and dryland Africa; from water harvesting in
Roman north Africa and south-west North America to shifting agriculture systems of the
forests of Asia and Africa; and from common fields of Europe to the *iriaichi* in Japan. It
has been rare, prior to the last decade or so, for the importance of these local institutions
to be recognized in agricultural and rural development. In both developing and industrialized countries, policy and practice has tended to be preoccupied with changing the
behaviour of individuals rather than of groups or communities, or indeed with changing
property regimes because traditional commons management is seen as destructive. At
the same time, modern agriculture has had an increasingly destructive effect on both the
environment and rural communities.[9]

A search through the writings of farmers and commentators, from ancient to contemporary times, soon reveals a very strong sense of connectedness between people and
the land. The Roman writer, Marcus Cato, on the opening page of his book *Di Agri
Cultura*, written 2200 years ago, celebrated the high regard in which farmers were held:
'when our ancestors... would praise a worthy man their praise took this form: "good
husbandman", "good farmer"; one so praised was thought to have received the greatest commendation'. He also said: 'a good piece of land will please you more at each visit'.
It is revealing that Roman agricultural writers like Cato, Varro and Columella spoke
of agriculture as two things: *agri* and *cultura*. The fields and the culture. It is only very
recently that we have filleted out the culture and replaced it with commodity.[10]

It is in China, though, that there is the greatest and most continuous record of agriculture's fundamental ties to communities and culture. Li Wenhua dates the earliest records
of integrated crop, tree, livestock and fish farming to the Shang-West Zhou Dynasties
of BC 1600–800. Later Mensius said in BC 400, 'if a family owns a certain piece of land
with mulberry trees around it, a house for breeding silkworms, domesticated animals
raised in its yard for meat, and crop fields cultivated and managed properly for cereals,
it will be prosperous and will not suffer starvation'. In one of the earliest recognitions of
the need for the sustainable use of natural resources, he also said, 'if the forests are timely
felled, then an abundant supply of timber and firewood is ensured, if the fishing net
with relatively big holes is timely cast into the pond, then there will be no shortage of
fish and turtle for use'.

Still later, other treatises such as the collectively written *Li Shi Chun Qiu* (BC 239)
and the *Qi Min Yao Shu* by Jia Sixia (600 AD) celebrated the fundamental value of agriculture to communities and economies, and documented the best approaches for sustaining food production without damage to the environment. These included rotation
methods and green manures for soil fertility, the rules and norms for collective management of resources, the raising of fish in rice fields, and the use of manures. As Li Wenhua
says, 'these present a picture of a prosperous, diversified rural economy and a vivid sketch
of pastoral peace'.[11]

But it was to be Cartesian reductionism and the enlightenment that changed things
many centuries later, largely casting aside the assumed folklore and superstitions of age-
old thinking. A revolution in science occurred in the late 16th and 17th centuries,

largely due to the observations, theories and experiments of the likes of Francis Bacon, Galileo Galilei, René Descartes and Isaac Newton, which brought forth mechanistic reductionism, experimental inquiry and positivist science.[12] These methods brought great progress, and continue to be enormously important. But an unfortunate side-effect has been a sadly enduring split in at least some of our minds between humans and the rest of nature.

As I discuss later, wilderness writers, landscape painters, ecologists and farmers of the 19th and 20th centuries sought to reverse, or at least temper, the dominance of the new thinking. But it has been a Sisyphean struggle, until perhaps recently when the mountain top has at least become more visible. It is, though, in the indigenous groups of the world that we find remnants of nature–people connectivity. One of the most comprehensive collections on the diversity of human cultures and their connectedness with nature and the land is Darrell Posey's 700 page volume, *Cultural and Spiritual Values of Biodiversity*. Containing contributions from nearly three hundred authors from across the world, these highlight 'the central importance of cultural and spiritual values in an appreciation and preservation of all life'. These voices of the earth demonstrate the widespread intimate connectivity that people have to nature, and their mutual respect and understanding.[13]

In Australia, Henrietta Fourmile of the Polidingi Tribe says:

> Not only is it the land and soil that forms our connections with the earth but also our entire life-cycle touches most of our surroundings. The fact that our people hunt and gather these particular species on the land means emphasis is placed on maintaining their presence in the future... What is sometimes called 'wildlife' in Australia isn't wild; rather it's something that we have always maintained and will continue gathering.

Pera of the Bakalaharil tribe in Botswana points to their attitudes in using and sustaining wild resources:

> Some of our food is from the wild – like fruits and some of our meat... We are happy to conserve, but some conservationists come and say that preservation means that we cannot use the animals at all. To us, preservation means to use, but with love, so that you can use again tomorrow and the following year.[14]

Johan Mathis Turi of the Saami reflects on the mutual shaping in the Norwegian arctic:

> The reindeer is the centre of nature as a whole and I feel I hunt whatever nature gives. Our lives have remained around the reindeer and this is how we have managed the new times so well. It is difficult for me to pick out specific details or particular incidences as explanations for what has happened because my daily life, my nature, is so comprehensive. It includes everything. We say 'lotwantua', which means everything is included.

A similar perspective is put by Gamaillie Kilukishah, an Inuit from northern Canada who in translation by Meeka Mike says:

You must be in constant contact with the land and the animals and the plants... When Gamaillie was growing up, he was taught to respect animals in such a way as to survive from them. At the same time, he was taught to treat them as kindly as you would another fellow person.

This Inuit perspective is common across the Arctic. Fikret Berkes documents the careful management by the Cree of the Canadian eastern sub-Arctic populations of beaver, caribou and fish. None of the species used by the Cree have become locally extinct since the glaciers departed the region some 4000–5000 years ago. Berkes says 'hunters are experts on the natural history of a number of species, and on food chain and habitat relationships'. The management of beaver is particularly clever. Cree communities appoint stewards, or beaver bosses, who oversee the codes and rules for hunting, and are also a chief source of knowledge about past hunting patterns and current beaver abundance. The trick to management is balance. If there are too many beavers, and the willow and aspen decrease until a threshold is passed, beaver numbers crash, and the whole system takes many years to recover. Cree management involves hunting once every four years to prevent such an ecosystem flip. Berkes indicates the subtle way the Cree see this balance: 'these adjustments are articulated in terms of the principle that it is the animals (and not the hunter) which are in control of the hunt'. Thus there is reciprocity between animal and hunter, and these connections echo similar rules for social relations. For the Cree, there is no fundamental difference between people and animals.[15]

Some believe that the ruin of common resources is inevitable – an unavoidable tragedy, as Garrett Hardin put it more than 30 years ago.[16] Each person feels compelled to put another cow on the common, as each derives all the benefit from the additional animal, but distributes the costs amongst all the other common users. In the contemporary context, each polluter continues to add greenhouse gases to the atmosphere, as they get all the immediate benefit of not having to pay the cost of abating the pollution, or of adopting clean practices. The costs, though, are spread amongst us all – including future generations who will have to pay for climate change. Other theorists have been equally pessimistic. Mansur Olson was convinced that unless there is coercion or individual inducements, then 'rational, self-interested individuals will not act to achieve their common or group interest'.[17] This indicates a problem with free-riders – individuals who take the benefit, but do not invest anything in return. The temptation, some would say, is always to free ride. The logic has been so compelling that the state has stepped in, developing policies directly or indirectly to privatize common property systems. Although this has been going on for centuries, it has accelerated in the late 20th century's experiment with modernism. Yet productive commons persist in many parts of the world.

In some places, the loss of local institutions has led to further natural resource degradation. In India, management systems for common property resources have been undermined, a critical factor in the increased over-exploitation, poor upkeep and physical degradation observed over the past half century. As local institutions have disappeared, so the state has felt obliged to take responsibility for natural resources, largely because of a mistaken assumption that they are mismanaged by local people. This solution is rarely beneficial for environments or for poor people. A key question, therefore, centres on how can we avoid this double tragedy of the commons, in which both nature and community are damaged? It is in precisely this area that there have been so many heroic transformations, and why there is increasing hope now for a new future for agricultural and food systems.[18]

On shaping and being shaped

Some may feel there is little value in connecting us to the land and nature. Is it not just something for indigenous people or remote tribes? What possible meaning or value can come from an abstract idea like connectedness to nature? First, even in our modern times, we as predominantly urban-based societies never seem to get enough of nature. People in cities and towns are wistful about lost rural idylls. They visit, on Sunday afternoons, or for occasional weekends, but on returning home, often feel that they should have stayed. Membership of environmental organizations in industrialized countries has never been higher, and is growing. In many developing countries, city people do not just go to rural areas for the experience – they return to their home farms. If you ask urban dwellers in cities from Nairobi to Dakar, 'where do you live?' they likely as not will give the name of their rural village or settlement rather than the city. Their family still farms; they earn in the city to invest in the farm and its community. Here the connectedness is tangible.

Yet an intimate connection to nature is both a basic right and a basic need. When it is taken away, we deny it was ever important, or simply substitute occasional visits and themed experiences. But it is still there, and it is valuable. Is it any wonder to discover that the gentle opportunities afforded by urban community gardens have brought meaning and peace to many people with mental health problems? For all of our time, we have shaped nature, and it has shaped us, and we are an emergent property of this relationship. We cannot suddenly act as if we are separated. If we do so, we simply recreate the wasteland inside ourselves.

The world we see, through our window, or from space above, is shaped by us. From a distance, it is of course larger than any obvious shaping. But this is not our scale. Our scale is more local, though the effects are often greater. What we see around us has been created by us. Agricultural landscapes are obviously created, whether rice terraces upon an Asian hillside, or prairie farms in the North American plains, or rolling European patchwork fields. But even most 'natural' or 'wild' landscapes are also creations of this interaction. Few forests are truly pristine wildernesses. Most arise from some human shaping, even the Amazon rainforests and the northern tundra. Strangely, most contemporary debates on human–nature interactions focus on how nature has been shaped by us, without fully accepting the second part of the equation – that we too must be shaped by this connection, by nature itself.

We are also shaped by our systems of food production, as they in turn shape nature, and rely upon its resources for success. We are shaped by what we know about these systems – whether we approve or disapprove, whether the food system is local or distant. We are, of course, fundamentally shaped by the food itself. Without food, we are clearly nothing. It is not a lifestyle add-on or a fashion statement. The choices we make about food affect both us intrinsically and nature extrinsically. We make one set of choices, and we end up with a diet-related disease and a damaged environment. We make another set, and we eat healthily, and sustain nature through sustainable systems of food production. In truth, it is not such a simple dichotomy as this. But once we accept the idea of the fundamental nature of this connection, then we start to see options for personal, collective and global recovery.

The connection is philosophical, spiritual and physical. We are buying a system of production when we purchase its food. In effect, we eat the view, and consume the landscape. Clearly, the more we consume of one thing, the more that it is likely to be produced. But if the system of production has negative side-effects, and cares not about the resources upon which it relies, then we have taken a path leading ultimately to disaster. On the other hand, if our choices mean more food comes from systems of agricultural production that increase the stock of nature, that improve the environment whilst at the same time producing the food, then this is a different path – a path towards sustainability. We must now shape this new path. We will, by walking it, also change ourselves. We will adapt and evolve, and new connections will be established.

Nature is amended and reshaped through our connections – both for the bad and for the good. But I am worried too, as the worst kind of reshaping occurs when nature is destroyed, or ignored, and then recreated in a themed context. Do not worry about the losses, we might be saying, we can make it better than the original. When nature is themed, the outcome is grim: plants and trees are made from plastic, sand is laid down by the millions of tonnes to create new beaches, and rocks are sprayed with cement to look more 'natural'.[19] But this should not diminish the value of nature as an escape, ultimately a mystery, and an otherness from the life in the city. It is an imagined world, as well as a real world full of great meaning and significance.

This disconnected dualism

Is nature part of us, and we a part of a grander scheme? Or are we, as humans, somehow separate? These are questions that have exercised philosophers, scientists and theologists through the ages, and particularly since the enlightenment period, when Newton's mechanics and Descartes' 'nature as machine' helped to set out a new way of thinking for Europeans. The result has been the gradual erosion of connections to nature, and the emergence in many people's minds of two separate entities – people and nature.

In recent years, with growing concerns for sustainability, the environment, and biodiversity, many different typologies have been developed to categorize shades of deep to shallow green thinking. Arne Naess sees shallow ecology, for example, as an approach centred on efficiency of resource use, whereas deep ecology transcends conservation in favour of biocentric values. Other typologies include Donald Worster's imperial and arcadian ecology, and the resource and holistic schools of conservation. For some, there is an even more fundamental schism – whether nature exists independently of us, or whether it is characterized as post-modern or as part of a post-modern condition. Nature to scientific ecologists exists. To post-modernists, though, it is all a cultural construction. The truth is, surely, that nature does exist, but that we socially construct its meaning to us. Such meanings and values change over time, and between different groups of people.[20]

There are many dangers in the persistent dualism that separates humans from nature. It appears to suggest that we can be objective and independent observers – rather than part of the system and inevitably bound up in it. Everything we know about the world we know because we interact with it, or it with us. Thus if each of our views is unique, we should listen to the accounts of others and observe carefully their actions. Another problem is that nature is seen as having boundaries – the edges of parks or protected areas. At the landscape level, this creates difficulties, as the whole is always more important than each part, and diversity is an important outcome.[21]

This leads inevitably to the idea of enclaves – social enclaves such as reservations, barrios or Chinatowns, and natural enclaves like national parks, wildernesses, sites of special scientific interest, protected areas or zoos. Enclave thinking leads us away from accepting the connectivity of nature and people. It appears to suggest that biodiversity and conservation can be in one place, and productive agricultural activities in another.[22] So is it acceptable to cause damage in most social and natural landscapes, provided you leave a few tasty morsels at the edges? Surely not. These enclaves will always be under threat at the borders, or simply be too small to be ecologically or socially viable. They also act as a sop to those with a conscience – we can justify the wider destruction if we fashion a small space for natural history to persist.

By continuing to separate humans and nature, the dualism also appears to suggest that we can invent simple technologies that can intervene to reverse damage caused by this very dualism. The greater vision, and the more difficult to define, involves looking at the whole, and seeking ways to redesign it. The Cartesian either/or between humans and nature remains a strange concept to many human cultures. It is only modernist thinking that has separated humans from nature in the first place, putting us up as distant controllers. Most peoples do not externalize nature in this way. From the Ashéninha of Peru to the forest dwellers of former Zaire, people see themselves as just one part of a larger whole. Their relationships with nature are dialectical and holistic, based on 'both/with' rather than 'either/or'.[23]

For the Arakmbut of the Peruvian rainforest, Andrew Gray says: 'no species is isolated, each is part of a living collectivity binding human, animal and spirit'. Mythologies and rituals express and embed these inter-relationships, both at the practical level, such as through the number of animals a hunter may kill, and how the meat should be shared, and at the spiritual, in which 'the distinction between animal, human and spirit becomes blurred'. One of the best known of these visible and invisible connections is the Australian Aboriginal peoples' Dreamtimes. Aboriginal people have inhabited Australia for 30,000 years or more, during which time some 250 different language groups developed intimate relations with their own landscapes. David Bennett says:

> Aboriginal peoples hold that there is a direct connection between themselves and their ancestral beings, and because they hold that their country and their ancestral beings are inseparable, they hold that there is a direct connection between themselves and their country.[24]

These connections are woven into The Dreamtime, or The Dreaming, which in turn shapes and fixes the norms, values and ideals of people into the landscape.

Each Aboriginal group has its own stories about creation of their land by their ancestors and these stories connect people with today's land. Such land is non-transferable.

It is not a commodity, so cannot be traded. Events took place here, and people invested their lives and built enduring connections – so no one owns it, or rather, everyone does. As Bennett also says, 'those who use the land have a collective responsibility to protect, sustainably manage and maintain their "country"'. How sad that those who came later showed so little of this responsibility, and little collective desire to protect what was already present.

Wilderness ideas

The idea of the wilderness struck a chord in the mid-19th century, with the influential writers Henry David Thoreau and John Muir setting out a new philosophy for our relations with nature. This grew out of a recognition of the value of wildlands for people's wellbeing. Without them, we are nothing; with them, we have life. Thoreau famously said in 1851 'in wildness is the preservation of the world'. Muir in turn indicated that: 'wildness is a necessity; and mountain parks and reservations are useful not only as fountains of timber and irrigating rivers, but as fountains of life'. But as Roderick Nash, Max Oelschlaeger, Simon Schama and many other recent commentators have pointed out, these concerns for wilderness represented much more than a defence of unencroached lands.[25] It involved the construction of a deeper idea. An imagination of something that never really existed, but which proved to be hugely successful in reawakening in North American and European consciences the fundamental value of nature.

Debates have since raged over whether 'discovered' landscapes were 'virgin' lands or 'widowed' ones, left behind after the death of indigenous peoples. Did wildernesses exist, or did we create them? Donald Worster, environmental historian, points out for North America that 'neither adjective will quite do, for the continent was far too big and diverse to be so simply gendered and personalised'.[26] In other words, just because they constructed this idea does not mean to say it was an error. Nonetheless, they were wrong to imply that the wildernesses in, say, Yosemite were untouched by human hand, as these landscapes and habitats had been deliberately constructed by Ahwahneechee and other native Americans and their management practices to enhance valued fauna and flora.

Henry David Thoreau developed his idea of people and their cultures as being intricately embedded in nature as a fundamental critique of mechanical ideas that had separated nature from its observers. His was an organic view of the connections between people and nature.[27] In his *Natural History*, Thoreau celebrates learning by 'direct intercourse and sympathy' and advocates a scientific wisdom that arises from local knowledge accumulated from experience combined with the science of induction and deduction. But he still invokes the core idea of wilderness untouched by humans – even though his Massachusetts had been colonized just two centuries earlier and had a long history of 'taming' both nature and local native Americans.

Nature is something to which we can escape as individuals. He celebrates the rhythms of walking and careful observation. In other words, Thoreau 'looked with awe at the ground... Here was no man's gardens, but the unhandselled globe. It was not lawn, nor pasture, not mead, nor woodland, not lea, nor arable, not wasteland. It was the fresh and natural surface of the planet Earth, as it was made forever and ever.' The important thing to note here is that the elegiac narrative of connections and intrinsic value had a huge influence on readers, and perhaps it is a small price to pay that he focused on the 'unhandselled globe' and the 'fresh and natural' to the exclusion of other constructed natures. For these woods were of course shaped in some way by previous peoples – they are an outcome of both people and nature, not a remnant of that which had always remained untouched until he happened along.[28]

The questions, 'is a landscape wild, or is it managed?' are perhaps the wrong ones to ask, as it encourages unnecessary and lengthy argument. What is more important is the notion of human intervention in a nature of which we are part. Sometimes such intervention means doing nothing at all, so leaving a whole landscape in a 'wild' state, or perhaps it means just protecting the last remaining tree in an urban neighbourhood or hedgerow on a field boundary. Preferably, intervention should mean sensitive management, with a light touch on the landscape. Or it may mean heavy reshaping of the land, for the good or the bad.

So it does not matter whether untouched and pristine wildernesses actually exist. Nature exists without us; and with us is shaped and reshaped. Most of what exists today does so because it has been influenced explicitly or implicitly by the hands of humans, mainly because our reach has spread as our numbers have grown, and as the effects of our consumption patterns have compounded the effect. But there are still places that seem truly wild, and these exist at very different scales and touch us in different ways. Some are on a continental scale, such as the Antarctic. Others are entirely local, a woodland amidst farmed fields, a saltmarsh along an estuary, a mysterious urban garden, all touched with private and special meanings.

This suggests that wild nature and wilderness can be discovered and touched on a personal scale. If we find a moment's peace on an hour's walk across a meadow by the river, does it matter that this is a shaped nature, and not a wild one? Wilderness is an idea, and it is a deep and appealing one. Some shaping of landscape can be so subtle that we would hardly notice. Nigel Cooper asks how natural is a nature reserve, and identifies a range of places where conceptions of nature are located in the British landscape, including biodiversity reserves, wilderness areas, historic countryside parks and what he calls 'companion places'. In our almost entirely farmed landscape, where nature is as much a product of agriculture as it is an input, the efforts to recognize and conserve biodiversity and wildness are varied. All of these are as much treasured by the people who make or experience them as those who gaze upon the wildest forests, savannahs or mountains.[29]

In all of these situations, we are a part, connected, and so affecting nature and land, and being affected by it. This is a fundamentally different position to one which suggests that wilderness is untouched, pristine, and so somehow better because it is separated from humans – who, irony of ironies, promptly want to go there in large numbers precisely because it appears separate. But an historical understanding of what has happened to produce the landscape or nature we see before us matters enormously when we use an idea to form a vision that clashes with the truth. An idea that this place is wild, and so these local people should be removed. Another idea that this place is ripe for development, and so a group of people should be dispossessed. The term wilderness has come to mean many things, usually implying an absence of people and presence of wild animals, but also containing something to do with the feelings and emotions provoked in people. Roderick Nash takes a particularly Eurocentric perspective in saying, 'any place in which a person feels stripped of guidance, lost and perplexed may be called a wilderness', though this definition may also be true of some harsh urban landscapes.[30] The important thing is not defining what it really is, but what we think it is, and then telling stories about it.

Language and memory on the frontier

Many stories about nature and our earth are embedded in local languages. Language and land are part of people's identities, and both are under threat. There are 5000–7000 oral languages spoken today, only about a half of which have more than 10,000 speakers each.[31] The rest, about 3400 languages, are spoken by only 8 million people, about a tenth of 1 per cent of the world's population. The top ten spoken languages now comprise about half of the world's population. A great deal of linguistic diversity is thus maintained by a large number of small and dwindling communities. They, like their local ecologies and cultural traditions, are under threat. Here there is a vicious circle. As languages come under threat, so do the stories people tell about their environments. Local knowledge does not easily translate into majority languages, and moreover, as Luisa Maffi states, 'along with the dominant language usually comes a dominant cultural framework which begins to take over'.[32]

Thus we increasingly lack the capacity to describe changes to the environment and nature, even if we were able to observe them. Slowly, it all slips away. Gary Nabhan and colleagues describe how the children of the Tohono O'odham (formerly known as the Papago) of the Sonoran Desert in the south-west US are losing both a connection with the desert and with their language and culture. Even though they hear the language spoken at home, they are not exposed to traditional story-telling, and are no longer able to name common plants and animals in O'odham – though they could easily name large animals of the African savannah seen on television. Nabhan called this process of erosion the 'extinction of experience'.[33]

These losses, too, are hastened by land degradation or removal for other purposes. Also in the Sonoran Desert, F. S. Molina found that his own people, the Yaqui or Yoeme, were unable to perform traditional rituals because of the disappearance of many local plants. Land is being settled by non-Yoeme, and converted to other uses.[34] Biodiversity slips away, and only the local indigenous people notice. But they are powerless in the global scheme of things. Their intimate spiritual and physical connection with nature is under threat, yet we on the outside may never notice it go. 'Yaquis have always believed that a close communication exists among all the inhabitants of the Sonoran Desert world in which they live: plants, animals, birds, fishes, even rocks and springs. All of these come together as part of one living community which Yaquis call the huya ania, the wilderness world.'[35] These problems are all connected. Luisa Maffi adds, 'The Yoeme elders' inability to correctly perform rituals due to environmental degradation thus contributes to precipitating language and knowledge loss and creates a vicious circle that in turn affects the local ecosystem'.[36]

The concept of the frontier suggests to me a place where people test out existing ideas on a new environment. As a result, both change. William Cronon and fellow historians indicate that self-shaping occurs rapidly on the frontier. The different identities of groups arriving from distant places, and those of people already present, clash and blend, merge and stand apart. Of the American frontier, they say: 'Self-shaping was a part of the very earliest frontier encounters and continues as a central challenge of regional life right down to the present'. People on the western frontier, as they pushed into what they saw as a 'wilderness' and 'free-land', had 'borrowed most of their cultural values... from Europe and older settlements back east'. They reshaped nature and themselves. They also, of course, imposed a new landscape on the old. Through conquest, the original owners were removed and corralled. New stories and mythologies emerged to give greater justification to these acts. One set of ideas about a landscape was replaced by another.[37]

The pioneering frontier historian, Frederick Jackson Turner, though promoting many ideas and views long since shown to be wrong and even downright racist, rightly indicated that the frontier repeated itself.[38] The frontier, where shaping of nature and self-shaping of societies, combined with a destruction of existing relationships and cultures, expands today at a pace beyond the appreciation of the majority. Most shaping does not bring benefits to us all, as the interwoven rug of nature and people is steadily pulled from beneath our feet. I am not concerned here with defining exactly what is a frontier, or indeed where they exist. Its use as an idea lies in the notion that one set of values about a land comes rapidly to be imposed on another. In modern times, the frontier is characterized both by the expansion of modern, industrialized agriculture, or by the loss of local associations and connectivity to the land. The problem is that those pushing out the frontier see it as progress; those exposed to the invasion see mainly destruction and loss. Of course, this applies too to the contemporary expansion of sustainable agriculture. When William Bradford stepped off the Mayflower, he saw a 'hideous and desolate wilderness'.[39] The pioneers at the frontier were not only carving out new lives, but battling it out with the wild country for survival. As Nash put it 'countless diaries, addresses and memorials of the frontier period represented wilderness as an "enemy" which had to be "conquered", "subdued" and "vanquished" by a "pioneer army". The same phraseology persisted into the present century.'[40]

In practice, of course, there is always mingling at the frontier, and what we see is a function of both sides' capacity to shape and reshape. Those coming along at the frontier bring connections to old cultures, but also new ideas of how to make improvements. Recipients at the frontier find new opportunities to trade, interact and learn. Out of these new connections can come new forms of cross-cultural dialogue. In the early north-east US, for example, where the received story is one of misunderstanding and conquest, the British and French learnt Iroquois languages, protocols and metaphors so as to aid trust and trading.[41] But it is also true that in the end there are clear winners and losers. As land beyond the frontier is seen as 'free', so it is taken, and this inevitably means conflict and violence. Cronon and colleagues say: 'Sometimes, it was perpetrated by individuals, and sometimes by the military power of the state. Always, it drew dark lines on a landscape whose newly created borders were defeated with bullets, blades and blood.'[42] Today, such frontier experiences are played out in the rainforests, swamps, hills and mountains of Latin America, Africa and Asia, and in the landscapes overwhelmed by modern agricultural technologies and narratives. What is gained is one thing – more food. What is lost has been too often invisible. Yet what are equally important are the cognitive frontiers inside us. We each have a journey to travel if we are to find new ways of protecting our world, and at the same time produce the food we need.

It does matter who tells the story

Who gets to tell the stories matters greatly. Every piece of land or landscape contains as many meanings and constructions as the people who have interacted with it. A modern, industrialized landscape, let's call it a monoscape, has few meanings. By contrast, a diverscape has many. Thus a single story of the land is not the only story – though many would have us believe it to be true. When the Europeans first brought their visions to the Pacific and Australasia, they saw the landscape, and met the people. But they did not give them great value – beyond, that is, curiosity and museum value. They sought to save them, convert them, enslave them. They imposed their stories on the landscape – even though Aboriginal peoples in Australia had walked the land for at least 1500 generations, and had accumulated extraordinary knowledge, understanding and compelling stories over timescales beyond any persisting European culture.[43] As Paul Carter describes, Captain Cook and the 'first arrivers' and narrators saw an empty space which could be settled and civilized.[44] The Australian landscape was awaiting history, and new stories could be created and imposed. They named all that they saw – in four months over 100 bays, capes and isles. Carter says 'for Cook, knowing and naming were identical'. Once these discovered places had been named for the first time, so they were known. The landscape begins its process of being reshaped. Cook sees, on deep black soils, 'as fine a meadow as ever was seen'. A meadow rather like those of home, and echoing John Muir's observation of 'wild' meadows in Yosemite that were actually created by controlled fires set by native Americans.

The naming of the new, which was actually old, with the old from elsewhere continued apace for decades, as explorers forced their way into the interior, aiming, as Carter put it, to 'dignify even hints of the habitable with significant names... Possession of the country depended ... to some extent, on civilising the landscape, bringing it into orderly being.' The new story is told and written, and the old slips away without notice. At the time, few bothered to find out something of the local stories of landscape, of the songlines stretching across both thousands of years and thousands of kilometres. Songlines wrap nature and the landscape inextricably into culture, identity and community. Take one away, and the whole falls apart.

Two hundred and twenty-nine years after Cook's landfall, I am standing with Phil and Suzie Grice on their Western Australian wool and cereal farm. They have an ecologically literate view of the landscape. They had seen what has happened through modern farming, and where it had led their family and neighbours. In a brief two centuries, modern farming and land management methods brought substantial economic benefit, but great harm, too, to the environment and land. Phil says: 'For two generations, the previous owner and his father pushed back the frontier, removing nature and replacing it with fields. Now, I'm replanting native vegetation as fast as I can and afford.' The farm is in Lower Balgarup catchment, 260km south-west of Perth, set in a landscape of ancient and deeply-weathered soils. But in the blink of an eye, it has changed. In the 40 years to 1990, 85 per cent of all the natural vegetation in the catchment was removed, with a profound impact on both hydrology and local biodiversity. Soils and water have become salinized, and farming itself threatened. The cost of expansion of the farming frontier has been destruction of the very resource on which farmers relied.[45]

Eighteen farmers set up the Lower Balgarup catchment group in 1990, covering an area of some 14,000 hectares. It is one of 400 Landcare groups in Western Australia. One of the first actions of the group was to survey the area of land degradation, as no one quite knew the extent of the problem. They were shocked to find more than 600 hectares of land affected by dryland salinity and waterlogging. Since then, Phil and his neighbours have planted 200,000 trees, constructed 100km of new fencing to protect creeks, and another 70km of drains and banks, and put down land to perennial grasses. The trees and grasses help to pump groundwater by evapotranspiration, so reducing salinity. But the task for the whole landscape is still massive. There are 19 million hectares of wheat and wool country in Western Australia, and already nearly 2 million hectares have been lost to dryland salinity. By 2010, another 3 million hectares are expected to have been lost and 40 rural towns in the wheat belt to have become vulnerable. This ancient landscape, where the rocks of the Yildirim Block underlying the catchment are 2.5 billion years old, needs thorough redesigning. Can these farmers, with their changed ways of thinking, now construct a new story?

Of course, what Cook saw, and later Muir in the Sierra Nevada, was conditioned by what they knew. If you believe in wildernesses, then you will see one and name it so. If you know a meadow as part of a pastoral scene, so you will see one more readily. If you see native vegetation simply taking up space where fields could be, then you remove it. However, it would be a mistake to believe that the effect of rewriting the landscape is only a one-way process. As Bernard Smith indicates with regard to the 'discovery' of Pacific peoples in the 18th century, their impact on Europe was perhaps as great as the impact of European culture and diseases on the Pacific.[46] When the Tahitian, Omani, arrived in England in 1774 with Captain Furneaux, according to Smith he 'created a sensation... He mingled in fashionable circles with a natural grace and became a lion of London society.' More importantly, his presence provoked new domestic criticism of Empire and its 'pilfered wealth', and even of the shortcomings of English society. A decade later, the son of the chief of one the Palau Islands accompanied Henry Wilson back to England, again to much public acclaim and self-criticism.

Nonetheless, there persisted a subtle misrepresentation of the story through landscape painting that, according to Smith, sought to 'evoke in new settlers an emotional engagement with the land that they had alienated from its aboriginal occupants'. The noble Pacific Islander, in traditional dress, or engaged in traditional ceremony or dance, or the boat full of arriving heroes sensitively stepping onto the beach hides the real story. Landings were more often accompanied by guns and violence, and long-term damage to societies and nature. Such systematic disenfranchisement has clearly been more common than sensitive interaction. George Miles similarly draws attention to the lack of voice given to Native Americans by incomers. Even though they had told their stories for centuries, suddenly they were silent, nobly silent to some, but more often, sadly even to the likes of Mark Twain, they were 'silent, sneaking, treacherous looking'.[47]

Part of the problem was that most Native Americans had a predominantly verbal culture, without alphabets. The Cherokee alphabet, for example, was only constructed by a young Cherokee, Sequoyah, in the early 19th century. It led to the printing of the first Native American newspaper in 1828, which was so successful in telling its story that the authorities of Georgia arrested its editors and confiscated the press six years later. It then reappeared as the Cherokee Advocate in 1843, from the Cherokee national capital of Talhequah, lasted until 1854, was closed down again, reappeared again in the mid 1870s, and then endured to 1906, when its 800–1000 Cherokee-only readers finally lost their only national language paper. During the 18th and 19th centuries, according to Miles, nearly every native American community embraced opportunities to write and read their own languages: 'from the Micmacs of Newfoundland to the Sioux of the plains, from the Apaches, Navajos and Yaquis of southwest and the Luiseños of California to the Aleuts and Eskimos of the Atlantic'.

It is, of course, easier to lose, intentionally or by accident, stories handed down by word of mouth. Once they have gone, there is no one to oppose those who would dominate with their own narrative. Then we forget why one thing is present in a landscape, why it used to be valuable, and what reasons we may have for looking after it.

Concluding comments

In this chapter, my aim has been to set the scene for a sustainable agricultural revolution by indicating that agricultural and food systems and the landscapes they shape are a common heritage to us all. For all our human history, we have been shaped by nature whilst shaping it in return. In recent times, that shaping has been destructive, with food seen as a commodity and no longer part of culture. In our modern and industrial age, we are losing our languages, memories and stories about land and nature. These disconnections matter, as they serve to promote a persistent dualism – that nature is separate from people, that nature can be conserved in wildernesses, and that economies can succeed without regard to the fundamental significance of agricultural and food systems.

Notes

1   Data on food production analysed from FAO's *FAOSTAT* database (www.fao.org). In the past 15 years, aggregate production in Europe has been largely stable due to supply management policies, whereas in the US it has grown by 35 per cent.

2   World population was 3.08 billion in 1960; 3.69 billion in 1970; 4.44 billion in 1980; and 5.27 billion in 1990. The annual growth rate of world population was 2.1 per cent in the late 1960s; had fallen to 1.3 per cent during the late 1990s; and is projected by the United Nations Population Fund, 1999, to fall further to 1 per cent by 2015, to 0.7 per cent by 2030, and to just 0.3 per cent by 2050.

3   For details of food policy analyses and challenges, see especially materials from the International Food Policy research Institute (www.ifpri.org) and FAO (www.fao.org). For specific papers, see Pinstrup-Andersen and Cohen, 1999; Pinstrup-Andersen et al, 1999; Delgado et al, 1999; FAO, 2000a.

4   In the US, $25 billion is spent each year by federal and state organizations to pro-
     vide extra food for this 12 per cent of the national population who are food inse-
     cure. On diets and obesity, see Eisinger, 1998; WHO, 1998, 2001; Lang et al, 1999;
     FAO, 2000a, 2000b; Lang et al, 2001

5   For more on the nutrition transition, see Popkin, 1998. During the period to 2020,
     the urban population in developing countries is expected to double to nearly 3.5
     billion, whilst rural numbers will only grow by 300 million to 3 billion. The num-
     bers of urban people will, for the first time in human history, have exceeded those
     in rural areas. Such a change will also affect food consumption. As rural people move
     to urban areas, and as urban people's disposable incomes increase, so they tend to go
     through the nutrition transition – particularly from rice to wheat, and from coarse
     grains to wheat and rice. They also tend to eat more livestock products, processed
     foods, and fruit and vegetables.

6   The annual demand for cereals is predicted by IFPRI to grow from 1400 million
     tons in 1995 to 2120 million tonnes by 2020. Of this 2.12 billion tonnes, 48 per
     cent will be for food and 21 per cent for animal feed in developing countries, and 8
     per cent for food and 23 per cent for animal feed in industrialized countries. Meat demand
     is expected to double by 2020 in developing countries to 190 million tonnes per
     year, and increase by a quarter in industrialized countries to 120 million tonnes.

7   On average, intensive livestock fed a diet of grain and silage only produce one mega-
     joule of meat for every three megajoules of grain eaten. There is another problem.
     As we eat more meat, so cereals are increasingly diverted for livestock feed, and those
     in food poverty stay in poverty. Today, 72 per cent of all cereals consumed in indus-
     trialized countries is for livestock feed. In developing countries, the pattern is
     inverted, with 74 per cent of all cereal still being directly consumed by humans. On
     the livestock revolution, see Rosegrant et al, 1997; Delgado et al, 1999. Also see
     White, 2000; Seidl, 2000.

8   I would like to acknowledge Linda Hasselstrom for her fine essay 'Addicted to Work'
     (1997), in which idea of converting human history to generations is developed. My
     estimates are slightly different from hers, as I use the dates of 7 million years before
     present (BP) for human divergence from apes, 12,000 BP for the start of agricul-
     ture, and a figure of 20 years for the average generation length. For more, see Dia-
     mond, 1997.

9   For more on collective action in agriculture, and the effects of modern agriculture,
     see Balfour, 1943; Huxley, 1960; Palmer, 1976; Picardi and Siefert, 1976; Jodha,
     1990; Ostrom, 1990; Bromley, 1992; Pretty, 1995a, b, 1998; Berkes and Folke,
     1998; Kothari et al, 1998. For a comprehensive review of how social systems have
     developed management practices based on ecological knowledge for dealing with the
     dynamics of local ecosystems, see Berkes and Folke, 1998. Petr Kropotkin was one
     of the first writers to give collective action prominence in his 1898 book, *Mutual
     Aid*. He drew attention to the 'immense importance which the mutual-support
     instincts, inherited by mankind for its extremely long evolution, play even now in
     our modern society, which is supposed to rest on the principle "every one for him-
     self, and the State for all"' (pxv). Kropotkin drew upon the history of guilds and
     unions in many countries, including craft guilds of mediaeval cities, brotherhood
     groups of Scandinavia, *artéls* and *druzhestva* of Russia, *amhari* of Georgia, communes

of France and the *Geburschaflen* of Germany. 'Organisations came into existence wherever a group of people – fishermen, hunters, travelling merchants, builders or settled craftsmen – came together for a common pursuit' (p171).

10 Cato, M. P. (1979) 'Di Agri Cultura', in Hooper W. D. (revised Ash, H. B.). Marcus Porcius Cato *On Agriculture*, and Marcus Terentius Varro *On Agriculture*. Harvard University Press, Cambridge, Massachusetts.

11 See Li Wenhua (2000) *Agro-Ecological Farming Systems in China*. For a classic text on Chinese agriculture, see F. H. King (1991) *Farmers of Forty Centuries*. Here he introduces the idea of permanent agriculture. For a review of history of innovation in China, see Temple (1986) *China: Land of Discovery and Invention*.

12 The dates of birth and death for these four are: Francis Bacon (1561–1626), Galileo Galilei (1524–1642), René Descartes (1596–1650) and Isaac Newton (1642–1727). Though the enlightenment provided the boost for modern science's disconnection, it is important to note it has not affected all sciences in the same way. There is great diversity within scientific disciplines. There is also great significance in the institutional location of and pressures on scientific research.

13 The voices of the earth quotes are all from Senanayake, in Posey (1999), pp125–152.

14 These principles are common in Brazil and Ecuador, where Daniel Mataho Cabixi of the Paraci people and Cristina Gualinga of the Quicha talk both about the fundamental connections and difficulties in modern times. First Cabixi: 'We have our mythological hero who is called Wasari... Wasari allocated territory to the different Paraci groups ... and taught them the technologies of hunting and preparing and consuming natural resources. Wasari further established political and economic principles revealing how to deal with other human beings and nature... Our traditional territory of twelve million hectares ... has been reduced to twelve hundred hectares today. Now the Paraci have to face a number of serious limitations in order to survive.' Says Gualinga: 'Nature, what you call biodiversity, is the primary thing that is in the jungle, in the river, everywhere. It is part of human life. Nature helps us to be free, but if we trouble it, nature becomes angry. All living things are equal parts of nature and we have to care for each other' (in Posey, 1999).

15 For the story of the Cree, see Berkes (1998).

16 See Hardin (1968) 'The Tragedy of the Commons'.

17 See Olson (1965).

18 For the study of 82 villages in India, see Jodha (1990, 1991). For more on the effects of local institutions on natural resources, see Scoones (1994); Pretty and Pimbert (1995); Leach and Mearns (1996); Pretty and Shah (1997); Ghimire and Pimbert (1997); Singh and Ballabh (1997).

19 For the dangers of theming our urban and rural spaces and the attempted manufacture of community, see Goin (1992); Garreau (1992); Barker (1998).

20 For more on the different types of thinking about the environment and types of sustainability, including postmodern views, see Hutcheon (1989); Naess (1992); Worster (1993); Benton (1994); Soulé and Lease (1995); Rolston (1997); Barrett and Grizzle (1999); Dobson (1999); Cooper (2000a). It is interesting to note that the term 'ecology' has now come to mean much more that a scientific discipline describing natural processes, but often now a noun mean environment as a whole. For a good

collection of the writings on the philosophies of environmentalism, see Sessions, 1995, *Deep Ecology for the 21st Century.*

21 For more on the landscape scale, see Foreman (1997); Klijn and Vos (2000); Cooper (2000a).

22 See Cronon et al (1992); Deutsch (1992); Brunkhorst et al (1997).

23 See Benton (1994); Gray (1999), p63.

24 See Bennett (1999), p104.

25 For Thoreau quote, see Nash (1973), p84 – quoted in turn from a speech by Thoreau on 23 April, 1851 to the Concord Lyceum. For Muir quote, see Oelschlaeger (1991). See also Nash (1973); Oelschlaeger (1991); Schama (1996). See also Vandergeest and DuPuis (1996).

26 Worster (1993), p5.

27 For a good review of Thoreau, see Oelschlaeger (1991), pp133–171.

28 Quotes are from Thoreau's *A Winter Walk*, p167; and Thoreau's *Maine Woods*, pp93–95. Also, see *The Writings of H. D. Thoreau Volumes 1–6* (Princeton University Press, 1981–2000).

29 See Cooper (2000b).

30 See Nash (1973), p3. For a discussion of the static and dynamic nature of locality and our desire, or otherwise, to conserve it, see Scruton, 1998, in *Town and Country*. Common Ground, a UK charity, make this point in the recent book on Community Orchards: 'defining beauty as mountains, and richness as rarity, has not only devalued the remainder, but it has diminished people's confidence to speak out for ordinary things... everyday places are as vulnerable as the special'. In the commonplace and the everyday, we form deeper cultural relations with nature and the land. See Common Ground's *The Common Ground Book of Orchards*.

31 Of the 5000–7000 oral languages persisting worldwide, 32 per cent are in Asia, 30 per cent in Africa, 19 per cent in the Pacific, 15 per cent in the Americas and 3 per cent in Europe. See Grimes (1996).

32 Maffi (1999), p31.

33 For more on the Tohono O'odham, see the work of Ofelia Zepeda at the University of Arizona on reinvigorating the language and its links to the land (www.u.arizona. edu/~mizuki/wain/wain0.html). The Tohono O'Odham abandoned the more common term for their culture, Papago, in the 1980s, as it means 'bean eaters'.

34 Molina (1998), p31.

35 Evers and Molina (1987).

36 For a discussion of the use of language and rhetoric to describe the transformation of the western interior of the US, see Lewis (1988), in Cosgrove and Daniels (eds). For an analysis of the linkage between language and understanding of the land amongst the Innu of Canada, see Samson (2002). Even when taught in school, young Innu are stripped of the experience of being on the land – they are taught a hunting language in a setting that presumes agriculture as the dominant activity rather than hunting.

37 See Cronon et al (1992), p4, p18.

38 On the work of Frederick Jackson Turner (1920), William Cronon and colleagues suggest that 'it would be a shame to lose the power of this insight just because Turner surrounded it with a lot of erroneous, misleading and wrong-headed baggage' (p6).

39 In Nash (1973), p23.
40 In Nash (1973), p27.
41 See Miles (1992); Jennings (1984).
42 See Cronon et al (1992), p15.
43 Assuming a date of 30,000–35,000 years before present for human arrival on the Australian continent – see Diamond (1997). See also Smith (1985); Carter (1987).
44 See Carter (1987), p9, p41, p54.
45 For more details on the problems of dryland salinity, see Australian National Drylands Salinity Programme (www.lwrrdc.gov.au/ndsp/index.htm). See also Pannell (2001) 'Salinity policy'.
46 See Smith (1985), pix; Smith (1987), p80.
47 Mark Twain is quoted in Miles (1992), p52. For later quote, see pp65–66.

References

Balfour, E. B. (1943) *The Living Soil*, Faber and Faber, London

Barker, P. (1998) 'Edge city' in Barnett, A. and Scruton, R. (eds) *Town and Country*, Jonathan Cape, London

Barrett, C. and Grizzle, R. (1999) 'A holistic approach to sustainability based on pluralism stewardship', *Environmental Ethics*, Spring, pp23–42

Bennett, D. (1999) 'Stepping from the diagram: Australian Aboriginal cultural and spiritual values relating to biodiversity' in Posey, D. (ed) *Cultural and Spiritual Values of Biodiversity*, IT Publications and UNEP, London

Benton, T. (1994) 'Biology and social theory in the environmental debate' in Redclift, M. and Benton, T. (eds) *Social Theory and the Global Environment*, Routledge, London

Berkes, F. (1998) 'Indigenous knowledge and resource management systems in the Canadian subarctic' in Berkes, F. and Folke, C. (eds) *Linking Social and Ecological Systems*, Cambridge University Press, Cambridge

Berkes, F. and Folke, C. (eds) (1998) *Linking Social and Ecological Systems*, Cambridge University Press, Cambridge

Bromley, D. W. (ed) (1992) *Making the Commons Work*, Institute for Contemporary Studies Press, San Francisco, California

Brunkhorst, D., Bridgewater, P. and Parker, P. (1997) 'The UNESCO biosphere reserve program comes of age: Learning by doing; landscape models for sustainable conservation and resource use' in Hale, P. and Lamb, D. (eds) *Conservation Outside Nature Areas*, University of Queensland, Queensland, pp176–182

Carter, P. (1987) *The Road to Botany Bay*, Faber and Faber, London

Cato, M. P. (1979) 'Di agri cultura' in Hooper, W. D. (revised Ash, H. B.) Marcus Porcius Cato *On Agriculture*, and Marcus Terentius Varro *On Agriculture*, Harvard University Press, Cambridge, Massachusetts

Common Ground (2000) *The Common Ground Book of Orchards*, Common Ground, London

Cooper, N. S. (2000a) 'Speaking and listening to nature: Ethics within ecology', *Biodiversity and Conservation*, vol 9, pp1009–1027

Cooper, N. S. (2000b) 'How natural is a nature reserve? An ideological study of British nature conservation landscapes', *Biodiversity and Conservation*, vol 9, pp1131–1152

Cronon, W., Miles, G. and Gitlin, J. (eds) (1992) *Under an Open Sky. Rethinking America's Western Past*, W W Norton and Co, New York

Delgado, C., Rosegrant, M., Steinfield, H., Ehui, S. and Courbois, C. (1999) 'Livestock to 2020: The next food revolution', IFPRI Brief 61. International Food Policy Research Institute, Washington, DC

Deutsch, S. (1992) 'Landscape of enclaves' in Cronon, W., Miles, G. and Gitlin, J. (eds) (1992) *Under an Open Sky. Rethinking America's Western Past*, W W Norton and Co, New York

Diamond, J. (1997) *Guns, Germs and Steel*, Vintage, London

Dobson, A. (ed) (1999) *Fairness and Futurity*, Oxford University Press, Oxford

Eisinger, P. K. (1998) *Towards an End of Hunger in America*, Brookings Institution Press, Washington, DC

Evers, L. and Molina, F. S. (1987) *Maso/Bwikam/Yaqui Deer Songs: A Native American Poetry*, Sun Track and University of Arizona Press, Tucson, Arizona

FAO (2000a) *Agriculture: Towards 2015/30*. Global Perspective Studies Unit, FAO, Rome

FAO (2000b) 'Food Security Programme', www.fao.org/sd/FSdirect/FSPintro.htm

Foreman, R. T. (1997) *Land Mosaics: The Ecology of Landscapes and Regions*, Cambridge University Press, Cambridge

Garreau, J. (1992) *Edge City – Life on the New Frontier*, Anchor Books, New York

Ghimire, K. and Pimbert, M. (1997) *Social Change and Conservation*, Earthscan, London

Goin, P. (1992) *Humanature*, University of Texas Press, Harrisonberg, Virginia

Gray, A. (1999) 'Indigenous peoples, their environments and territories' in Posey, D. (ed) (1999) *Cultural and Spiritual Values of Biodiversity*, IT Publications and UNEP, London

Grimes, B. (1996) *Ethnologue: Languages of the World*, 12th edition. Summer Institute of Linguistics, Dallas, Texas

Hardin, G. (1968) 'The tragedy of the commons', *Science*, vol 162, pp1243–1248

Hasselstrom, L. (1997) 'Addicted to work' in Vitek, W. and Jackson, W. (eds) *Rooted in the Land: Essays on Community and Place*, Yale University Press, New Haven and London

Hutcheon, L. (1989) *The Politics of Postmodernism*, Routledge, London

Huxley, E. (1960) *A New Earth: An Experiment in Colonialism*, Chatto and Windus, London

Jennings, F. (1984) *The Ambiguous Iroquois Empire*, W W Norton and Co, New York

Jodha, N. S. (1990) 'Common property resources and rural poor in dry regions of India', *Economic and Political Weekly*, vol 21, pp1169–1181

Jodha, N. S. (1991) 'Rural common property resources: A growing problem', *Sustainable Agriculture Programme Gatekeeper Series No 24*, IIED, London

King, F. H. (1911) *Farmers of Forty Centuries: Permanent Agriculture in China, Korea and Japan*, Rodale Press, Pennsylvania

Klijn, J. and Vos, W. (eds) (2000) *From Landscape Ecology to Landscape Science*, Kluwer Academic Publishers, Dordrecht

Kothari, A., Pathak, N., Anuradha, R. V. and Taneja, B. (1998) *Communities and Conservation: Natural Resource Management in South and Central Asia*, Sage, New Delhi

Kropotkin, P. (1898) *Mutual Aid*, Extending Horizon Books, Boston (1955 edition)

Lang, T., Barling, D. and Caraher, M. (2001) 'Food, social policy and the environment: Towards a new model', *Social Policy and Administration*, vol 35, no 5, pp538–558

Lang, T., Heasman, M. and Pitt, J. (1999) *Food, Globalisation and a New Public Health Agenda*, International Forum on Globalization, San Francisco

Leach, M. and Mearns, R. (1996) *The Lie of the Land*, Routledge, London

Lewis, G. M. (1988) 'Rhetoric of the western interior: Modes of environmental description in American promotional literature of the 19th century' in Cosgrove, D. and Daniels, S. (1988) *The Iconography of Landscape*, Cambridge University Press, London

Li Wenhua (2000) *Agro-Ecological Farming Systems in China*, Man and the Biosphere Series Volume 26. UNESCO, Paris

Maffi, L. (1999) 'Linguistic diversity' in Posey, D. (ed) (1999) *Cultural and Spiritual Values of Biodiversity*, IT Publications and UNEP, London

Miles, G. (1992) 'To hear an old voice' in Cronon, W., Miles, G. and Gitlin, J. (eds) (1992) *Under an Open Sky. Rethinking America's Western Past*, W W Norton and Co, New York, pp52–70

Molina, F. S. (1998) 'The wilderness world is respected greatly: The Yoeme (Yaqui) truth from the Yoeme communities of Arizona and Sonora, Mexico' in Maffi, L. (ed) *Language, Knowledge and the Environment*, Smithsonian Institution Press, Washington, DC

Naess, A. (1992) 'Deep ecology and ultimate premises', *Society and Nature*, vol 1, no 2, pp108–119

Nash, R. (1973) *Wilderness and the American Mind*, Yale University Press, New Haven

Oelschlaeger, M. (1991) *The Idea of Wilderness*, Yale University Press, New Haven

Olson, M. (1965) *The Logic of Collective Action: Public Goods and the Theory of Groups*, Harvard Press, London

Ostrom, E. (1990) *Governing the Commons: The Evolution of Institutions for Collective Action*, Cambridge University Press, New York

Palmer, I. (1976) *The New Rice in Asia: Conclusions from Four Country Studies*, UNRISD, Geneva

Pannell, D. (2001) 'Salinity policy: A tale of fallacies, misconceptions and hidden assumptions', *Agricultural Science*, vol 14, no 1, pp35–37

Picardi, A. C. and Siefert, W. W. (1976) 'A tragedy of the commons in the Sahel', *Techology Review*, vol 14, pp35–54

Pinstrup-Andersen, P. and Cohen, M. (1999) 'World food needs and the challenge to sustainable agriculture', Paper for Conference on Sustainable Agriculture: New Paradigms and Old Practices? Bellagio Conference Centre, Italy, 26–30 April 1999

Pinstrup-Andersen, P., Pandya-Lorch, R. and Rosegrant, M. (1999) *World Food Prospects: Critical Issues for the Early 21st Century*, IFPRI, Washington, DC

Popkin, B. (1998) 'The nutrition transition and its health implications in lower income countries', *Public Health Nutrition*, vol 1, no 1, pp5–21

Posey, D. (ed) (1999) *Cultural and Spiritual Values of Biodiversity*, IT Publications and UNEP, London

Pretty, J. N. (1995a) *Regenerating Agriculture: Policies and Practice for Sustainability and Self-Reliance*, Earthscan, London; National Academy Press, Washington, DC; ActionAid, Bangalore

Pretty, J. N. (1995b) 'Participatory learning for sustainable agriculture', *World Development*, vol 23, no 8, pp1247–1263

Pretty, J. N. (1998) *The Living Land: Agriculture, Food and Community Regeneration in Rural Europe*, Earthscan, London

Pretty, J. N. and Shah, P. (1997) 'Making soil and water conservation sustainable: From coercion and control to partnerships and participation', *Land Degradation and Development*, vol 8, pp39–58

Pretty, J. N. and Pimbert, M. (1995) 'Beyond conservation ideology and the wilderness myth', *Natural Resources Forum*, vol 19, no 1, pp5–14

Rolston, H. (1997) 'Nature is for real: Is nature a social construct?' in Chappell, T. (ed) *Philosophy of the Environment*, Edinburgh University Press, Edinburgh

Rosegrant, M. W., Leach, N. and Gerpacio, R. (1997) 'Alternative futures for world cereal and meat consumption', *Proceedings of Nutrition Science*, vol 58, pp219–234

Samson, C. (2002) *Innu Naskapi-Montagnais. A Way of Life that Does Not Exist. Canada and the 'Extinguishment of the Innu'*, ISER Press, St Johns, Newfoundland

Schama, S. (1996) *Landscape and Memory*, Fontana Press, London

Scoones, I. (1994) *Living with Uncertainty: New Directions in Pastoral Development in Africa*, IT Publications, London

Seidl, A. (2000) 'Economic issues and the diet and the distribution of environmental impact', *Ecological Economics*, vol 34, no 1, pp5–8

Sessions, G. (ed) (1995) *Deep Ecology for the 21st Century*, Shambhala Publications, Boston

Singh, K. and Ballabh, V. (1997) *Cooperative Management of Natural Resources*, Sage, New Delhi

Smith, B. (1985) *European Vision and the South Pacific*, 2nd edition, Yale University Press, New Haven and London

Smith, B. (1987) *Imagining the Pacific*, Yale University Press, New Haven

Soulé, M. E. and Lease, G. (eds) (1995) *Reinventing Nature? Responses to Postmodern Deconstruction*, Island Press, Washington, DC

Temple, R. K. G. (1986) *China: Land of Discovery and Invention*, Patrick Stephenson, Wellingborough

Thoreau, H. D. (1837–1853) *The Writings of H. D. Thoreau, Volumes 1–6* (published 1981–2000), Princeton University Press, Princeton, New Jersey

Thoreau, H. D. (1906) *The Writings of H. D. Thoreau (including 'A Winter Walk')*, Houghton Mifflin, Boston

Thoreau, H. D. (1987) *Maine Woods*, Harper and Row, New York

Turner, F. J. (1920) *The Frontier in American History*, Holt, Rinehart and Winston, New York

Vandergeest, P. and DuPuis, E. M. (1996) 'Introduction' in DuPuis, E. M. and Vandergeest, P. (eds) *Creating Countryside: The Politics of Rural and Environmental Discourse*, Temple University Press, Philadelphia

White, T. (2000) 'Diet and the distribution of environmental impact', *Ecological Economics*, vol 34, no 1, pp145–153

World Health Organization (WHO) (1998) 'Obesity. Preventing and managing the global epidemic', WHO Technical Report 894. WHO, Geneva

WHO (2001) *Food and Health in Europe. A Basis for Action*, Regional Office for Europe, WHO, Copenhagen

Worster, D. (1993) *The Wealth of Nature: Environmental History and the Ecological Imagination*, Oxford University Press, New York

19

The Farm as Natural Habitat

Dana L. Jackson

'We should be having our summer board meeting on a farm. It's really beautiful at my place now.' When Dan French brought this up at the annual meeting of the Land Stewardship Project's board of directors, everyone nodded in agreement. Why didn't we think about scheduling the meeting there? I had spent a few days at the French farm several summers ago, sitting under a tent listening to instructors in the holistic management course held there and looking at Muriel's flower garden and the black-and-white dairy cows in the pasture beyond. A light breeze brought us the fragrance of green alfalfa from the barn where Dan's son was unloading bales. The class walked down to the creek, where we talked about the water cycle and how to judge water quality by observing the kinds of insects and fish in the water. The gravel on the creek bottom sparkled in the clear water. There was a flurry of birds and birdsong in the taller grass of a pasture section that hadn't been grazed for awhile.

When you drive up to the French farm, it looks like an interesting place, with its barns and outbuildings, vegetable and flower gardens, and shady picnic area. It doesn't look like those farmsteads one often sees in the Corn Belt, where the house and a machinery shed or two seem to just stick up out of a corn field, as if the owners had planted every inch they could on the place. Actually, if you drive on Interstate 35 between Saint Paul and Des Moines, you don't even see many houses. The landscape in July seems to be covered just with corn, a seemingly endless monotony of green stalks broken occasionally by shorter bushy soybeans.

It's hard to imagine what it must have looked like when Europeans first settled the Midwest, when it was a wilderness covered with prairie, forests, clear streams, and herds of buffalo. Too quickly it became dominated by agricultural uses interrupted by a few patches of prairie or woods around lakes or rivers that harboured remnants of natural habitat. Some prairie plants survived in pastures and meadows until they were replaced by fields of corn and soybeans in the last part of the 20th century. Then animals were moved into barns and feedlots, fences came down, and habitat edges disappeared. It only took about 150 years to reduce biological diversity on this landscape to a numbing sameness.

It is no surprise that people passionate about wildlife and the preservation of natural habitats have concentrated on protecting other places, those dramatic expanses of land where more of the original landscape remains, such as the Boundary Waters Canoe Area

Note: Reprinted from *The Farm as Natural Habitat*, by Jackson, D. L. and Jackson, L. L. (eds), copyright © (2002) by Island Press. Reproduced by permission of Island Press, Washington, DC

in northern Minnesota, the rugged mountains of Colorado and Montana, and roadless areas in Alaska. Such conservationists have accepted the agricultural Midwest, especially the Corn Belt, as a sacrifice area, like an open pit iron mine, or an oil field, where we mine the rich soil and create toxic wastes to extract basic raw materials. But the environmental impacts of this kind of mining are not confined to farming country. No nature preserves within its watersheds or wildlife area downstream on the Mississippi River can be adequately protected from farming practices that simplify ecosystems to a few manageable species and replace ecosystem services with industrial processes.

People who live in rural areas or urban residents who drive through them may not know that they are seeing a biologically impoverished landscape, because they have no knowledge of its diversity before modern agriculture. Others may know or imagine what the land looked like with different kinds of crops, meadows and livestock in pastures, but they accept its simplification because they are convinced that the main trends in agriculture can't be overcome. Agribusiness has successfully persuaded farmers, politicians, civic leaders, and even conservationists to believe that agricultural modernization leads to specialization and industrialization, and that financially viable alternatives are unavailable even though such modernization reduces the rural quality of life and harms the environment.

In this chapter, I will introduce an alternative vision for agriculture that defies the trends considered inevitable. It is a vision inspired by Aldo Leopold's writing that farming and natural areas should be interspersed, not separated, and by the farmer-members of the Land Stewardship Project, whose ways of managing farms have created a natural habitat for them, for their crops and livestock, and for the native plants and animals of the area. I will also describe two sustainable farming practices that currently are improving biological diversity on rural landscapes and showing the real possibility of this vision. Subsequent chapters describe these farmers and many others in more detail and discuss issues relevant to the vision and its feasibility. But first, let's look at the practices of mainstream industrial farming that render the countryside an ecological sacrifice zone.

Rural lands as industrial zones

The loss of biological diversity was not the only environmental consequence of creating the Corn Belt. Soil erosion, depletion of water resources, contamination of groundwater and surface water from fertilizers and pesticides (Soule and Piper, 1991), and a steady silt load in rivers are some of the consequences of so much tilled land. The sediment load in the Minnesota River at Mankato is equal to a ten-tonne dump truck load moving by approximately every five and a half minutes (Minnesota Pollution Control Agency, 1994).

The most serious environmental consequences are yet to come because of the growing consolidation in the livestock industry fueled by the abundance of cheap corn and soybeans. Each year an increasing number of poultry and hogs raised in the Corn Belt are not dispersed across the countryside on independent farms but are instead concentrated in large operations. Hundreds of thousands of chickens and tens of thousands of hogs are confined in buildings, creating huge quantities of manure that pose serious environmental risks to ground and surface water. Hydrogen sulphide fumes in the stench emitted from the operations have sickened neighbours. People don't want to live close to these hog factories or visit relatives close to hog factories. The once rich prairies that became bucolic communities are now industrial zones, suitable for 'neither man nor beast'.

Dairy farmers also feed the bounteous harvest of the Corn Belt to cattle confined in barns and milked three times a day. Dairy operations with 1000–2000 cows are replacing traditional family-sized farms with 100 or fewer cows. They manage large quantities of manure the same way as hog factories do and present the same risks to water quality. Travellers through Wisconsin's wooded hill lands graced with small dairy farms in the valleys may be unaware of how this landscape will change if consolidation continues in the dairy industry and four dairy farmers go out of business each day in the state as they did between 1992 and 1997 (USDA, 1997). Where large-scale dairies replace small ones, the scenes of black-and-white cows grazing on green pastures and moving in line to and from red barns are being replaced by fields of corn and soybeans with nary a cow in sight.

Factory livestock operations have popped up like mushrooms across the entire Midwest and Great Plains. They have also grown rapidly in southern states and are emerging everywhere state laws are weak and local communities naively believe the industry's forecasts for economic development. California led the way with its 1000 cow dairies and became the leading milk producer in the country; as a result, departments of agriculture in traditional dairy states are promoting California-style dairying. Agricultural economists encourage farmers to expand their operations to be efficient and convince them that all dairy cows and pigs, like poultry, are going to be raised in large-scale confinement operations in the future. It's inevitable.

This mantra of 'it's inevitable' is happily chanted by the corporate processors of pork that benefit from large supplies of cheap hogs, and, sadly, this mantra is repeated by many farmers. Some of them borrowed heavily to expand and build hog confinement buildings, and when pork prices plunged to an historic low in the 1990s, they went bankrupt. The huge packing plants that encouraged industrial production prospered and consolidated into even larger corporations through mergers (Heffernan, 1999).

We are seeing rural landscapes all across the US changing for the worse because farmers believe that further industrialization in livestock agriculture is inevitable and that they must 'get big or get out'. Some farmers incur staggering debt to increase the size of their operations, some form family corporations to share the costs of expansion, others invest in new buildings and technology to become contract producers for corporations, and some just leave farming and sell out to neighbours who want to expand. A house in the country isn't so romantic any more, because it might very well be within odour range of one of these hog expansions. Hay meadows and pastures with wildflowers and grassland birds are few and far between, and many streams running through fields have been cleared of trees and wildlife. If a family can't earn a living on the land, and it isn't a beautiful or healthful place to live, they might as well move to town. The land serves utilitarian purposes only, sacrificing natural values that once made it a home, not only for humans, but also for all kinds of creatures.

The disappearance of diversity in farming country has occurred steadily, mostly without notice or comment. Politicians and policy makers, the US Department of Agriculture, land grant universities, and many farmers and rural people accept the loss of biological diversity on the land as a necessary cost of efficient high production. There is some nostalgia in older people for a favourite fishing or swimming hole on the creek of the farm on which they grew up, but farming is a business and you can't be sentimental about it. Most travellers aren't aware that many of the monotonous fields they see along the highways harboured wildlife in prairie pastures and hayfields as recently as the 1960s. They only know that if they want to see woods and prairies and wildlife, they must head for a publicly owned park or wildlife area where agriculture is not practiced.

Aldo Leopold and a different vision for agriculture

Aldo Leopold, the Midwest's most famous conservationist, disapproved of the separation of natural areas from farming. To him it didn't make sense to protect forests in a special area and accept the absence of trees on agricultural land, when the farm was then left without the conservation benefit of erosion control and wind breaks. 'Doesn't conservation imply a certain interspersion of land uses, a certain pepper-and-salt pattern in the warp and woof of the land use fabric?' he asked (Leopold, 1991). Leopold believed that conservation efforts on certain parts of the land would fail if other parts were ruthlessly exploited. He wrote in the essay 'Round River':

> Conservation is a state of harmony between men and land. By land is meant all of the things on, over, or in the earth. Harmony with land is like harmony with a friend; you cannot cherish his right hand and chop off his left. That is to say, you cannot love game and hate predators; you cannot conserve the waters and waste the ranges; you cannot build the forest and mine the farm. The land is one organism (Leopold, 1966).

Although Leopold knew that agriculture was becoming more industrialized and wrote about the dangers of a farm becoming a factory, he could not have imagined the enormous livestock factories in production today. The transformation of so many meadows, prairies, and wetlands into corn, beans and hogs in Iowa, the state of his birth, and conversion of family-sized dairy farms into milk factories and corn fields in his adopted state of Wisconsin would astonish and grieve him. However, if someone told him about the zone of hypoxia in the Gulf of Mexico, 7000 square miles depleted of marine life because of excess nutrients flowing down the Mississippi River from the Corn Belt, I doubt if he would be surprised.

It is understandable that people accept these trends as the destiny of agriculture if they cannot clearly see alternatives. But there is an alternative – another trend – that could produce a landscape of farms which are natural habitats rather than ecological sacrifice areas.

A strong minority of modern farmers, like Dan and Muriel French, have not turned their farms into factories nor abandoned their chosen profession but are instead leading agriculture in an entirely different direction. Their creative initiatives to make farming more economically sound and environmentally friendly are producing benefits for them, for society at large, and for the land. The trends of these models are toward independent farms supporting families and communities while restoring biological diversity and health to the land.

Using an ecological approach to management decisions, these farmers are restoring a relationship between farming and the natural world that improves the sustainability of both. This relationship makes the farm a natural habitat. It is a natural habitat for humans in that it is a healthful and aesthetic place to live and earn a living. The farm is a natural habitat for the crops and livestock because they are able to use ecosystem services for fertility and pest control rather than fossil fuel and man-made chemicals. And the farm is a natural habitat for native plants and animals, a refuge that encourages biological diversity along streams, in pastures, and along uncultivated edges.

Farming practices for natural habitat

Farmers themselves don't talk about turning their farms into natural habitats. It happens as a result of the way that they choose to farm. Many farmers became interested in changing their practices in the 1980s, particularly during a period of low prices, high production costs and minuscule profits. A number of newly formed farming organizations around the country helped them lower their use of purchased inputs, such as chemical fertilizer and pesticides, and develop more environmentally friendly practices. For example, the Land Stewardship Project (LSP) in Minnesota began to hold workshops and field days about the practice of *management intensive rotational grazing*. This involves dividing a pasture into sections or 'paddocks' with electric fences and allowing the animals to graze each area intensively for a short period of time before moving them on to another area. In conventional grazing, livestock roam freely in an open pasture, often overgrazing some areas and causing erosion.

Management intensive rotational grazing roughly mimics grazing patterns of migrating buffalo herds that preceded European settlement on the plains and prairies, but domestic livestock return to graze an area much sooner than did buffalo. The length of time that animals graze a particular paddock usually depends upon the rate of recovery of the forage after grazing and its nutritional value, which requires farmers or ranchers to become attentive observers of their pastures and all that is growing there.

A group of farmers wanted to know how they could tell whether the switch to management intensive rotational grazing was making their farms more sustainable. In response, LSP established a biological, social and financial monitoring project, conducting research on six diversified livestock and dairy farms that used management intensive rotational grazing. The project team that worked together for three years included university researchers and state agency staff in addition to the six farmers and LSP staff. To conduct biological monitoring, researchers helped the six farmers collect biological, physical and chemical soil quality data from 60 plots and make observations about pasture vegetative species and ground cover. They sampled wells and kept precipitation records. The farmers learned to survey their land for breeding birds, frogs and toads, and they helped fisheries scientists survey streams passing through four of the team farms and through one paired farm to analyse the effects of management intensive rotational grazing on stream banks and stream invertebrate and fish populations.

These farms were seen as natural habitats, not as ecological sacrifice areas. The farmers wanted to find out if the soil and the water quality in streams on their farms were improving, just as they wanted to know if their financial bottom lines were improving by cutting production costs. They were not accepting the 'inevitable', that they must get big or get out.

The farmers in the monitoring project, and many others who have been constituents of the LSP, practice holistic management, a decision making process based on goal setting, planning and monitoring. This process was developed by Alan Savory, who founded the Center for Holistic Resource Management in Albuquerque, New Mexico, in 1984. LSP staff taught many holistic management courses throughout the Upper Midwest. They developed a research project to monitor the effectiveness of management decisions made by the six farmers who had taken the course and were making the switch from conventional grazing to management intensive rotational grazing.

Holistic management contains four elements that distinguish it from conventional farm management and provide managers with strong incentives to make environmentally sound decisions. First, as part of the goal-setting process, it directs managers to develop a long-term vision for how they want the landscape to look far into the future. Second, the model teaches basic recognition of ecosystem processes that farms are dependent upon: the water cycle, the mineral cycle, plant succession and energy flow (Savory, 1998). Farmers strive to understand these processes and harmonize their farming practices with them. For example, farmers can rely on nitrogen fixation in legumes and the recycling of nutrients in manure to provide fertility for fields. Third, holistic management places a high value on biological diversity both in crop systems and in areas on the land not used for farming. And last, practitioners consider the effect of any proposed action or choice of enterprise upon quality of life for the community as well as for themselves. They understand that their land is part of a larger whole and how they manage it will affect the landscape around them and the lives of people in the community. Holistic management has become an effective tool for those who want to be good stewards of the land and earn a living on it at the same time.

Though holistic management has been used on all kinds of farming operations, it was developed in connection with rotational grazing. Farmers in the Upper Midwest often began using holistic management and management intensive rotational grazing approaches simultaneously. Cattle grazing on public lands in western states has been considered such a disaster by environmentalists that many have a negative view of grazing anywhere. However, at the landscape level in the Midwest and in parts of the Great Plains and the South management intensive rotational grazing provides visible environmental improvement in farming, especially where field crops have been converted to permanent pastures and livestock eat more grass than grain. Fewer acres of corn and soybeans also mean fewer applications of chemical pesticides, herbicides and fertilizer, which decreases the potential for contamination of surface and groundwater. When corn and soybeans are replaced by perennial grasses, there is less soil erosion (Cambardella and Elliot, 1992; Rayburn, 1993).

Dairy farmers have widely adapted management intensive rotational grazing. Between 1993 and 1997, the number of Wisconsin dairy farmers using variations of this grazing method increased by 60 per cent (ATFFI, 1996). Milk cows on most conventional dairy farms are confined in 'loafing barns' or corrals between milkings and are never allowed out to graze. On very large operations of 500–1500 or more cows, feed is brought to the cows and all of their manure is pumped out of manure pits or scraped and hauled out of the barns to be spread on fields. Conventional dairy farmers work hard to produce the corn and alfalfa to feed the dairy herd, and capital costs for equipment and barns are high. In contrast, grass-based dairy farmers usually move cattle daily but claim that their work load and costs of production are much less because the cattle walk around in the paddocks, get most of their own food, and disperse their own manure (ATFFI, 1996). With more feed produced in pastures, a farmer uses less machinery and fossil fuel (Rayburn, 1993). Some grass-based farmers 'don't have much iron', as they say, because they've sold most of the machinery they formerly needed for large fields of corn. With fewer acres planted for feed, they can share machinery with neighbours, employ custom harvesters to bring in their crops or even buy feed from other farmers. For these dairy farmers, management intensive rotational grazing is a farming practice that benefits them as much as it benefits the land and the water.

Poultry and hog farmers also use management intensive rotational grazing. Hogs can be put on pasture to graze, at least for part of their food, and spread their own manure in the grass. Hogs can spend most of their time outdoors and farrow in pastures. Farmers in the Upper Midwest often combine outdoor and indoor production systems by bringing hogs into open-ended metal hoop buildings covered with canvas for the winter. Hogs bed in deep straw or corn stalks, which composts with their manure, warming the hogs in the process and producing nearly composted, dry fertilizer for the fields when the barns are cleaned. Manure is not a toxic waste in management intensive rotational grazing or hoop house production systems, and the cost to the farmer of handling it and the public for regulating it is little or nothing. In fact, overall production costs are so much lower that farmers can make a profit as long as they have fair access to markets (Dansingburg and Gunnink, 1995) or sell cooperatively with other farmers or directly to consumers. If market prices are too low, farmers can use these hoop houses for other purposes, such as storing hay or machinery, which gives them a flexibility that producers trying to pay off the debt for a high-tech, single-use confinement facility don't have. Using management intensive rotational grazing and deep-bedded straw systems in hoop houses, farmers can take advantage of ecosystem services in providing animal feed and managing manure. These systems are efficient alternatives to the industrial production models for livestock and can compatibly exist alongside or as part of natural ecosystems.

The benefits of diversity

Diversified farms producing feed for their own livestock may rotate crops of alfalfa or other legumes, corn, soybeans and small grains such as barley or oats, in contrast to conventional cash grain farms that rotate only corn and soybeans or grow corn with no rotation. For example, Jaime DeRosier employs a complex rotation of hay, wheat, barley, vetch, flax, buckwheat, corn and soybeans on his large organic farm in north-western Minnesota (DeRosier, 1998). The Fred Kirschenmann farm in North Dakota rotates up to ten different grain or hay crops in three different rotations (Anonymous, 2000). In all parts of the country, farmers are also planting several different kinds of grasses and legumes in their pasture mixes, planting fields in strips of several crops, intercropping one species with another (such as field peas with small grains), and using cover crops between plantings of major crops. In California, orchards, vineyards, and specialty crop farms have added cover crops and farmscape plantings to attract pollinators and other beneficial insects (CAFF, 2000).

The benefits of biodiversity in agriculture were effectively laid out in a report with that title by a task force of the Council for Agriculture Science and Technology, co-chaired by ecologist G. David Tilman and geneticist Donald N. Duvick (CAST, 1999). The report stresses the dependency of modern agriculture upon biological diversity and advocates greater attention to preserving diversity both in domesticated crops and livestock, and in the natural landscape.

The *Benefits of Biodiversity* also discusses the dependence of modern agriculture upon ecosystem services, such as pollination, generation of soils and renewal of their fertility, pest control, and decomposition of wastes. It acknowledges the importance of preserving biodiversity by protecting natural areas and proposes that we substantially increase the worldwide network of biodiversity reserves and preserve large blocks of land in native ecosystems.

This report was not produced by CAST for the purpose of rerouting agriculture from the direction trends are leading. However, if followed, just one recommendation would lead us toward a landscape of farms that are natural habitats:

> Increase the capacity of rural landscapes to sustain biodiversity and ecosystem services by maintaining hedgerows/windbreaks; leaving tracts of land in native habitat; planting a diversity of crops; decreasing the amount of tillage; encouraging pastoral activities and mixed-species forestry; using diverse, native grasslands; matching livestock to the production environment; and using integrated pest management techniques.

The six farmers who participated in the LSP monitoring project use many of these practices and have created more natural diversity on their land. Just by converting cropland to pasture they created new habitat for soil microbes, insects, birds, reptiles, amphibians and small mammals. Species that would have been adversely affected by chemical pesticides and fertilizers used on crops found a more favourable environment in the pastures.

Because of the emphasis on diversity and biological monitoring in holistic management, farmers in the project became advocates of diversity and astute observers of wildlife. A newsletter distributed to monitoring team members contained the following notes in a column called 'Farmer Observations':

> Mike saw first red clover blossoms on June 6. Mike saw a hummingbird on clover in his extended rest pad. He suggests that each farmer photograph their rest areas and notice the smell intensified by flowering plants. Ralph saw two baby bobolinks on July 14. He noticed the young are bunching up and may move soon (Land Stewardship Project, 1995).

These farmers are not conventional in any sense of the word. Mike and Jennifer Rupprecht pay meticulous attention to erosion control and species diversity in their pasture, getting excited when they find native prairie species on their land. Ralph Lentz likes to show people the prairie grasses in his pastures and to talk about how he has used managed intensive grazing to improve the stability of stream banks on his land (DeVore, 1998). Dave and Florence Minar began working with a local monitoring team, after the original LSP monitoring project concluded, in the area of Sand Creek, the tributary that dumps the most sediment into the Minnesota River. Art Thicke is ecstatic when he talks about the birds he sees while moving cattle – birds that weren't there when those pastures were planted to corn and soybeans (King and DeVore, 1999).

The increase of grassland birds wasn't just a phenomenon on Art's farm or on the other five farms in the monitoring project. Other farmers in the Upper Midwest report that they see more grassland birds such as bobolinks *(Dolichonyx oryzivorus)* and dickcissels *(Spiza americana)* since they replaced row crops with grass pastures. The Agriculture Ecosystems Research Project in the agronomy department at the University of Wisconsin has been comparing continuously grazed dairy pastures with rotationally grazed pastures, and preliminary results show that many more birds and more different species use rotational pastures than use continuous pastures (Paine, 1996). The increased acres of permanent grass in pasture, combined with conservation reserve land that has been in grass for several years, has created large areas of habitat for game birds also. Additional habitat is created where trees are allowed to grow again along drainages in pastures that were formerly tilled fields.

The farmers actively engaged in the Land Stewardship Project's monitoring project, and many others practicing monitoring as a result of studying holistic management, are protecting or restoring diverse colours and textures in the 'warp and woof of the land use fabric'. To nurture the diversity of wildlife they have come to appreciate, and the wildlife they have begun to understand as indicators of ecosystem health, these farmers are developing and protecting more habitat niches in wood lots, along roadsides, on orchard and pasture edges, and along streams and ponds. They are leaving areas in their pastures ungrazed during nesting season for grassland birds and removing low areas in fields from cultivation to restore wetlands.

The important point is not that these farmers have become naturalists. The natural habitat they are creating on their land is not because they set out to entice native plant and animal species to reinhabit their farms. Their management decisions and farming practices are turning their farms into a natural habitat for humans, crops and livestock, *and* wild plants and animals too. Then, as they make the connections between biological diversity, the economic health of the farm, and the quality of their lives, farmers have begun consciously to make decisions to encourage even more biological diversity on their farms. Such farms should be the model for agriculture in the 21st century. To make that happen, a large group of constituents are needed who understand the possibilities for farms to be natural habitats and to transform rural landscapes.

Building a constituency

Aldo Leopold wrote that no government conservation programmes with their subsidies for farmers could cause landowners to take good care of the land unless they felt an ethical responsibility for it. The ultimate responsibility for conservation was the farmer's (Leopold, 1991). From the latest agricultural census, we can see that less than 2 per cent of the US population are farmers (USDA, 1997), and not all of them are the family farmers Leopold had in mind but include large-scale farmers managing thousands of acres, often on behalf of investors or on contract with corporations. There aren't enough private landowners on farms to rescue the agricultural landscape from ruin, even if those that exist possess a strong land ethic. We would be foolish to depend upon giant producers and processors such as Tyson, IBP and Smithfield corporations to exercise a land ethic. Whose responsibility is it then? It is a public responsibility. Good farming produces public goods, and the public must support good farming. Instead of accepting industrial agriculture as a necessary evil and counting on regulations to soften its negative environmental and social consequences, the public (particularly conservationists and environmentalists) should use their dollars and their votes and their influence to bring about agroecological restoration.

If asked whether it is all right to consider agricultural land as an ecological sacrifice area, most conservationists would loudly say no. But without thinking about it, many have acquiesced to the inevitability of farms becoming corporate factories when they have been involved in state or national processes to establish regulations for feedlots. Activist organizations have worked for strong regulations of non-point source water pollution and confined animal feeding operations, and their chief opponents have often been farmers, or farm organizations, which has caused them to develop antagonism for farmers. Many haven't had the opportunity to know farmers whose diversified livestock systems operate without need of regulations. If conservationists could get to know farmers who are stewards of the soil, water and the wild and learn about their management philosophy and the farming practices they use, perhaps they would see possibilities for making basic changes in US agriculture that would restore rural landscapes to greater biological diversity and environmental health.

Dave Palmquist, the interpretative naturalist at southeast Minnesota's Whitewater State Park, the most popular park in Minnesota with about one-third of a million visitors a year, knows a stewardship farm family. He has taken groups of campers ten miles away from the park to visit the 275-acre farm owned and operated by Mike and Jennifer Rupprecht, one of the six farms in LSP's monitoring project. His reason: 'There's an increasing understanding you can't save the world within state parks. The 65 little pieces of Minnesota (state parks) aren't going to do it. If you have to go outside your park to tell an important story that relates to the park area, do that.' Palmquist believes that visitors are impressed. 'It's clear to the visitors that these farmers embrace diversity and see themselves as being part of the bigger environment. The more diversity, the more bobolinks, bluebirds, etcetera, they have on their land, the better they feel. If they can make a living there, maintain a family farm, and be gentler on the environment, that's very exciting for them' (DeVore, 1996).

This kind of agroecological restoration is occurring on many farms today, illustrating that farms can be managed to give rural landscapes a mixture of agricultural and natural ecosystems that preserve much of local biodiversity and provide ecosystem services essential to agriculture. We need the heirs to Aldo Leopold's thought and inspiration and those who respect the work of modern ecologists such as David Tilman and naturalists like Dave Palmquist to help society see this vision of the farm as natural habitat and work to turn it into reality.

Conclusion

This vision does not promise that a landscape of such farms will reproduce the ecosystem that existed before white Europeans conquered the land, but neither will it be covered with factories. When farms are factories, they produce commodities and profit for agribusiness and charge external costs to the land and rural communities. When farms are natural habitats for humans, domesticated crops and livestock, and also for wild plants and animals, they produce food and multiple other benefits for society. And such farms can be the sources for further ecological restoration in the landscape.

No doubt interspersing a variety of uses on farms will mean different problems to overcome than those we now face, both ecologically and economically, because we still have a lot to learn about farming with the wild. Creating farms as natural habitats will require more sophisticated strategies for disease and pest suppression in crops and livestock. It will also require greater emphasis on diversification and resilience and less emphasis on simplification and short-term fixes. These are problems in farming that require ecological solutions.

Farming-system problems can be solved. The perhaps intractable problem is how to influence social evolution so that a land ethic, and not pure utilitarianism, guides land-use decisions. We need all people to look at farming with new eyes, to see the potential of the farm as natural habitat, and to refuse to accept the inevitability of farms becoming rural factories to serve the global economy. We must teach that 'the land is one organism'.

References

Agricultural Technology and Family Farm Institute (ATFFI) (1996) *Grazing in Dairy-land: The Use and Performance of Management Intensive Rotational Grazing among Wisconsin Dairy Farms*, Technical Report no 5, University of Wisconsin, College of Agriculture, Madison

Anonymous (2000) 'Farmer chosen as next Leopold Center director', *Leopold Letter*, vol 12, no 2, p6

Cambardella, C. A. and Elliot, E. T. (1992) 'Particulate soil organic matter changes across a grassland cultivation sequence', *Soil Science Society of America Journal*

Community Alliance for Family Farms (CAFF) (2000) *Farmer to Farmer*, May, June

Council for Agriculture Science and Technology (CAST) (1999) *Benefits of Biodiversity*, Task Force Report no 133, Ames, Iowa

Dansingburg, J. and Gunnink, D. (1995) *An Agriculture That Makes Sense: Making Money on Hogs*, Land Stewardship Project, White Bear Lake, Minnesota

DeRosier, J. (1998) *My Cover Crop Rotation Program*, Jaime DeRosier, Red Lake Falls, Minnesota

DeVore, B. (1996) 'An agrarian ecological tour', *The Land Stewardship Letter*, vol 14, pp2–3

DeVore, B. (1998) 'The stream team', *The Minnesota Volunteer*, vol 61, pp10–19

Heffernan, W. (1999) *Report to the Farmers Union: Consolidation in the Food and Agriculture System*, National Farmers Union, Ames, Iowa

King, T. and DeVore, B. (1999) 'Bringing the land back to life', *Sierra*, Jan/Feb, pp 36–391

Land Stewardship Project (1995) *Monitoring Project Monthly Newsletter*, June

Leopold, A. (1966) *A Sand County Almanac with Essays on Conservation from Round River*, Ballantine Books, New York

Leopold, A. (1991) 'The farmer as a conservationist', in Callicott, B. and Flader, S. (eds) *The River of the Mother of God and Other Essays by Aldo Leopold*, University of Wisconsin Press, Madison

Minnesota Pollution Control Agency (1994) *Executive Summary: Minnesota River Assessment Project Report*, Minnesota Pollution Control Agency, St Paul, Minnesota

Paine, L. (1996) 'Pasture songbirds', *Pasture Talk*, May, pp8–9

Rayburn, E. B. (1993) 'Potential ecological and environmental effects of pasture and BGH technology', in Liebhardt, W. C. (ed) *The Dairy Debate: Consequences of Bovine Growth Hormone and Rotational Grazing Technologies*, University of California, Davis

Savory, A. (1998) *Holistic Resource Management*, Island Press, Washington, DC

Soulé, J. D. and Piper, J. K. (1991) *Farming in Nature's Image: An Ecological Approach to Agriculture*, Island Press, Washington, DC

United States Department of Agriculture (USDA) (1997) *Census of Agriculture*, vol 1, United States Department of Agriculture, National Agricultural Statistics Service, Washington, DC

20

Diet and Health: Diseases and Food

Tim Lang and Michael Heasman

Let Reason rule in man, and he dares not trespass against his fellow-creature, but will do as he would be done unto. For Reason tells him, is thy neighbour hungry and naked today, do thou feed him and clothe him, it may be thy case tomorrow, and then he will be ready to help thee.

Gerrard Winstanley, English Leveller, 1609–1676[1]

Core arguments

The Productionist paradigm is critically flawed in respect of human health. Half a century ago it responded to issues then seen as critical but which now require radical revision. While successfully raising the caloric value of the world food supply, it has failed to address the issue of quality, and as a result, there is now a worldwide legacy of externalized ill-health costs. The world's human health profile is now very mixed. Within the same populations, in both developed and developing countries, there exists diet-related disease due both to under- and over-consumption. The pattern of diet that 30 years ago was associated with the affluent West is increasingly appearing in the developing countries, in a phenomenon known as the 'nutrition transition': while the incidence of certain diet-related diseases has decreased, such as heart disease in the West, others are increasing, particularly diabetes and obesity worldwide, and heart disease in the developing world. Massive global inequities in income and expectations contribute to this double burden of disease, and current policies are failing to address it.

Introduction

One of the key Food Wars is over the impact of the modern diet on human health. In the last quarter of the 20th century, nutrition moved from the sidelines of public health

---

Note: Reprinted from *Food Wars*, by Lang, T. and Heasman, M., copyright © (2004) Earthscan, London

to being central to the marketing of foodstuffs, and major public health campaigns urged consumers to improve their diets.

This human health dimension is central to our critique of the Productionist paradigm in two respects. First, even though global food production has increased to meet caloric needs, its nutritional content may be less than desirable. Second, food distribution remains deficient: nearly a billion people remain malnourished. In this chapter, we explore the relationship between diet and the range of disease and illnesses that are associated with food choices. We discuss, too, the existence of gross inequalities within and between countries in the form of food poverty amidst food abundance and wealth.

In late 2002 and 2003, a wave of new public health reports reminded the world that diet is a major factor in the causes of death and morbidity. Although deeply unpalatable to some sections of the food industry, these reports were sober reminders of the enormity and scale of the public health crisis. The joint WHO and FAO's 2003 report on diet, nutrition and the prevention of chronic diseases drew attention to high prevalence of diseases which could be prevented by better nutrition, including:[2]

- obesity;
- diabetes;
- cardiovascular diseases;
- cancers;
- osteoporosis and bone fractures;
- dental disease.

Of course, these diseases are not solely exacerbated by poor diet but also by lack of physical activity. In truth, this report was only reiterating the story of nutrition's impact on public health that had been rehearsed for many years, and the evidence for which was judged to be remarkably sound, but as Dr Gro-Harlem Brundtland, then the Director-General of the WHO, stated in the report: 'What is new is that we are laying down the foundation for a global policy response.' To this end, the WHO set up an international consultation dialogue to prepare its global strategy on diet, physical activity and health, scheduled to be launched in 2004. By international agency standards, this relatively speedy shift from evidence to policy making indicates the real urgency of the problem. The draft strategy was launched ahead of schedule in December 2003.[3]

Already by 2002, the WHO had produced a major review of the national burdens that such diseases cause. Of the top ten risk factors associated with non-communicable diseases, food and drink contribute to eight (with the two remaining – tobacco and unsafe sex – not associated with diet and food intake):[4]

- blood pressure;
- cholesterol;
- underweight;
- fruit and vegetable intake;
- high body mass index;
- physical inactivity;
- alcohol;
- unsafe water, sanitation and hygiene.

In the 2003 World Cancer Report, the most comprehensive global examination of the disease to date, the WHO stated that cancer rates could further increase by 50 per cent to 15 million new cases in 2020.[5] To stem the rise of this toll, the WHO and the International Agency for Research on Cancer (the IARC) argued that three issues in particular need to be tackled:

- Tobacco consumption (still the most important immediate avoidable risk to health).
- Healthy lifestyle and diet, in particular the frequent consumption of fruit and vegetables and the taking of physical activity; early detection and screening of diseases to allow prevention and cure.
- In addition to these UN reports, the International Association for the Study of Obesity (the IASO) revised its figures of the global obesity pandemic: it estimates that 1.7 billion people are overweight or obese, a 50 per cent increase on previous estimates. The IASO's International Obesity Task Force stated that the revised figures meant that most governments were simply ignoring one of the biggest risks to world population health.[6]

These reports testify to an extensive body of research and evidence from diverse sources around the world of the link between food availability, consumption styles and specific patterns of disease and illness. Table 20.1 confirms some of the diet-related causes of death throughout the world. Good health and longevity were intended to result from ensured sufficiency of supply; at the beginning of the 21st century, far from diet-related ill health being banished from the policy agenda, it appears to be experiencing a renewed crisis.

Under the old Productionist paradigm, the main focus was under-nutrition. Yet at the end of the 20th century, with diseases such as heart disease, cancers, diabetes and obesity rampant worldwide, not just in the affluent West, a new focus must be placed on diet and inappropriate eating. In this chapter, we begin to explore wider societal changes which impose progress in this regard through demographic shifts, maldistribution of and poor access to food, and spiralling health-care costs. These factors add weight to our argument that the Productionist paradigm is beyond its own sell-by date.

Policy making is failing to address the causes of these food-related health problems and too often resorts to only palliative measures. This is partly because the Productionist paradigm's approach to health narrows the framework for considering alternative solutions: by being centred on striving to increase output, it has taken only a medicalized, rather than a socially determined, view of health.

Could the proponents of Productionism have anticipated the scale of these most recent health concerns? To some extent, they could not. Even excessive intake of fats as causing ill health might have been something of a shock for the Productionist paradigm, as it was almost heretical to argue that too much of a nutrient could be harmful to health.[7] Part of the problem here was the essential paternalism of the paradigm which assumed total knowledge of all variables needed to make good food policy: governments and companies could be trusted to look after the public health; the consumers' role was to select products to create their own balanced diets. Recent history, however, has shown that governments and the food supply chain failed to adapt to new scientific knowledge in relation to food and health. Nationally and internationally, the influence of health scientists on public policy has been minimal. Consumerism triumphed.

Table 20.1 *Some major diet-related diseases*

Problem	Extent/comment
Low birthweights	30 million infants born in developing countries each year with low birthweight: by 2000, 11.9 per cent of all newborns in developing countries (11.7 million infants)
Child under-nutrition	150 million underweight pre-school children: in 2000, 32.5 per cent of children under 5 years in developing countries stunted, amounting to 182 million pre-school children. Problem linked to mental impairment. Vitamin A deficiency affects 140–250 million schoolchildren; in 1995, 11.6 million deaths among children under 5 years old in developing countries
Anaemia	Prevalent in schoolchildren; maternal anaemiapandemic in some countries
Adult chronic diseases	These include adult-onset diabetes, heart disease and hypertension, all accentuated by early childhood under-nutrition
Obesity	A risk factor for some chronic diseases (see above), especially adult-onset diabetes.[8] Overweight rising rapidly in all regions of the world
Underweight	In 2000, an estimated 26.7 per cent of pre-school children in developing countries.
Infectious diseases	Still the world's major killers but incidence worsened by poor nutrition; particularly affects developing countries
Vitamin A deficiency	Severe vitamin A deficiency on the decline in all regions, but sub-clinical vitamin A deficiency still affects between 140 and 250 million pre-school children in developing countries, and is associated with high rates of morbidity and mortality

Source: adapted from ACC/SCN 2000[9]

To some extent, too, the public health world has colluded in its own marginalization from 'live' policy making by its fixation with deficiency diseases: for example, on programmes of food fortification or on protein shortages which could be made up by increased meat and dairy production. Despite a successful worldwide campaign to increase intake of folic acid following the discovery of its connection with spina bifida (neural tube defect syndrome), the overall impact of nutritional science in policy making has been negligible. Its response to the current epidemic of heart disease has been 'health education' – advice, leaflets and exhortations to change behaviour – explaining it as caused by modern lifestyles, rather than by preventable dietary deficiencies. Almost as soon as the Productionist paradigm was put in place worldwide in the last half of the 20th century, global campaigns were needed to address the increase in degenerative diseases. However, the necessary policy instruments were not in place to tackle the health impact of long-term shifts in diet. The UN bodies which noted the evidence of new patterns of ill health were merely intergovernmental bodies who lacked any administrative power and influence to act on the global and national level. Commercial interests, on the other hand, had no such limits and could pursue their global ambitions, selling foods and a lifestyle around the world without regard to their consequences, and being able to defend their actions as being in the public interest.

Instead, the developed world now must confront one of the most challenging food and health disasters ever to face humankind: an epidemic of obesity with little prospect of an end in sight and the prospect of a new wave of diet-related diseases in its wake. It has little in its armoury with which to combat the causes of obesity, now affecting significant numbers of children and with even graver implications for future population health. Health education is ineffective; consumerism is part of the problem, but politically it is nearly sacrosanct.

Meanwhile, hunger and insufficiency continue, ironically, to prevail. As a 1995 FAO review of the global picture starkly put it: '[H]unger... persists in developing countries at a time when global food production has evolved to a stage when sufficient food is produced to meet the needs of every person on the planet.'[10] Over-consumption and under-consumption coexist. There is gross inequality of global distribution and availability of food energy. The same review asserted that Western Europe, for example, has in theory 3500 kilocalories (kcal) available per person per day and North America has 3600, while sub-Saharan Africa has 2100 and India has 2200. By 2015 the FAO calculates that 6 per cent of the world's population (462 million people) will be living in countries with under 2200kcal available per person. And by 2030, in the most optimistic scenario, in sub-Saharan Africa 15 per cent of the population will be undernourished. Numbers of the undernourished look set only to decline much more slowly than suggested by targets, for example those of the World Food Summit of 1996.[11]

The transnational nature of these patterns of diet-related disease demands public policy attention. The enormity of this human health problem cannot be over-emphasized. Diseases associated with deficient diet account for 60 per cent of years of life lost in the established market economies.[12]

Scientists categorize diseases into two broad groups: communicable (carried from person to person or via some intermediary factor; these include diseases such as malaria, food poisoning, SARS); non-communicable (acquired by lifestyle or other mismatch between humans and their environment, such as cardiovascular disease and cancers). Table 20.2 indicates that in the developed world, deaths through infectious and parasitic diseases are very low compared to developing countries, while diet-related non-communicable diseases like coronary heart disease and cancers are high in both developed and developing worlds. Degenerative disease rates are already high in the developing world. Figure 20.1 gives the leading causes of mortality by age to give another view of the global disease patterns.

The WHO and the FAO reports stress that world health in general is in transition with non-communicable diseases now taking a higher toll than communicable diseases. Figure 20.2 shows the WHO prognosis of how the rates of non-communicable disease are expected to rise. Factors in this health transition include diet, demographic change (such as an ageing population) and cultural factors related to globalization.

Table 20.1 *Number of deaths by WHO regions, estimates for 2002 (thousands)*

	Africa	Western Pacific	Europe	The Americas	Eastern Mediterranean	South-east Asia	Total world-wide
Infectious & parasitic diseases	5787	794	212	394	959	2968	11,114
Cardiovascular diseases	1136	3817	4857	1927	1080	3911	16,728
Cancers	410	2315	1822	1115	272	1160	7094
Respiratory infections[a]	1071	511	273	228	365	1393	3841
Perinatal and maternal causes	585	371	69	192	371	1183	2771
Injuries	747	1231	803	540	391	1267	4979

Note: a This does NOT include respiratory diseases; includes upper and lower respiratory infections and otitis media

Source: WHO, *Shaping the Future*, World Health Report, Geneva, 2003, calculated from Annex Table 2

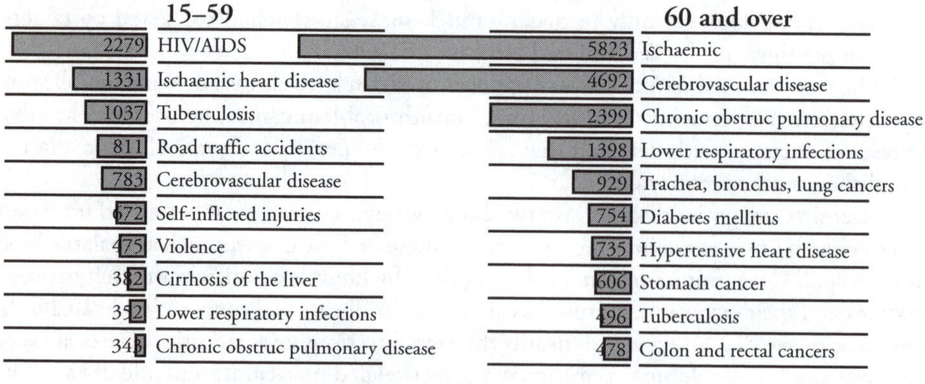

15–59		60 and over	
2279	HIV/AIDS	5823	Ischaemic
1331	Ischaemic heart disease	4692	Cerebrovascular disease
1037	Tuberculosis	2399	Chronic obstruc pulmonary disease
811	Road traffic accidents	1398	Lower respiratory infections
783	Cerebrovascular disease	929	Trachea, bronchus, lung cancers
672	Self-inflicted injuries	754	Diabetes mellitus
475	Violence	735	Hypertensive heart disease
382	Cirrhosis of the liver	606	Stomach cancer
352	Lower respiratory infections	496	Tuberculosis
348	Chronic obstruc pulmonary disease	478	Colon and rectal cancers

Source: WHO, *World Health Report 2003*

Figure 20.1 *Leading causes of mortality, by age, 2002*

The nutrition transition

In a series of papers, Professor Barry Popkin and his colleagues have argued that there is what they term a 'nutrition transition' occurring in the developing world, associated primarily with rising wealth.[13 14] The thesis, which has been extensively supported by country and regional studies,[15] argues simply that diet-related ill health previously asso-

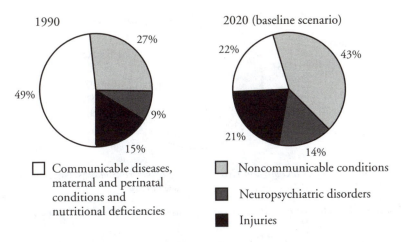

1990

27%

49%

9%

15%

2020 (baseline scenario)

22%

43%

21%

14%

☐ Communicable diseases, maternal and perinatal conditions and nutritional deficiencies

▨ Noncommunicable conditions

◼ Neuropsychiatric disorders

◼ Injuries

Source: WHO, *Evidence, Information and Policy*, 2000

Figure 20.2 *Anticipated shift in global burden of disease 1990–2020, by disease group in developing countries (WHO)*

ciated with the affluent West is now becoming increasingly manifest in developing countries.[16, 17] The 'nutrition transition' suggests shifts in diet from one pattern to another: for example, from a restricted diet to one that is high in saturated fat, sugar and refined foods, and low in fibre. This transition is associated with two other historic processes of change: the demographic and epidemiological transitions. Demographically, world populations have shifted from patterns of high fertility and high mortality to patterns of low fertility and low mortality. In the epidemiological transition, there is a shift from a pattern of disease characterized by infections, malnutrition and episodic famine to a pattern of disease with a high rate of the chronic and degenerative diseases. This change of disease pattern is associated with a shift from rural to urban and industrial lifestyle.

WHO researchers have noted that changes in dietary patterns can be driven not just by rising income and affluence but also by the immiseration that accompanies others' rising wealth;[18] low income countries are experiencing the effects of the transition but cannot afford to deal with them.[19] Popkin argues that, while the nutrition transition brings greater variety of foods to people who previously had narrow diets, the resulting health problems from the shift in diet should not be traded off against the culinary and experiential gains. Consumers might enjoy the new variety of foods that greater wealth offers but they are often unaware of the risk of disease that can follow. The implications of the nutrition transition now ought to exercise the minds of global as well national policy makers: certainly health policy specialists are concerned at the rise of degenerative diseases in low- and middle-income countries.[20, 21]

Nutrition may have recently become a key notion in modern dietary thinking but it only echoes the insights of an earlier generation of researchers which included nutrition and public health pioneers such as Professors Trowell and Burkitt, whose observations from the 1950s to the 1980s led them to question 'whether Western influence in Africa, Asia, Central and South America and the Far East is unnecessarily imposing our

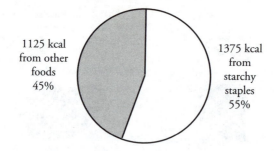

Source: National Survey of Income and Expenditure of Urban Households. Government of China, 1990; FAO, *State of Food Insecurity*, 2000; www.fao.org

Figure 20.3 *Diet of a well-nourished Chinese adult (2500kcal/person/day)*

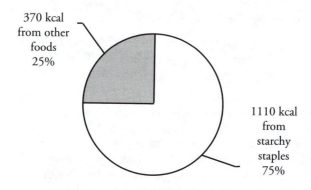

Source: National Survey of Income and Expenditure of Urban Households, Government of China, 1990; www.fao.org

Figure 20.4 *Diet of an undernourished Chinese adult (1480kcal/person/day)*

diseases on other populations who are presently relatively free of them'.[22] Trowell and Burkitt, both with long medical experience in Africa, could easily explain the variation in infectious diseases, but not the variation in rates of non-infectious diseases such as heart disease between countries at different economic levels of wealth and development. In Africa in the post-World War II period, they witnessed the rise of key indicators for diseases such as heart disease and high blood pressure in peoples who had previously had little experience of them.[23] The dietary transition is swift. An FAC study of very undernourished Chinese people (living on 1480kcal per day) shows that they derive three-quarters of their energy intake from starchy staples such as rice, while better-fed Chinese (living on 2500kcal per day) are able to reduce their energy intake from such staples and to diversify their food sources (see Figures 20.3 and 20.4 which compare the diets of undernourished and well-nourished people in China).

Popkin has shown how this same process occurs in both urban and rural populations in developing countries with rising incomes. Figures 20.5 and 20.6 show the relationship between per capita income and what predominantly rural and predominantly

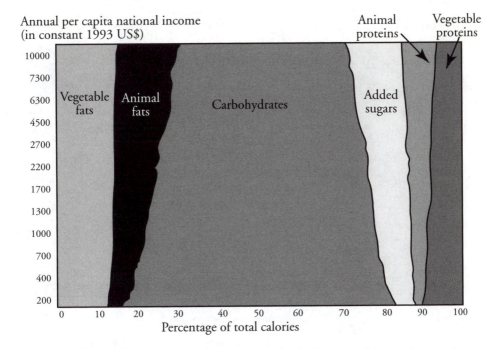

Annual per capita national income
(in constant 1993 US$)

Source: FAO/World Bank/Popkin, B (1998) 'The nutrition transition and its health implications on lower income countries', *Public Health Nutrition*, 1, 5–21

Figure 20.5 *Relationship between the proportion of energy from each food source and GNP per capita, with the proportion of the urban population at 25 per cent, 1990*

urban populations eat as both get wealthier:[24] both eat more meats and fats, and reduce carbohydrate, as a proportion of their overall diet. But there still remain differences between urban and rural populations, probably due to their different levels of activity, access to dietary ingredients and cultural mores.[25] The more urban population also consumes more added sugars as it gets wealthier, whereas the rural population consumes less. Popkin and his colleagues' point is that changing economic circumstances markedly shape the mix of nutrients in the diet and that lifestyle factors – such as the degree of urbanization[26] and changing labour patterns – have a major effect on health.

The transition is occurring in areas that usually receive little food policy attention. A study by the WHO has reported that in the Middle East changing diets and lifestyles are now resulting in changing patterns of both mortality and morbidity there too.[27] Dietary and health changes can be rapid. In Saudi Arabia, for instance, meat consumption doubled and fat consumption tripled between the mid-1970s and the early 1990s; in Jordan, there has, in the same timescale, been a sharp rise in deaths from cardiovascular disease. These problems compound older Middle-Eastern health problems such as protein-energy malnutrition, especially among children. In China, the national health profile began to follow a more Western pattern of diet-related disease as the population gradually urbanized,[28] coinciding with an increase in degenerative diseases. Consumption of legumes

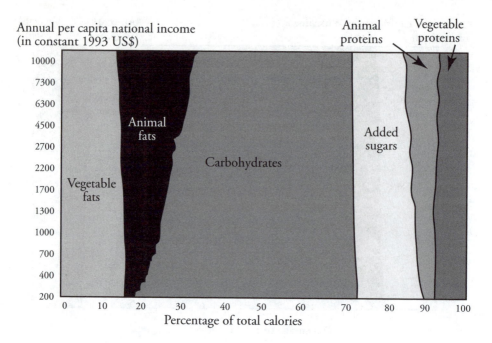

Source: FAO/World Bank/Popkin, B (1998) 'The nutrition transition and its health implications on lower income countries', *Public Health Nutrition*, 1, 5–21

Figure 20.6 *Relationship between the proportion of energy from each food source and GNP per capita, with the proportion of the urban population at 75 per cent, 1990*

such as soyabean was replaced by animal protein in the form of meat. One expert nutritional review of this problem concluded that exhorting the Chinese people to consume more soy when they were voting with their purses to eat more meat would be ineffective 'in the context of an increasingly free and global market'.[29] Such studies can suggest that the battle to prevent Western diseases in the developing world appears already to have been lost. If the nutrition transition is weakening health in China, the world's most populous and fastest economically growing nation, which has 22 per cent of the world's population but only 7 per cent of its land, what chance is there for diet-related health improvements throughout the developing world?

As populations become richer, they substitute cereal foods for higher-value protein foods such as milk, dairy products and meat, increased consumption of which is associated with Westernization of ill health. Relatively better-off populations also consume a greater number of non-staple foods and have a more varied, if not healthier, diet.[30] Thus we have the modern nutritional paradox: in the same low-income country there may be ill health caused by both malnutrition and over-nutrition; in the same rural area of a poor country both obesity and underweight can coexist.

In policy terms the challenge is whether India, China, Latin America or Africa, for example, can afford the technical fixes that the West can resort to in order to improve diet-related health:[31] coronary by-pass operations; continuous drug regimes; expensive

Table 20.3 *Types and effects of malnutrition*

Type of malnutrition	Nutritional effect	No. of people affected globally (× billion)
Hunger	deficiency of calories and protein	at least 1.2
Micronutrient deficiency	deficiency of vitamins and minerals	2.0–3.5
Over-consumption	excess of calories, often accompanied by deficiency of vitamins and minerals	at least 1.2–1.7

Source: Gardner and Halweil (2000), based on WHO, IFPRI, ACC/SCN data[33]

drugs and foods with presumed health-enhancing benefits;[32] and subscriptions to gyms and leisure centres. The affluent middle classes in the developed world might be able to afford such fixes but the vast numbers in the developing world certainly will not. Technical fixes are not societal solutions.

It could be argued that the increase in degenerative diseases is the inevitable downside of economic progress. The problem for policy-making is how to differentiate between protecting the already protective elements of traditional, indigenous diets such as legumes, fruit and vegetables, and opening up more varied food markets, which is deemed to be good economic policy. In practice, too few policy makers in the developing world have been prepared to fight to keep 'good' elements of national and local diets or to constrain the flow of Western-style foods and drinks into their countries lest they infringe support for trade liberalization. Thus, in stark terms, trade and economic policies have triumphed over health interests. US-style fast foods – the 'burgerization' of food cultures – have been hailed as modernity. We must now expose the production, marketing and prices of fast food,[34, 35] their nutritional value and their impact on health.[36]

Three categories of malnutrition: Underfed, overfed, and badly fed

More than 2 billion people in the world today have their lives blighted by nutritional inadequacy. On one hand, half of this number do not have enough to eat; on the other hand, a growing army of people exhibit the symptoms of overfeeding and obesity. In both cases, the international communities are floundering for solutions, and malnutrition results, as indicated by the following table.

Figure 20.7 highlights the role of the mother in infant health. Even before conception, the mother's own nutrition is vital.[37] It is now understood that children who are born with a low birthweight are at increased risk of developing heart disease and that good nourishment of the foetus is key. That nutrition affects disease patterns and life expectancy is now well documented.[38]

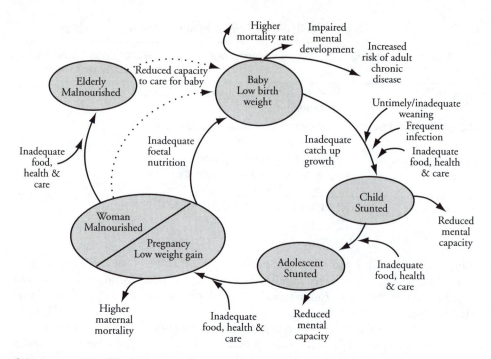

Source: ACC/SCN (2000) *Nutrition through the Life Cycle: 4th Report on the World Nutrition Situation.* UN Administrative Committee on Co-ordination Subcommittee on Nutrition, New York, p8

Figure 20.7 *Life cycle – the proposed causal links*

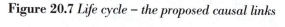

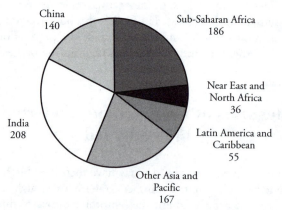

Source: State of Food Insecurity in the World 2000. Available at www.fao org/DOCREP/X8200E/x8200e03.htm#P0_0

Figure 20.8 *Number of undernourished by region, 1996–1998 (millions)*

One of the particularly tragic consequences of undernourishment is its impact on the world's children. UNICEF calculates that 800 million children worldwide suffer malnu

Table 20.4 *Projected trends in undernourishment by region, 1996–2030*

	1996–98	2015	2030	1996–98	2015	2030
	Per cent of population			Millions of people		
Sub-Saharan Africa	34	22	15	186	184	165
Near East/North Africa	10	8	6	36	38	35
Latin America and the Caribbean	11	7	5	55	45	32
China and India	16	7	3	348	195	98
Other Asia	19	10	5	166	114	70
Developing countries	18	10	6	791	576	400

Source: FAO (2000) *Agriculture: Towards 2015/30*, Technical interim Report, April, Rome, FAO, www.fao.org

trition at any given time (Figure 20.8 gives the FAO's estimated locations of these millions. Table 20.4 then gives the sobering projections for 2015 and 2030.) High proportions of Asian and African mothers are undernourished, largely due to seasonal food shortages, especially in Africa. About 243 million adults in developing countries are deemed to be severely undernourished.[39] This type of adult under-nutrition can impair work capacity and lower resistance to infection.

Against a rapid growth in world population, well-informed observers agree that greater food production is needed for the future.[40, 41, 42, 43] One estimate suggests that by 2020 there will be 1 billion young people growing up with impaired mental development due to poor nutrition. At a conservative estimate, this means there will be 40 million young people added to the total each year.[44]

The obesity epidemic

As early as 1948, there were medical international groups researching the incidence of obesity in various countries.[45] There were official reports at country level by the early 1980s,[46] and there has also been a commercial and consumer response to obesity for even longer.[47] But the grip of international obesity was in fact confirmed by the WHO's Task Force on Obesity in 1998. Today, overweight and obesity are key risk factors for chronic and non-communicable diseases.[48] In developing countries obesity is more common amongst people of higher socio-economic status and in those living in urban communities. In more affluent countries, it is associated with lower socio-economic status, especially amongst women and rural communities.[49] Historically and biologically, weight gain and fat storage have been indicators of health and prosperity. Only the rich could afford to get fat. By 2000, the WHO was expressing alarm that more than 300 million people were defined as obese, with 750 million overweight, ie pre-obese: over a billion people deemed overweight or obese globally.[50] But by 2003, this figure had been radically revised upwards when the International Association for the Study of Obesity (the IASO) calculated that up to 1.7 billion people were now overweight or obese. The new

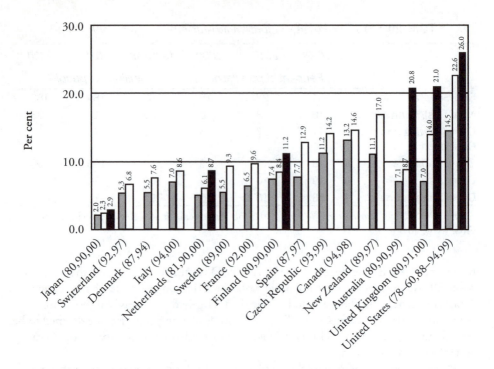

Source: OECD Health Data 2002, available at www.oecd.org/pdf/M00031000/M00031130.pdf

Figure 20.9 *Obesity in adult population across OECD countries*

figures were in part due to more accurate statistics but also to the recalculation of obesity benchmarks, which acknowledged rising obesity in Asia.[51]

Particularly worrying is that extreme degrees of obesity are rising even faster than the overall epidemic: in 2003, 6.3 per cent of US women, that is 1 in 16, were morbidly obese, with a body mass index of 40 or more.

Obesity is defined as an excessively high amount of body fat or adipose tissue in relation to lean body mass. Standards can be determined in several ways, notably by calculating population averages or by a mathematical formula known as 'body mass index' (BMI),[52] a simple index of weight-for-height: a person's weight (in kilos) divided by the square of the height in metres (kg/m^2). BMI provides, in the words of the WHO, 'the most useful, albeit crude, population-level measure of obesity'. A personal BMI of between 25 and 29.9 is considered overweight; 'obesity' means a BMI of 30 and above; a personal BMI of less than 17 is considered underweight. There is some argument about whether the definition of overweight (a BMI within the 25–29.9 range) should be lowered from 25 to 23, in which case tens of millions more people would be considered overweight, and such an unofficial re-classification has led to the disparity between current world obesity figures.

BMI levels are a useful predictor of risk from degenerative diseases. Unutilized food energy is stored as fat. Currently, the US National Institutes of Health consider that all adults (aged 18 years or older) who have a BMI of 25 or more are at risk from premature

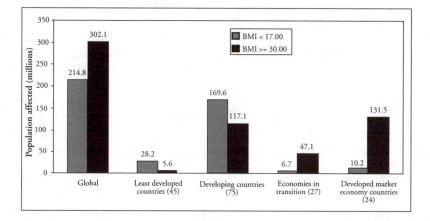

Source: WHO, *Nutrition for Health and Development: A Global Agenda for Combating Malnutrition.* 2000, available at www.who.inVnut/dbj3mLhtrn

Figure 20.10 *Global population affected by underweight and obesity in adults, by level of development, 2000*

death and disability as a consequence of overweight and obesity.[53] Men are at risk who have a waist measurement greater than 40 inches (102cm); women are at risk who have a waist measurement greater than 35 inches (88cm). Whilst height is obviously also taken into consideration, we should regard these measurements as key health benchmarks.

Figure 20.9 shows how, in a remarkably short time, the rate of obesity within countries is rising. In the UK, for instance, between 1980 and 2000, obesity trebled from 7 per cent of the population to 21 per cent.[54] Particularly alarming is that the 'North Americanization' of obesity is spreading down Latin America.[55] Figure 20.10 illustrates the level of obesity in comparison to underweight in developed and developing countries. In many countries, levels of obesity are double what they were 15 years ago.[56] In Peru, Tunisia, Colombia, Brazil, Costa Rica, Cuba, Chile, Mexico and Ghana, for example, overweight adults outnumber those who are thin. Even Ethiopia and India, traditionally beset by under-nutrition[57] and starvation now have the added burden of an emerging obesity problem. The trend to obesity is occurring in countries with different economic profiles, from the Asian 'Tiger' economies to the oil-rich Middle East.[58]

Rising obesity rates among children are particularly troubling to health professionals, as this trend suggests massive problems of degenerative disease for the future. In Jamaica and Chile, for instance, one in ten children is obese; in Japan, a country historically with a very low incidence of fat in its diet and with a low incidence of obesity, the frequency of obesity in school children has increased from 5 per cent to 9 per cent for girls and 10 per cent for boys in 1996.[59] (Table 20.5 summarizes the rapid rise in obesity as measured by comparing initial surveys with follow-up worldwide studies. The final column of the table shows how obesity is becoming out of control in developed and developing countries alike.) Even in Australia, obesity rose 3.4-fold for boys and 4.6-fold for girls between 1985 and 1995; in Egypt, 3.9-fold between 1978 and 1996; in Morocco, 2.5-fold in just five years, from 1987 to 1992; in Scotland by 2.3-fold for boys

Table 20.5 *Global increases in the prevalence of childhood obesity*[a]

Country	Index of measurement	Age of children	Date of first study (% obesity)	Date of second study (% obesity)	Growth of obesity incidence from first to second study
USA[b]	BMI=95th percentile	6–11 12–19	1971/74 (4%) 1971/74 (6%)	1999 (13%) 1999 (14%)	Up 3.3-fold Up 2.3-fold
England[b]	Age-adjusted BMI cut-off linked to adult value of 30kg/m²	4–11	1984 (0.6% boys; 1.3% girls)	1994 (1.7% boys; 2.6% girls)	Up 2.8-fold (boys) Up 2.0-fold (girls)
Scotland[c]	Age-adjusted BMI cut-off linked to adult value of 30kg/m²	4–11	1984 (0.9% boys; 1.8% girls)	1994 (2.1 % boys; 3.2% girls)	Up 2.3-fold (boys) Up 1.8-fold (girls)
China[d]	Age-adjusted BMI cut-off linked to adult value of 25kg/m²	6–9 10–18	1991 (10.5%) 1991 (4.5%)	1997(11.3%) 1997 (6.2%)	Up 1.1-fold Up 1.4-fold
Japan[e]	>120% of standard weight	10	1970 (<4% boys; =4% girls)	1996 (=10% boys; =9% girls)	Up 2.5 fold (boys) Up 2.3-fold (girls)
Egypt[f]	Weight-for height >2 SD from median	0–5	1978 (2.2%)	1996 (8.6%)	Up 3.9-fold
Australia[g]	Age-adjusted BMI cut-off linked to adult value of 30kg/m²	7–15	1985 (1.4% boys; 1.2% girls)	1995 (4.7% boys; 5.5% girls)	Up 3.4-fold (boys) Up 4.6-fold (girls)
Ghana[f]	Weight-for- height >2 SD from median	0–3	1988 (0.5%)	1993/94 (1.9%)	Up 3.8-fold
Morocco[f]	Weight-for- height >2 SD from median	0–5	1987 (2.7%)	1992 (6.8%)	Up 2.5-fold
Brazil[d]	Age-adjusted BMI cut-off linked to adult value of 25 kg/m²	6–9 10–18	1974 (4.9%) 1974 (3.7%)	1997 (174%) 1997 (2.6%)	Up 3.6-fold Up 3.4-fold
Chile[h]	Weight-for-height >2 SD from median	0–6	1985 (4.6%)	1995 (72%)	Up 1.6-fold
Costa Rica[f]	Weight-for-height >2 SD from median	0–6 (1982) 1–7 (1996)	1982 (2.3%)	1996 (6.2%)	Up 2.7-fold

Table 20.5 *Global increases in the prevalence of childhood obesity*[a]

Country	Index of measurement	Age of children	Date of first study (% obesity)	Date of second study (% obesity)	Growth of obesity incidence from first to second study
Haiti[6]	Weight-for-height >2 SD from median	0–5	1978 (0.8%)	1994/95 (2.8%)	Up 3.5-fold

Sources: a Ebbeling, C. B., Pawlak, D. B. and Ludwig, D. S. (2002) 'Childhood obesity: Public health crisis, common sense cure', *The Lancet*, vol 360, 10 August, pp473–482; b National Center for Health Statistics (1999) 'Prevalence of overweight among children and adolescents: United States', 1999–2000, available at www.cdc.gov/nchs/ products/pubs/pubd/hestats/overweight99. htm (accessed 29 January 2002); c Chinn, S. and Rona, R. J. (2001) 'Prevalence and trends in overweight and obesity in three cross-sectional studies of British children, 1974–94', *BMJ*, vol 322, pp24–26; d Wang, Y., Monteiro, C. and Popkin, B. M. (2002) 'Trends of obesity and underweight in older children and adolescents in the US, Brazil, China, and Russia', *American Journal of Clinical Nutrition*, vol 75, pp971–977; e Murata, M. (2000) 'Secular trends in growth and changes in eating patterns of Japanese children', *American Journal of Clinical Nutrition*, vol 72 (suppl), pp1379S–1383S; f deOnis, M. and Blossner, M. (2000) 'Prevalence and trends of overweight among preschool children in developing countries', *American Journal of Clinical Nutrition*, vol 72, pp1032–1039; g Magarey, A. M, Daniels, L. A. and Boulton,T. J. C. (2001) 'Prevalence of overweight and obesity in Australian children and adolescents: Reassessment of 1985 and 1995 data against new standard international definitions', *Medical Journal of Australia*, vol 174, pp561–564; h Filozof, C., Gonzalez, C., Sereday, M., Mazza, C. and Braguinsky, J. (2001) 'Obesity prevalence and trends in Latin American countries', *Obesity Review*, vol 2, pp99–106.

and 1.8-fold for girls between 1984 and 1994. A child's weight can be thrown off balance by a daily consumption of only one sugar-sweetened soft drink of 120kcals; over ten years, this intake would turn into 50kg of excess growth. Although their review also fully acknowledged the role of genetics, the authors pointed to pressures on children's diets from advertisements to help explain the rapidity of consumption and obesity changes.[60]

Health education seems to be powerless before this rising tide of obesity. On the island of Mauritius, for instance, a study which examined adults over a period of five years found that, despite a national programme promoting healthy eating and increased physical activity, obesity levels had increased dramatically:[61] men with a BMI above 25 increased from 26.1 per cent to 35.7 per cent and for women the figure grew from 37.9 per cent to 47.7 per cent during the five-year study. The government of Mauritius concluded that a National Nutrition Policy and National Plan of Action on Nutrition was needed.[62] Even in the US, the homeland of fast food, President George W Bush was so alarmed by the obesity crisis that in 2002 he launched a national debate. He has long had good reason for concern,[63] as even as far back as 1986, the economic costs of illness associated with overweight in the US were estimated to be $39 billion; today the estimated cost of obesity and overweight is about $117 billion.[64] The rise in US obesity is dramatic: between 1991 and 2001, adult obesity increased by 74 per cent. The percentage of US children and adolescents who are defined as overweight has more than doubled since the early 1970s, and about 13 per cent of children and adolescents are now seriously overweight.[65] These general US figures disguise marked differences between ethnic groups and income levels: according to the Centers for Disease Control and Prevention, 27 per cent of black and about 21 per cent of Hispanics of all ages are considered obese – that is, a third overweight – compared with a still worrying but lower 17 per cent among whites.[66] The poor are more obese than the more affluent within the US. The price of food is a key driver of obesity: saturated fats from dairy and meat and hydrogenated (trans) fats are relatively cheap.[67]

The connection between overweight and health risk is alarmingly highlighted by the following list of the physical ailments that an overweight population (with a BMI higher than 25) is at risk of:[68]

- high blood pressure, hypertension;
- high blood cholesterol, dyslipidemia;
- type-II (non-insulin-dependent) diabetes;
- insulin resistance, glucose intolerance;
- hyperinsulinaemia;
- coronary heart disease;
- angina pectoris;
- congestive heart failure;
- stroke;
- gallstones;
- cholescystitis and cholelithiasis;
- gout;
- osteoarthritis;
- obstructive sleep apnea and respiratory problems;
- some types of cancer (such as endometrial, breast, prostate and colon);
- complications of pregnancy;
- poor female reproductive health (such as menstrual irregularities, infertility and irregular ovulation);
- bladder control problems (such as stress incontinence);
- uric acid nephrolithiasis;
- psychological disorders (such as depression, eating disorders, distorted body image, and low self esteem).

There is a vocal position – particularly articulated in the US – arguing that the critique of obesity is an infringement of personal liberty and 'size-ist, making cultural value statements. If someone wants to be fat and is content and loved by others, goes this argument, what does it matter? The list of health problems given above is surely an answer to this position. The costs of what is presented as an 'individual' problem are, in fact, society wide. The ill-health that results is paid for either in direct costs or in a societal drag – lost opportunities, inequalities and lost efficiencies. This is why policy makers have to get to grips with obesity and the world's weight problem.

Both obesity and overweight are preventable. At present the debate about obesity is divided about which of three broad strategies of action is the best to address. One strand argues that it is a problem caused by over-consumption (diet and the types of food) and over-supply; another that it is lack of physical activity; and the third that there might be a matter of genetic predisposition. Certainly, the emphasis has to be on changing the environmental determinants that allow obesity to happen. A pioneering analysis by Australian researchers in the mid-1990s proposed that the obesity pandemic could only be explained in 'ecological' terms: Professors Garry Egger and Boyd Swinburn set out environmental determinants such as transport, pricing and supply; they claimed that environmental factors were so powerful in upsetting energy balances that obesity could be viewed as 'a normal response to an abnormal environment'.[69] So finely balanced are caloric intake and physical activity than even slight alterations in their levels can lead to weight gain. Swinburn and Egger assert that no amount of individual exhortation will reduce worldwide obesity;[70, 71] transport, neighbourhood layout, home environments, fiscal policies and other alterations of supply chains must be tackled instead.

Calculating the burden of diet-related disease

During the 1990s, world attention was given to calculating the costs of what has been called the 'burden of disease'. Five of the ten leading causes of death in the world's most economically advanced country, the US, were, by the 1980s, diet-related: coronary heart disease, some types of cancer, stroke, diabetes mellitus and atherosclerosis. Another three – cirrhosis of the liver, accidents and suicides – were associated with excessive alcohol intake.[72] Together these diseases were accounting for nearly 1.5 million of the 2.1 million annual deaths in the US. Only two categories in the top ten – chronic obstructive lung disease and pneumonia and influenza – had no food connection.

In a 1990s study published by the World Bank, 'The Global Burden of Disease',[73] the authors Murray and Lopez gave a detailed review of causes of mortality in eight regions of the world. Ischaemic heart disease accounted for 6.26 million deaths. Of these, 2.7 million were in established market economies and formerly socialist economies of Europe; 3.6 million were in developing countries (out of 50.5 million deaths from all causes in 1990). Stroke was the next most common cause of death (4.38 million deaths, almost 3 million in developing countries), closely followed by acute respiratory infections (4.3 million, 3.9 million in developing countries). Other leading causes of death include diarrhoeal disease (almost totally occurring in developing countries), chronic obstructive pulmonary disease, tuberculosis, measles, low birthweight, road-traffic accidents and lung cancer, with only diarrhoea and low birthweight having a diet-related aetiology. They also calculated that cancers caused about 6 million deaths in 1990. About 2.4 million cancer deaths occurred in established market economies and former socialist economies of Europe. By 1990, therefore, there were already 50 per cent more cancer deaths in less developed countries than in developed countries.

For their analysis, Murray and Lopez created a new index they called the DALY, standing for the 'disability adjusted life year'. A DALY is the sum of life years lost owing to premature death, and years lived with disability (adjusted for severity). It is thus a measure of both death and disability (both mortality and morbidity). The top ten DALYs in all developing regions combined already included ischaemic heart disease and cerebovascular disease. Murray and Lopez's report concluded: 'Clearly, the focus of research and debate about health policy in developing regions should address the current chal-

Table 20.6 *DALYs lost by cause for the developed and developing countries, 1990 and 2020*

Cause	Developed		Developing	
	1990 (%)	2020 (%)	1990 (%)	2020 (%)
Infectious diseases	7.8	4.3	48.7	22.2
Cardiovascular disease	20.4	22.0	8.3	13.8
Coronary heart disease	9.9	11.2	2.5	5.2
Stroke	5.9	6.2	2.4	4.2
Diabetes	1.9	1.5	0.7	0.7
Cancer	13.7	16.8	4.0	9.0
Neuropsychiatric disorders	22.0	21.8	9.0	13.7
Injuries	14.5	13.0	15.2	21.1

Source: Murray, C. J. L and Lopez, A. D. (1996) *The Global Burden of Disease: A Comprehensive Assessment of Mortality and Disability from Diseases. Injuries and Risk Factors in 1990 and Projected to 2020.* Harvard University Press on behalf of the World Bank and WHO, Cambridge, Massachusets

Table 20.7 *DALYs lost by selected causes for the EU and Australia around 1995*

Cause	EU %	Australia %
Smoking	9.0	9.5
Alcohol consumption	8.4	2.1
Diet and physical activity	8.3	16.4
Overweight	3.7	2.4
Low fruit and vegetable intake	3.5	2.7
High saturated fat intake	1.1	2.6
Physical inactivity	1.4	6.8

Sources: National Institute of Public Health, Stockholm (1997)[74, 75]

lenges presented by the epidemiological transition now, rather than several decades hence.' Table 20.6 gives their original breakdown for the world of the DALYs by main disease, present and anticipated.

The authors anticipated that the greatest increase in cardiovascular disease-related DALYs would occur in developing countries, up 8.3 per cent in 1990 to 13.8 per cent in 2020 – a rising burden of disease for those countries which could least afford it. The corresponding increase in developed countries' DALYs associated with non-communicable diseases was calculated to be only relatively slight, rising from 20.4 per cent to 22.0 per cent. (The developed world already had a high base rate of DALYs from diet-related disease.) (Interestingly, there is hardly any movement in diabetes figures for developing countries and a fall for developed countries, yet it should be noted that diabetes figures are in fact rising rapidly worldwide. The newness of this diet-related epidemic might have been too late for Murray and Lopez's 1990 data.)

One purpose of the DALY method is to enable policy makers to estimate the relative risk of major factors for health. Table 20.7 gives the Swedish National Institute of Public Health's summary of the calculated impacts of smoking, alcohol, diet and physical activity for key DALYs in the EU and Australia. Again, the diet-related disease toll is very high. Smoking, as was noted at the start of this chapter, is a major contributory factor in heart disease but the dietary factors, when separated, were almost as great.

The DALY approach was extended in the 'World Health Report 2002' which was based on a series of massive multi-country studies designed to test and refine the methodology. Figure 20.11 details risk factors by level of development. The results, however, merely deepened the insights from the earlier study. If anything, the burden of diet-related disease and of lack of physical activity received even higher profile. Special studies on the impact of lack of fruit and vegetables in the diet showed great impact. The WHO–FAO 2003 report underlined how a variety of diseases, from heart disease to diabetes, were all associated with the same dietary pattern: over-consumption, excess fat, under-consumption of fruit and vegetables and excess added sugar and salt.[76]

The financial costs

In 2001, the Commission on Macroeconomics and Health, created by the WHO, argued that there were mutual benefits to be had from improved health and for the economy, particularly for those in low-income countries.[77] Table 20.8 shows how general health care costs are rising rapidly in many developed economies; in the developing world, the costs of health care for degenerative diseases are now also looming as a serious concern. The growth of health expenditure is sometimes higher than the growth of gross domestic product (GDP). Table 20.9 gives a breakdown of the direct and indirect costs for a number of key diet-related diseases in the US; these costs are immense, even for such a rich society.

Health ministries, it appears, are locked in a model which tends to be curative rather than preventative. The UK health care system, for instance, costs £68 billion for around 60 million people, costs that are anticipated to rise to between £154 billion ($231 billion) and £184 billion ($276 billion) by 2022–2023 in 2002 prices.[79] In other words, at constant prices, UK health care costs are doubling.

In the context of diet-related disease, the direct and indirect financial tolls of ill health could offer opportunities for positive policy intervention through a health-enhancing food supply chain. An estimate for the UK by the Oxford University British Heart Foundation Health Promotion Research Group has calculated that coronary heart disease (CHD) – constituting about half of all cases of cardiovascular disease – costs the UK £10 billion per annum. These costs are made up of £1.6 billion in direct costs (primarily to the taxpayer through the costs of treatment by the NHS) and £8.4 billion in indirect costs to industry, and to society as a whole, through loss of productivity due to death and disability.[80] (This is probably an underestimate of the direct costs to the UK's National Health Service as these costs do not include the cancer treatment costs.)

A report chaired in 2002 by Derek Wanless, a former head of the NatWest Bank, for the Chancellor of the Exchequer produced not dissimilar calculations.[81] It estimated that costs for the health service will rise alarmingly if targets are not met to reduce CHD and cancers. CHD treatment costs (drugs like statins and surgical techniques like re-

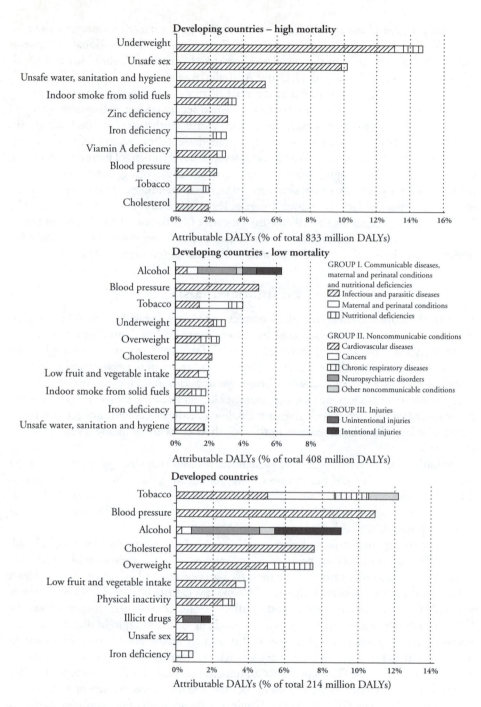

Source: The World Health Report 2002

Figure 20.11 *Burden of disease attributable to ten selected leading risk factors by level of development and type of affected outcome*

Table 20.8 *Growth of expenditure on health, 1990–2000*

	Real per capita growth rates, 1990–2000 (%)		Health spending as per cent of GDP		
	Health spending	GDP	1990	1998	2000
Australia	3.1	2.4	7.8	8.5	8.3
Austria	3.1	1.8	7.1	8.0	8.0
Belgium	3.5	1.8	7.4	8.5	8.7
Canada	1.8	1.7	9.0	9.1	9.1
Czech Republic	3.9	0.1	5.0	7.1	7.2
Denmark	1.7	1.9	8.5	8.4	8.3
Finland	0.1	1.8	79	6.9	6.6
France	2.3	1.4	8.6	9.3	9.5
Germany	2.2	0.2	8.7	10.6	10.6
Greece	2.8	1.9	7.5	8.7	8.3
Hungary[a]	2.0	2.7	71	6.9	6.8
Iceland	2.9	1.6	79	8.3	8.9
Ireland	6.6	6.4	6.6	6.8	6.7
Italy	1.4	1.4	8.0	7.7	8.1
Japan	3.9	1.1	5.9	7.1	7.8
Korea	7.4	5.1	4.8	5.1	5.9
Luxembourg[b]	3.7	4.5	6.1	5.8	6.0
Mexico	3.7	1.6	4.4	5.3	5.4
Netherlands	2.4	2.3	8.0	8.1	8.1
New Zealand	2.9	1.5	6.9	7.9	8.0
Norway	3.5	2.8	78	8.5	7.5
Poland[b]	4.8	3.5	5.3	6.4	6.2
Portugal	5.3	2.4	6.2	8.3	8.2
Slovak Republic	..	4.0	..	5.9	5.9
Spain	3.9	2.4	6.6	76	7.7
Switzerland	2.5	0.2	8.5	10.6	10.7
United Kingdom	3.8	1.9	6.0	6.8	7.3
United States	3.2	2.3	11.9	12.9	13.0
OECD Average[c,d]	3.3	2.2	72	8.0	8.0
EU Average	3.1	2.3	7.4	8.0	8.0

Source: OECD (2002) Health Data 2002, available at www.oecd.org/pdf/M00031000/ M0Q031130.pdf (p1). a Hungary: 1991–2000; b Luxembourg and Poland: 1990–1999; c OECD averages exclude the Slovak Republic because of missing 1990 estimates; d Unweighted averages.

Note: No recent estimates are available for Sweden and Turkey.

Table 20.8 *Economic costs of diet- and exercise-related health problems, US*

Disease	Direct costs US$ billion (medical expenditures)	Indirect costs US$ billion (productivity losses)	Total costs US$ billion
Heart disease	97.9	77.4	175.3
Stroke	28.3	15.0	43.3
Arthritis	20.9	62.9	83.8
Osteoporosis	n.a.	14.9	14.9
Breast cancer	8.3	7.8	16.1
Colon cancer	8.1	n.a.	8.1
Prostate cancer	5.9	n.a.	5.9
Gail bladder disease	6.7	0.6	7.3
Diabetes	45.0	55.0	100.0
Obesity	55.7	51.4	107.1
			Total = 561.8

Sources: National Institutes of Health (1998) and Wolf and Colditz (1998)[78]
Note: Costs are expressed in constant 1998 dollars, using the Consumer Price Index.

vascularization) would add an additional £2.4 billion per annum by 2010–11, doubling CHD expenditure. Such calculations remind us of the multi-headed nature of ill health. Smoking, diet, physical activity, genetics, environment and socio-economic background all have direct health outcomes. Wanless and his team were convinced by US scientific work that high cholesterol – 'which is mainly due to diet' – accounts for 43 per cent of CHD incidence, compared to 20 per cent for smoking. This sort of evidence shows that the poor diet has such far-reaching financial implications that it warrants higher political attention. This case was confirmed by a second Wanless study arguing for the economic value of facing the public health costs of poor diet, lifestyle and education.[82] However, for the last quarter of a century policy attention has been directed to cutting costs, not by altering the food supply chain, but by such policies as contracting out services and by privatization. In the UK, less than £5 million a year is spent on food-related health education. Meanwhile, drug companies and surgeons only offer expensive but highly sophisticated solutions when the patient is already sick.

Indeed, drug treatments can be hugely expensive. A trial on over 20,000 UK people with high risks for heart disease showed that giving patients a type of drug known as statins reduced the risk of a first coronary attack by 25 per cent but would cost £1 ($1.5 or 1.5) per patient per day.[83] Currently, 1.8 million people are prescribed statins, costing UK£750 million a year. Taking statins for three years can reduce the risk of a heart attack by up to a third.

Coronary heart disease (CHD)

Since 1999, the WHO has attributed 30 per cent of all annual global deaths – that is, of 15 million people – to cardiovascular disease.[84, 85] The majority of those deaths are in

Table 20.10 *Age-standardized deaths per 100,000 population from CHD selected countries, 1968–1996: men*

Men	1968	1978	1988	1998
Finland	718	664	477	340
UK	517	546	434	297
Austria	327	349	262	226
US	694	504	292	224
Australia	674	409	315	202
Canada	543	457	296	200
Italy[a]	230	249	172	150
Belgium[a]	345	313	184	147
Spain	99	165	146	125
France	152	154	118	92
Japan	92	74	52	58

Note: a latest statistics for 1994

low- and middle-income countries. In 1998, 86 per cent of DALYs were lost to cardio-vascular disease worldwide.

The main risk factors for heart disease are high blood pressure, smoking and lipid concentrations (cholesterol levels). Others include age, sex, family history and the presence of diabetes. WHO recommendations for reducing CVD include:[86]

- regular physical activity;
- linoleic acid;
- fish and fish oils;
- vegetables and fruits, including berries;
- potassium;
- low to moderate alcohol intake.

The WHO judges that there is convincing evidence for the increasing risks from:

- myristic and palmitic acid;
- trans-fatty acids;
- high sodium intake;
- overweight;
- high alcohol intake.

In regard to CHD, public health policy has tended to focus on two things: health education as prevention, and improved medical treatment through drug, hospital and surgical care. It has also urged behavioural change, in particular a reduction of total fat intake and especially of saturated fats (mainly from animal meat and dairy fats). This health promotion policy has had an effect: rates of heart disease are declining in most affluent

Table 20.11 *Age-standardized deaths per 100,000 population from CHD selected countries, 1968–1996: women*

Women	1968	1978	1988	1996
UK	175	182	156	107
Finland	204	177	141	93
US	273	185	119	92
Austria	120	119	84	81
Australia	268	186	117	73
Canada	198	155	100	72
Belgium[a]	111	100	61	46
Italy[a]	87	82	51	43
Spain	33	46	39	34
France	49	44	30	22
Japan	45	99	21	21

Source: British Heart Foundation from WHO country statistics
Note: a latest statistics for 1994

Western countries, after years of steady increase since the immediate post-World War II period (see Tables 20.10 and 20.11).[87]

The global picture is more complex, however.[88] For example, the steep rise in CHD in the newly independent countries of Eastern Europe (such as Belarus, Azerbaijan and Hungary) is worrying. Leaving the strictures of the Soviet era means only that already high rates of CHD have risen further. Even in countries considered to have a healthy diet, like Greece and Japan, social change is being accompanied by changing patterns of diet-related disease: Greece's CHD and obesity rates are rising as it changes to a more Northern European diet high in animal fats, following entry to the European Union and increased tourism. Death rates from CHD may have dropped in the US and Finland, but it should be remembered that their morbidity and costs are still high, as was shown by the Global Burden of Disease studies.

This complexity keeps epidemiologists busy around the world, but the rapidity of change should bring little surprise. In 1981 Trowell summarized the emergence of CHD amongst East Africans: in the 1930s, he reported, autopsies had shown zero CHD in East Africa, and only one case among 2994 autopsies conducted in Makere University Medical School over the period 1931–1946. However, by the 1960s CHD in this region was emerging as a major rather than peripheral health problem.[89]

In China, between 1991 and 1995,[90] CHD accounted for 15 per cent of all deaths. Cholesterol levels here, compared to those found in Western populations, were low but were increasing rapidly among the urban populations where a more affluent lifestyle was being adopted. Daily intake of meat, eggs and cooking oil had increased while intake of legumes and cereals had decreased. A reduction in the consumption of Western fast foods was also recommended as were increasing physical activity levels, an urging which could be applied to many urbanizing developing countries.

Food-related cancers

Since the 1980s, dietary factors have been thought to account for around 30 per cent of cancers in Western countries, making diet second only to tobacco as a preventable cause of cancer;[91] in developing countries diet accounted for around 20 per cent.[92] Table 20.12 gives the 1997 review of food-cancer research by the World Cancer Research Fund. An updated report is due out in 2006.

The annual WHO World Health Report has shown that cancers are increasing world-wide,[93] and the 2003 World Cancer Report suggested that, like obesity, rising cancer rates are preventable. By virtue of steadily ageing populations, cancer could further increase by 50 per cent to 15 million new cases a year by 2020. In 2000, 6.2 million people died of cancer worldwide (12.5 per cent of all deaths), but 22.4 million were living with cancer. In the South, cancers of the oesophagus, liver and cervix are more common, while in the North, there is a predominance of cancers of the lung, colon, pancreas and breast.

The most significant cause of death among men is lung cancer and among women breast cancer, but certain lifestyle changes, such as to diet or smoking habits, would alter these patterns. Some cancers are closely associated with diets centred on well-cooked red meats, animal proteins and saturated fats in large quantities, with a daily routine that takes in little physical activity.[94] Indeed, many cancers could be prevented by modifying dietary habits to include more fruits, vegetables, high-fibre cereals, fats and oils derived from vegetables, nuts, seeds and fish, and by limiting the intake of animal fats derived from meat, milk and dairy products.[95, 96] A number of published studies show that an increase in anti-oxidant nutrients such as beta-carotene, vitamins C and E, zinc and selenium could also decrease the risk of certain cancers and there seems to be strong evidence that eating a diet rich in fresh fruit and vegetables will reduce the risk of stomach cancer.[97] Yet the nutrition transition is being driven in a different direction – towards a diet actually higher in processed foods and animal fats, key food industries within the Productionist paradigm.

Diabetes

The incidence of Type II diabetes is, alarmingly, on the increase. This form of diabetes was formerly known as non-insulin-dependent diabetes mellitus, occurring when the body is unable to respond to the insulin produced by the pancreas; it accounts for around 90 per cent of cases worldwide. In Type I diabetes (formerly known as insulin-dependent), the pancreas fails to produce the insulin which is essential for survival; this form develops most frequently in children and adolescents, but is now being increasingly noted later in life.[98] It is anticipated that cases of Type II diabetes will rise coming years (see Table 20.13): the WHO anticipates a doubling in the number of cases from 150 million in 1997 to 300 million in 2025, with the greatest number of new cases being in China and India.[99]

Diabetes is the fourth main cause of death in most developed countries. Research demonstrates the association between excessive weight gain, central adiposity (fat around the waist) and the development of Type II diabetes. Diabetics are two to four times more likely to develop cardiovascular diseases than others, and a stroke is twice as common in people with diabetes and high blood pressure as it is for those with high blood pressure alone.

Table 20.12 *Cancers preventable by dietary means*

	Global ranking (incidence)	Global incidence (1000s)	Dietary factors (convincing or probable)	Non-dietary risk factors (established)	Preventable by diet			
					Low estimate High estimate (%)	High estimate Low estimate (%)	Low estimate (1000s)	High estimate (1000s)
Mouth And pharynx	5	575	Vegetables & fruits[a] alcohol[a]	Smoking[a] betel[a]	33	50	190	288
Nasopharynx			salted fish[b]	EBV[b]				
Larynx	14	190	Vegetables & fruits alcohol	Smoking	33	50	63	95
Oesophagus	8	480	Vegetables & fruits deficient diets alcohol	Smoking Barrett's oesophagus	50	75	240	360
Lung	1	1320	Vegetables & fruits	Smoking occupation	20	33	254	436
Stomach	2	1015	Vegetables & fruits refrigeration salt salted foods	*H. pylori*	66	75	670	761
Pancreas	13	200	Vegetables & fruits meat, animal fat	Smoking	33	50	66	100
Gallbladder	–	c	–					
Liver	6	540	Alcohol contaminated food	MBV and HCV	33	66	178	356
Colon, rectum	4	875	Vegetables	Smoking	66	75	578	656

Table 20.12 *Cancers preventable by dietary means*

	Global ranking (incidence)	Global incidence (1000s)	Dietary factors (convincing or probable)	Non-dietary risk factors (established)	Preventable by diet — Low estimate (%)	High estimate (%)	Low estimate (1000s)	High estimate (1000s)
			Physical activity, Meat, Alcohol	Genes, Ulcerative colitis, *S. sinensis*, NSAIDs				
Breast	3	910	Vegetables, Rapid early growth, Early menarche, Obesity, Alcohol	Reproductive genes, Radiation	33	50	300	455
Ovary	15	190	–	Genes	10	20	19	38
Endometrium	16	170	Obesity	Reproductive, OCs, Oestrogens	25	50	43	85
Cervix	7	525	Vegetables & fruits	Reproductive, HPV, Smoking	10	20	53	105
Prostate	9	400	Meat or meat fat or dairy fat		10	20	40	80
Thyroid	–	100[d]	Iodine deficiency	Radiation	10	20	10	20
Kidney	17	165	Obesity	Smoking	25	33	41	54

Table 20.12 *Cancers preventable by dietary means*

	Global ranking (incidence)	Global incidence (1000s)	Dietary factors (convincing or probable)	Non-dietary risk factors (established)	Preventable by diet			
					Low estimate (%)	High estimate (%)	Low estimate (1000s)	High estimate (1000s)
Bladder	11	310		Phenacetin Smoking Occupation S. haematobium	10	20	31	62
Other	2,355	– –			10	10	236	236
Total (1996)	10,320						3022	4187
							29.3%	40.6%

Source: Table 9.1.2 in World Cancer Research Fund (1997), *Food, Nutrition and the Prevention of Cancer: A Global Perspective*. Washington, DC: World Cancer Research Fund/American Institute for Cancer Research. Reproduced by permission.

Notes: Included as 'dietary factors' in this table are various foods, nutrients, alcoholic drinks, body weight and physical activity. The panel has estimated the extent to which specific cancers or cancer in general are preventable by the dietary and associated factors described in this report. The figures suggested are ranges consistent with current scientific knowledge, and take established non-dietary risk factors, notably the use of tobacco, specific infections and occupational exposures to carcinogens, into account. The arrows represent either decreasing risk or increasing risk (T).

Figures on global ranking and incidence: Parkin et al (1993); WHO (1997)

a mouth and pharynx; also chewing tobacco; b nasopharynx; c reliable worldwide data are not collected by IARC for this site; d conservative estimate based on the IARC (1993)

Table 20.13 *Prevalence of diabetes worldwide*

	2000	2030	Projected growth
Africa	7020,553	18,244,638	160
Mediterranean	15,189,760	43,483,842	186
Americas	33,014,823	66,828,417	102
European	33,380,754	48,411,977	45
SE Asia	45,810,544	122,023,693	166
Western Pacific	36,138,079	71,685,158	98
Total	171,000,000	366,000,000	114

Source: WHO (2004) *Diabetes Action Programme*. Geneva, World Health Organization, available at www.who. int/diabetes/facts/worid_figures/en/

In 2000, India recorded 32.7 million diabetics, China 22.6 million and the US 15.3 million, while Brazil recorded only 3.3 million and Italy 3.1 million. In 2000, the five countries with the highest diabetes prevalence in the adult population only were Papua New Guinea (15.5 per cent), Mauritius (15 per cent), Bahrain (14.8 per cent), Mexico (14.2 per cent) and Trinidad & Tobago (14.1 per cent).[100, 101] Such disparate statistics reflect a transition from traditional diet and from an activity-based lifestyle to a more sedentary one. By 2025, the prevalence of diabetes is anticipated to triple in Africa, the Eastern Mediterranean, the Middle East and South Asia. It is expected to double in the Americas and the Western Pacific and to almost double in Europe. In India, incidence is much higher in urban than rural populations:[102] in urban Chennai (Madras), for example, cases of diabetes rose by 40 per cent in 1988–1994. Incidence is rising among male urban dwellers of South India compared to the rural male population. In addition to Diabetes mellitus, the prevalence of non-insulin-dependent diabetes (NIDDM) increased dramatically within the urban populations of India within just a decade.[103] In Thailand, also, NIDDM is more pronounced amongst females in the urban population than it is in the rural population,[104] whilst in the rural environment, incidence of NIDDM amongst males is higher.

In the UK, Professor David Barker and colleagues have shown that adult diabetes is associated with low birthweight,[105] while studies in India suggest that poor interuterine growth, combined with obesity later in life is associated with insulin resistance, diabetes and increased cardiovascular risk.[106] Once again, a single disease seems attributable to a pattern of poor nutrition related to the lifecycle, and is one whose costs are externalized onto society as a whole and health care in particular. Devastating complications of diabetes, such as blindness, kidney failure and heart disease, are imposing a huge financial burden: in some countries 5–10 per cent of national health budgets.

Food safety and foodborne diseases

Whilst attention to such non-communicable diseases is of vital importance, food safety, foodborne diseases and other communicable diseases remain uppermost within food and public health policy, partly due to consumer campaigns about risks and to heightened media awareness of poor food processing standards. Food safety problems include risks from:[107]

- veterinary drug and pesticide residues;
- food additives;
- pathogens (ie illness-causing bacteria, viruses, parasites, fungi and their toxins);
- environmental toxins such as heavy metals (eg lead and mercury);
- persistent organic pollutants such as dioxins;
- unconventional agents such as prions associated with BSE.

In particular, companies have had to respond to new public awareness about food safety issues, and new regimes of traceability have been implemented to enable companies to track food ingredients in order to eliminate subsequent legal or insurance liability consequences. In this respect, food companies are anxious to present themselves as guardians of the public health.[108] The attention food safety receives is predictably higher in affluent countries when, on the evidence, the burden of ill health is far greater in the developing world, due to lack of investment and infrastructure, including drains, housing, water supplies and food control systems. The World Health Report 2002 pointed out that, in developing countries, water supply and general sanitation remain the fourth highest health-risk factor, after underweight, unsafe sex and blood pressure.[109] In developing countries which are building their food export markets, there is too often a bipolar structure, with higher standards for foods for export to affluent countries than for domestic markets. There ought to be a cascading down into internal markets of these higher standards.[110]

Environmental risks to health are a significant problem on the global scale and, in Western countries in the 1990s, new strains of deadly bacteria such as *E. coli 0157* captured policy attention, an estimated 30 per cent of people having suffered a bout of foodborne disease annually. The US, for instance, reports an annual 76 million cases, resulting in 325,000 hospitalizations and 5000 deaths.[111] The WHO estimates that 2.1 million children die every year from the diarrhoeal diseases caused by contaminated water and food,[112, 113] asserting that each year worldwide there are 'thousands of millions' of cases of foodborne disease.[114]

In early industrializing countries, a grand era of engineering made dramatic health improvements in public health. Part of that investment included the introduction of effective monitoring and hygiene practice systems, such as the establishment of local authority laboratories and training, the packaging of foods and processes such as milk pasteurization. Today, public health proponents are actively trying to promote a 'second wave' of food safety intervention but this time using a risk-reduction management system known as Hazards Analysis Critical Control Point (HACCP), an approach designed to build safety awareness and control of potential points of hygiene breakdown into food handling and management systems. HACCP also encourages the creation of a 'paper' trail to enable tracking along the production process, essential in order to obviate errors and enable learning. Breakdowns in food safety have in the past led to major political and business crises, with governments under attack and new bodies responsible for food safety being set up in many countries. As food supply chains become more complex and as the scale of production, distribution and mass catering increases, so the chances for problems associated with food contamination rise; mass production breakdowns in food safety spread contamination and pathogens widely. An outbreak of *Salmonellosis* in the US in 1994, for example, affected an estimated 224,000 people.[115] *Listeria monocytogenes* has a fatality rate of 30 per cent, a fact that seriously dented UK public confidence in the 'cook-chill' and 'oven-ready' foods of the late 1980s.

Cross-border trade in agricultural and food products, as well as international pacts have brought food safety to the fore.[116] The Director-General of the WHO, in a speech on food safety to the UN Codex Alimentarius Commission, said: 'globalisation of the world's food supply also means globalisation of public health concerns'.[117] Crises over BSE, *Salmonellosis* and *E. coli.* for example, had had a significant political impact throughout both the UK and EU, for instance,[118] and many countries have experienced a fast rise in incidences of *Salmonellosis* and *Campylobacter* infections since the 1980s, both bacteria being associated with meat and meat products. Despite countries such as Denmark and Sweden having strict policies governing the extermination of flocks and herds found to be carrying *Salmonella,* the incidence continues through the contamination of feedstuffs, and in Denmark in 1998 the percentage of positive flocks with *Campylobacter* was 47.1 per cent.

Thus, in many developed countries with good monitoring systems, the incidence of foodborne disease has in fact risen during the era of the Productionist paradigm: in West Germany cases of infectious *S. enteritis* rose from 11 per 100,000 head of population in 1963 to 193 per 100,000 in 1990;[119] in England and Wales formal notifications of the same disease rose from 14,253 cases in 1982 to 86,528 in 2000. These cases resulted in millions of days lost from work but, fortunately, relatively few deaths.

Bacteria fill gaps left by nature, evolving new strains; but they are constantly evolving even as science combats existing strains. The new food processes and systems of distribution ushered in by the food technology revolution of the second half of the 20th century provided many opportunities for bacteria to develop and colonize new niches. The incidence of *Salmonella* in the UK, for example, first rose, and then, following good monitoring, hygiene intervention and political pressure, fell right back – in two decades.

Table 20.14 gives a list from the WHO of some of the pathogenic organisms that are associated with food and food hygiene: viruses, bacteria, trematodes (flukeworms), cestodes (tapeworms) and nematodes (roundworms), the last three all small worms that can be found either in soil, fish or meats. The first two are concerns in global food trade particularly. In the case of bacteria such as *Listeria monocytogenes*, only 657 cases were reported throughout the European Union in 1998;[120] in the same period, deaths from cardiovascular disease in the EU totalled 1.5 million, 42 per cent of all deaths,[121] while, in 1990, diarrhoeal diseases accounted for 11,000 years of life (DALYs) lost out of a total of 22.7 million in Europe; in the same year, cardiovascular disease accounted for 7 million, diabetes for 371,000 and cancer of the colon and rectum for 593,000,[122] and five times as many years of life were lost due to drug addiction than to diarrhoeal diseases.

Despite a low health burden in the developed world, the financial costs of food poisoning can be significant. Estimates in the US suggest that the diseases caused by major pathogens cost up to $35 billion each year in medical costs and lost productivity.[123] Policy makers must be concerned about both foodborne illness and degenerative diseases, the latter of which do not as yet receive sufficient political attention.

Table 20.14 *Some pathogenic organisms associated with public health, which may be transmitted through food*

Bacteria	Protozoa
Bacillus cereus	Cryptosporidium spp
Brucella spp	Entamoeba histolyuca
Campylobacter jejunl and coli	Giardia lamblia
Clostridium botulinum	Toxoplasma gondii
Clostridium perfringens	
Escherichia coli	
(pathogenic strains)	**Trematodes** (flukeworms)
Listeria monocytogenes	Fascioia hepatica
Mycobacterium bovis	Opistorchis felineus
Salmonella typhi and paratyphi	
Salmonella (non-typhi) spp	**Cestodes** (tapeworms)
Shigella spp	Diphyllobotrium latum
Staphylococcus aureus	Echinococcus spp
Vibrio cholerae	Taenia solium and saginata
Vibrio parahaemoiyticus	
Vibrio fulnificus	**Nematodes** (roundworms)
Yersinia enterocolitica	Anisakis spp
	Ascaris lumbricoides
Viruses	Trichinella spiralis
Hepatitis A	Trichuris trichiura
Norwalk agents	
Poliovirus	
Rotavirus	

Source: WHO European Centre for Health and Environment, Rome, 2000

Food poverty in the Western world

Most public health concern about food poverty rightly centres on the developing world, but it is also important to recognize that the impact of food poverty is significant in the developed world. The new era of globalization has unleashed a reconfiguration of social divisions both between and within countries; these social divisions are particularly marked in societies such as the UK and the US which have pursued neo-liberal economic policies. Indeed, one review of EU food and health policies estimated that food poverty was far higher in the UK than any other EU country,[124] where inequalities in income and health widened under the Conservative government of 1979–1997. The proportion of people earning less than half the average income grew[125] and the bottom tenth of society experienced a real, not just relative, decline in income and an increase in social health distinctions. This was a the converse of the post-World War II years of Keynesian social democratic policies during which inequalities narrowed: lower UK socio-economic groups now experience a greater incidence of premature and low birthweight babies, and of heart disease, stroke and some cancers in adults. Risk factors such as bottle-feeding, smoking, physical inactivity, obesity, hypertension, and poor diet were clustered in the lower socio-economic groups[126] whose diet traditionally derives from cheap energy forms such as meat products, full-cream milk, fats, sugars, preserves, potatoes and cereals with little reliance on vegetables, fruit, and wholemeal bread. Essential nutrients such as calcium, iron, magnesium, folate and vitamin C are more likely to be ingested by the higher socio-economic groups:[127, 128] their greater purchasing power creates a market for healthier foods such as skimmed milk, wholemeal bread, fruit and other low-fat options. Similarly, in the US, hunger has been a persistent cause of concern for decades and rising during the 1990s when the Census Bureau calculated that 11 million Americans lived in households which were 'food insecure' with a further 23 million living in households which were 'food insecure without hunger' (in other words at risk of hunger).[129] Other US surveys of the time estimated that at least 4 million children aged under 12 were hungry and an additional 9.6 million were at risk of hunger during at least one month of the year. Despite political criticisms of these surveys, further research suggested that even self-reported hunger, at least by adults, is a valid indication of low intakes of required nutrients. It should be noted that, ironically, the US spent over US$25 billion on federal and state programmes to provide extra food for its 25 million citizens in need of nutritional support.[130]

Implications for policy

This chapter has sketched the bare bones of a highly complex global picture of diet-related health. Over the last half-century, epidemiologists have generated many facts, figures and arguments about the role of food in the creation and prevention of ill health, linking what humans eat with their patterns of disease. They raise a number of important questions: how much of a risk does poor diet pose? What proportion of the known incidence of key diseases like cancer, heart disease, diabetes and microbiological poisoning can be attributed to the food supply? What levels of certainty can be applied to the many studies that have been produced? Is diet a bigger factor than, say, tobacco or genetics? For policy makers, the uncomfortable fact is that the pattern of diet-related diseases summarized in this chapter appears to be closely associated with the Productionist paradigm. Whilst the paradigm had as its objective the need to produce enough to feed people, its harvest of ill health was mainly sown in the name of economic development. Yet the public health message is clear: if diet is inappropriate or inadequate, population ill health will follow. Diet is one of the most alterable factors in human health, but despite strong evidence for intervention, public policy has only implemented lesser measures such as labelling and health education while the supply chain remains legitimized to produce the ingredients of heart disease, cancer, obesity and their diet-related degenerative diseases.

In making these tough assertions, we are aware that to piece together all food research evidence is immensely complex: more research is always needed; scientific understanding inevitably advances and is refined along the way. But surely, there is enough evidence for action. Certainly there is no shortage of reports and studies with which to inform policy. Calling for more research ought not to be an excuse for policy inaction. Policy procrastination is merely poor political prioritization.

Policy attention needs to shift from the overwhelming focus, enshrined in the Productionist paradigm, on under-consumption and under-supply to a new focus on the relationship between the over-supply of certain foodstuffs, excessive marketing and mal-consumption, and do so simultaneously within and between countries. Historically, there has been too much focus on public education as the main driver of health delivery; the diet and health messages, while welcome, have not always had the widespread or long-lasting effect that current data suggests is needed. While there have been reductions, for example, in coronary heart disease mortality rates in affluent societies, this is not universally true, and health education as framed in the West may not be universally appropriate. The food supply chain itself must be re-framed and must target wider, more health-appropriate goals.

Even rich countries are struggling to provide and fund equitable solutions to problems caused by diet: drugs and surgery, designer health foods, scientific research and public health education. But for developing countries, the majority of humanity, who have even fewer resources and weaker health care infrastructure, the picture is even more desperate. At the heart of the food policy challenge is the need to reinforce the notion of entitlement to food. While the 1948 Universal Declaration of Human Rights asserted the right to food for health for all, even into the new millennium the call is still not being adequately met, and, for humanity's sake, it must now be pursued with more vigour.

Notes

1 Cited in Hill, C. (1975) *The World Turned Upside Down: Radical Ideas During the English Revolution*, Penguin, Harmondsworth, p391
2 WHO & FAO (2003) *Diet, Nutrition and the Prevention of Chronic Diseases. Technical Report Series 916*, World Health Organization, Geneva; Food and Agriculture Organization, Rome
3 WHO (2003) *Draft Global Strategy on Diet, Physical Activity and Health*, World Health Organization, Geneva. Available at www.who.int/hpr/gs.strategy.document
4 WHO (2002) *World Health Report 2002*, World Health Organization, Geneva
5 WHO (2003) *Word Cancer Report*, World Health Organization/International Agency for Research on Cancer, Geneva
6 IOTF (2003) *Call for International Obesity Review as Overweight Numbers Reach 1.7 Billion*, press release, International Obesity Task Force/International Association for the Study of Obesity, London
7 Ewin, J. (2001) *Fine Wines and Fish Oils: The Life of Hugh Macdonald Sinclair*, Oxford University Press, Oxford

8   WHO (1998) *Obesity: Preventing and Managing the Global Epidemic: Report of a WHO Consultation on Obesity*, World Health Organization, Geneva

9   Commission on the Nutrition Challenges of the twenty-first Century (2000). *Ending Malnutrition by 2020: An Agenda for Change in the Millennium.* Final Report to the ACC/SCN, *Food and Nutrition Bulletin*, vol 21, no 3, September. United Nations University Press, New York, p19ff

10   Alexandratos, N. (ed) (1995) *World Agriculture: Towards 2010: a FAO study*, John Wiley & Son, Chichester

11   Burinsma, J. (ed) (2003) *World Agriculture: Towards 2015/2030*, Food and Agriculture Organization, Rome; Earthscan, London, p5

12   Murray, C. J. L. and Lopez, A, D. (1996) *Global Burden of Disease*, World Health Organization, Geneva

13   Popkin, B. M. (1999) 'Urbanization, lifestyle changes and the nutrition transition', *World Development*, vol 27, no 11, pp1905–1916

14   Cabellero, B. and Popkin, B. (eds) (2002) *The Nutrition Transition*, Elsevier, New York

15   Popkin, B. M. (2001) 'An overview on the nutrition transition and its health implication: The Bellagio meeting', *Public Health Nutrition*, vol 5 (1A), pp93–103

16   Popkin, B. M. (1994) 'The nutrition transition in low-income countries: An emerging crisis', *Nutrition Reviews*, vol 52, pp285–298

17   Drewnoski, A. and Popkin, B. (1997) 'The nutrition transition: New trends in the Globalisation', *Nutrition Reviews*, vol 155, pp31–43

18   Pena, M. and Bacallao, J. (eds) (2000) *Obesity and Poverty: A New Public Health Challenge*, Pan American Health Organization, (WHO), Washington, DC

19   Popkin, B. M. (1997) 'The nutrition transition in new income countries: An emerging crisis', *Nutrition Reviews*, vol 52, no 19, pp285–298

20   WHO (2002) *The World Health Report 2002: Reducing Risks, Promoting Healthy Life*, WHO, Geneva

21   Lenfant, C. (2001) 'Can we prevent cardiovascular diseases in low- and middle-income countries?' *Bulletin of the World Health Organization*, vol 9, no 10, pp980–982

22   Robson, J. (1981) 'Foreword' in Trowell, H. and Burkitt, D. (eds) *Western Diseases: Their Emergence and Prevention*, Edward Arnold, London

23   Trowell, H. and Burkitt, D. (eds) (1981) *Western Diseases: Their Emergence and Prevention*, Edward Arnold, London

24   Popkin, B. M. (1998) 'The nutrition transition and its health implications in 'lower income Countries', *Public Health Nutrition*, vol 1, pp5–21

25   Cabellero, B. and Popkin, B. (eds) (2002) *The Nutrition Transition.* Elsevier, New York

26   IFPRI (2002) *Living in the City: Challenges and Options for the Urban Poor*, International Food Policy Research Institute, Washington, DC

27   Verster, A. (1996) 'Nutrition in transition: The case of the Eastern Mediterranean Region' in Pietinen, P., Nishlda, C. and Khaltaev, N. (eds) *Nutrition and Quality of Life: Health Issues for the 21st century*, World Health Organization, Geneva, pp57–65

28   Chen, J., Campbell, T. C., Li, J. and Peto, R. (1990) *Diet, Lifestyle and Mortality in China: Study of the Characteristics of 65 Counties*, Oxford University Press, Oxford

29   Geissler, C. (1999) 'China; the soybean-pork dilemma', *Proceedings of the Nutrition Society*, vol 58, pp345–353

30 Dowler, E. and Pryer, J. (1998) 'Relationship of diet and nutritional status', in *Encyclopaedia of Nutrition*, Academic Press, New York

31 Lang, T. (1997) 'The public health impact of globalisation of food trade', in Shetty, P. and McPherson, K. (eds) *Diet. Nutrition and Chronic Disease: Lessons from Contrasting Worlds*, John Wiley and Sons, Chichester

32 Pan Lang, T. (2001) 'Trade, public health and food', in McKee, M., Garner, P. and Stott, R. (eds) *International Co-operation in Health*, Oxford University Press, Oxford

33 Heasman, M. and Mellentin, J. (1999) 'Responding to the functional food revolution', *Consumer Policy Review*, vol 19, pp152–159

34 Schlosser, E. (2001) *Fast Food Nation: The Dark Side of the All-American Meal*, HarperCollins, New York

35 Vidal, J. (1997) *McLibel: Burger Culture on Trial*, Macmillan, London

36 Gardner, G. and Harwell, B. (2000) 'Underfed and overfed: The global epidemic of malnutrition', *Worldwatch paper* 150, Worldwatch Institute, Washington, DC

37 Barker, D. J. P. (ed) (1992) *Fetal and Infant Origins of Adult Disease*, British Medical Journal, London

38 Barker, D. J. P. (2001) 'Cutting edge', *THES*, 1 June, p22 Commission on the Nutrition Challenges of the 21st Century (2000) 'Ending malnutrition by 2020: An agenda for change in the millennium. Final report to the ACC/SCN', *Food and Nutrition Bulletin*, vol 21, p3, Supplement, September. United Nations University Press, New York, p19

39 Pinstrup-Anderson, P. (2001) *Achieving Sustainable Food Security for All: Required Policy Action*, International Food Policy Research Institute, Washington, DC

40 See the Projections in FAO (2000) *The State of Food Insecurity in the World*, Food and Agriculture Organization, Rome, pv, 6

41 Smil, V. (2000) *Feeding the World: A Challenge for the 21st Century*, MIT Press, Cambridge, Massachusetts

42 Dyson, T. (1996) *Population and Food: Global Trends and Future Prospects*, Routledge, London

43 ACC/SCN (2000) *Nutrition through the Life Cycle: 4th Report on The World Nutrition Situation*, United Nations Administrative Committee on Co-ordination Sub-Committee on Nutrition (ACC/SCN), New York, p8

44 International Association for the Study of Obesity (2003) *Obesity Newsletter*, October, IASO, London

45 Royal College of Physicians of London (1983) 'Obesity. A report of the Royal College of Physicians', *Journal of the Royal College of Physicians of London*, vol 17, no 1, pp5–65

46 Stearns, P. (1997) *Fat History: Bodies and Beauty in the Modern West*, New York University Press, New York

47 WHO (1998) *Obesity: Preventing and Managing the Global Epidemic: Report of a WHO Consultation on Obesity*, World Health Organization WHO/NUT/NCD/98.1, Geneva, pi

48 Pena, M. and Bacallao, J. (eds) (2000) *Obesity and Poverty: A New Public Health Challenge*, Pan American Health Organization (WHO), Washington, DC

49 WHO (2000) *Nutrition for Health and Development: A Global Agenda For Combating Malnutrition*, World Health Organization, Geneva. Available at www.who.int/nut/dbjDmi.htm

50 IOTF (2003) 'Call for international obesity review as overweight numbers reach 1.7 billion', press release. International Obesity Task Force/International Association for the Study of Obesity, London

51 Centre for Disease Control (2002), www.cdc.gov/nccdphp/dnpa/obesity/basics.htm

52 National Institutes of Health (1998) *Clinical Guidelines on the Identification. Evaluation. and Treatment of Overweight and Obesity in Adults*, Department of Health and Human Services, Bethesda, Maryland; National Institutes of Health; National Heart, Lung and Blood Institute www.nhlbi.nih.gov/guidelines/obesity/ob_home.htm

53 OECD Health data, www.oecd.org/pdf/M00031000/M00031130.pdfpg. 5

54 Pena, M. and Bacallao, J. (eds) (2000) *Obesity and Poverty: A New Public Health Challenge*, PAHO Scientific Publication, Washington, DC, no 576

55 WHO (1998) *Obesity: Preventing and Managing the Global Epidemic: Report of a WHO Consultation on Obesity*, World Health Organization WHO/NUT/NCD/98.1, Geneva

56 Vepa, S. et al (2002) *Food Insecurity Atlas of India*, MS Swaminathan Research Foundation and UN World Food Programme, Chennai

57 Verster, A. (1996) 'Nutrition in transition: The case of the Eastern Mediterranean Region' in Pietinen, P., Nishida, C. and Khaltaev, N. (eds) *Nutrition and Quality of Life: Health Issues for the 21st century*, World Health Organization, Geneva, pp57–65

58 Murata, M. (2000) 'Secular trends in growth and changes in eating patterns of Japanese children', *American Journal of Clinical Nutrition*, vol 72 (suppl), pp1379S–1383S

59 Ebbeling, C. B., Pawlak, D. B. and Ludwig, D. S. (2002) 'Childhood obesity: Public health crisis, common sense cure', *The Lancet*, vol 360, 10 August, pp437–482

60 Hodge, A. M., Dowse, G. K., Gareeboo, H. and Tuomilehto, J. et al (1996) 'Incidence, increasing prevalence and predictors of change in obesity and fat distribution over 5 years in the rapidly developing population of Mauritius', *International Journal of Obesity & Related Metabolic Disorders*, vol 20, no 2, pp137–146

61 Chitson, P. (1995) 'Integrated intervention programmes for combating diet-related chronic diseases' in Pietinen, P., Nishida, C. and Khaltaev, N. (eds) Proceedings of the 2nd WHO Symposium on Health Issues for the 21st Century: Nutrition and Quality of Life. Kobe, Japan, 24–26 November 1993, World Health Organization, Geneva, pp269–287

62 Kuczmarski, R. et al (1994) 'Increasing prevalence of overweight among US adults', *JAMA*, vol 272, no 3, pp205–211

63 CDC (2002) *Physical Activity and Good Nutrition: Essential Elements to Prevent Chronic Diseases and Obesity 2002*, Department of Health and Human Services, Centers for Disease Control and Prevention, Atlanta

64 Centre for Disease Control (2002) www.cdc.gov/nccdphp/dnpa/obesity/basics.htm

65 Barboza, D. (2000) 'Rampant obesity, a debilitating reality for the urban poor', *New York Times*, 26 December, D5

66 Nestle, M. (2002) *Food Politics*, University of California Press, Berkeley, California

67 Stunkard, A. J. and Wadden, T. A. (eds) (1993) *Obesity: Theory and Therapy*, Raven Press, New York

68 Egger, G. and Swinburn, B. (1997) 'An 'ecological' approach to the obesity pandemic', *British Medical Journal*, vol 315, pp477–480

69 Swinburn, B., Egger, G. and Raza, F. (1999) 'Dissecting obesogenic environments: The development and application of a framework for identifying and prioritizing environmental interventions for obesity', *Preventive Medicine*, vol 29, pp563–570

70 Swinburn, B. and Egger, G. (2002) 'Preventive strategies against weight gain and obesity', *Obesity Reviews*, vol 13, pp289–301

71 US Department of Health and Human Services (1988) *The Surgeon-General's Report on Nutrition and Health*, Report 88–50210, DHHS, Washington, DC, pp2–6

72 Murray, C. J. L. and Lopez, A. D. (1997) 'Mortality by cause for eight regions of the world: Global burden of disease study', *The Lancet*, vol 349, 3 May, pp1269–1276, 1347–1352, 1436–1442, 1498–1504

73 Murray, C. J. L. and Lopez, A. D. (eds) (1996) *The Global Burden of Disease: A Comprehensive Assessment of Mortality and Disability from Diseases. Injuries and Risk Factors in 1990 and Projected to 2020*, Harvard School of Public Health on behalf of the World Health Organization and the World Bank, Cambridge, Massachusetts

74 National Institute of Public Health, Stockholm (1997) *Determinants of the Burden of Disease in the European Union*, NIPH, Stockholm; Mathers, E., Vos, T. and Stevenson, C. (1999) *The Burden of Disease and Injury in Australia*, AIHW, Canberra, 1999 (Catalogue No PHE-17)

75 WHO and FAO (2003) *Diet, Nutrition and the Prevention of Chronic Diseases. Technical Report Series 916*, World Health Organization; Rome, Food and Agriculture Organization, Geneva

76 Commission on Macroeconomics and Health (2001) *Macroeconomics and Health; Investing in Health for Economic Development*, World Health Organization, Geneva

77 Cited in Kenkel, D. S. and Manning, W. (1999) 'Economic evaluation of nutrition policy or there's no such thing as a free lunch', *Food Policy*, vol 24, pp145–162

78 Wanless, D. (2002) *Securing Our Future Health: Taking a Long-Term View. Final Report*, H M Treasury, London, April

79 Maniadakis, N. and Rayner, M. (1998) *Coronary Heart Disease Statistics: Economics Supplement*, British Heart Foundation, London, www.heartstats.org

80 Wanless, D. (2002) *Securing Our Future Health: Taking a Long-Term View. Final Report*, H M Treasury, London, April

81 Wanless, D. (2004) *Securing Good Health for the Whole Population, Final Report*, H M Treasury, London, 25 February

82 Heart Protection Study Collaborative Group (2002) 'MRC/BHF Heart Protection Study of cholesterol lowering with simvastatin in 20,536 high-risk individuals: A randomised placebo-controlled trial', *The Lancet*, vol 360, pp7–22

83 WHO (1999) *World Health Report 1999*, World Health Organization, Geneva

84 WHO (2002) *World Health Report 2002*, World Health Organization, Geneva

85 WHO (2003) *Diet. Nutrition and the Prevention of Chronic Diseases*, Technical Series 916. World Health Organization Geneva

86 Dobson, A. J, Evans, A., Ferrario, M., Kuulasmaa, K. A. et al (1998) 'Changes in estimated coronary risk in the 1980s: Data from 38 populations in the WHO MONICA Project. World Health Organization. Monitoring trends and determinants in cardiovascular diseases', *Annals of Medicine*, vol 30, no 2, pp199–205

87 Rayner, M. (2000) 'Impact of nutrition on health', in Sussex, J. (ed) *Improving Population Health in Industrialised Nations*, Office of Health Economics, London, pp24–40

88 Trowell, H. (1981) 'Hypertension, obesity, diabetes mellitus and coronary heart disease', in Trowell, H. and Burkitt, D. (eds) *Western Diseases: Their Emergence and Prevention*, Edward Arnold, London

89 Zhou, B. (1998) 'Diet and cardiovascular disease in China in diet', in Shetty, P. and Gopalan, C. (eds) *Nutrition and Chronic Disease: An Asian Perspective*, Smith-Gordon, London

90 Doll, R. and Peto, R. (1981) 'The causes of cancer: Quantitative estimates of avoidable risks of cancer in the United States today', *Journal of the National Cancer Institute*, vol 66, pp1191–1308

91 Miller, A. B. (2001) *Diet in cancer prevention*, World Health Organization, Geneva, available at www.who.int/ncd/

92 WHO (1999) *World Health Report 1999*, World Health Organization, Geneva

93 Zheng, W., Sellers, T.A., Doyle, T. J., Kushi, L. H., Potter, J. D. and Folsom, A. R. (1995) 'Retinol, antioxidant vitamins, cancers of the upper digestive tract in a prospective cohort study of postmenopausal women', *American Journal of Epidemiology*, vol 142, pp955–960

94 World Cancer Research Fund/American Institute for Cancer Research (1997) *Food, Nutrition and the Prevention of Cancer: A Global Perspective*, AICR, Washington, DC

95 WHO (2003) *World Cancer Report*. World Health Organization/International Agency for Research on Cancer, Geneva, pp62–67

96 World Cancer Research Fund/American Institute for Cancer Research (1997) *Food, Nutrition and the Prevention of Cancer: A Global Perspective*, AICR, Washington, DC

97 WHO (2002) *Diabetes Mellitus Factsheet 138*, April update, World Health Organization, Geneva

98 WHO and FAO (2003) *Diet, Nutrition and the Prevention of Chronic Diseases. Technical Report Series 916*, World Health Organization, Geneva; Food and Agriculture Organization, Rome, p72

99 International Diabetes Federation (2000) *Diabetes Atlas 2000*, International Diabetes Federation, Brussels

100 International Diabetes Federation, www.idf.org

101 Yajnik, C. S. (1998) 'Diabetes in Indians: Small at birth or big as adults or both?', in Shetty, P. and Gopalan, C. (eds) *Nutrition and Chronic Disease: An Asian Perspective*, Smith-Gordon, London

102 Ramachandran, A. (1998) 'Epidemiology of non-insulin-dependent diabetes mellitus in India', in Shetty, P. and Gopalan, C. (eds) *Diet, Nutrition and Chronic Disease: An Asian Perspective*, Smith-Gordon, London

103 Vannasaeng, S. (1998) 'Current status and measures of control for diabetes mellitus in Thailand', in Shetty, P. and Gopalan, C. (eds) *Diet, Nutrition and Chronic Disease: An Asian Perspective*, Smith-Gordon, London

104 Barker, D. J. P. (ed) (1992) *Fetal and Infant Origins of Adult Disease*, British Medical Journal, London

105 Yajnik, C. S. (1998) 'Diabetes in Indians: Small at birth or big as adults or both?', in Shetty, P. and Gopalan, C. (eds) *Diet, Nutrition and Chronic Disease: An Asian Perspective*, Smith-Gordon, London

106 Buzby, J. (2001) 'Effects of food safety perceptions on food demand and global trade' in Regmi, A. (ed) *Changing Structure of Global Food Consumption and Trade*,

US Department of Agriculture, Washington, DC, Agriculture and Trade Report WRS-01-1

107 Nestle, M. (2003) *Safe Food: Bacteria. Biotechnology and Bioterrorism*, University of California Press, Berkeley, California

108 WHO (2002) *World Health Report 2002*, World Health Organization, Geneva

109 Barling, D. and Lang, T. (2003) *Codex. The European Union and Developing Countries: An Analysis of Developments in International Food Standards Setting*, Report to the Department for International Development. City University Institute of Health Sciences, London

110 WHO (2002) *Food Safety and Foodborne Illness*, WHO Information Fact Sheet 237, World Health Organization, Geneva, January

111 WHO (2002) *Food Safety – A Worldwide Public Health Issue*, World Health Organization, Geneva www.who.int/fsf/fctshtfs.htm

112 Buzby, J. (2001) 'Effects of food safety perceptions on food demand and global trade', in Regmi, A (ed) *Changing Structure of Global Food Consumption and Trade*, US Department of Agriculture, Washington, DC, Agriculture and Trade Report WRS-01-1

113 Brundtland, G. H. (2001) *Speech to 24th Session of Codex Alimentarius Commission*, Geneva, 2 July, WHO, Geneva

114 WHO (2002) *Food Safety and Foodborne Illness*, WHO Information Fact Sheet 237, World Health Organization, Geneva, January

115 WHO (2002) *Food Safety*, Agenda item 12.3, 53rd World Health Assembly, WHO, 20 May

116 Brundtland, G. H. (2001) *Speech to 24th Session of Codex Alimentarius Commission*, Geneva 2 July, World Health Organization, Geneva

117 Phillips, The Lord of Worth Matravers, Bridgeman, J. and Ferguson-Smith, M. (2000) *The BSE Inquiry: Report: Evidence and Supporting Papers of the Inquiry into the Emergence and Identification of Bovine Spongiform Encephalopathy (BSE) and Variant Creutzfeldt-Jakob Disease (VCJD) and the Action Taken in Response to It up to 20 March 1996*, 16 vols, The Stationery Office, London

118 WHO (2002) *Food Safety – A Worldwide Public Health Issue*, Geneva, available at www.who.int/tsf/fctshfis.htm

119 DG SANCO (2000) *Trends and Sources of Zoonotic Agents in Animals. Feedstuffs. Food and Man in the European Union in 1998*, European Commission Part 1, Brussels. Prepared by the Community Reference Laboratory on the Epidemiology of Zoonoses, BgVV, Berlin, Germany

120 Rayner, M. and Peterson, S. (2000) *European Cardiovascular Disease Statistics 2000*, BHF Health Promotion Research Group, University of Oxford, Oxford

121 Data from Murray and Lopez, National Institute of Public Health (Sweden) and WHO, compiled in Rayner, M. and Peterson, S. (2000) *European Cardiovascular Disease Statistics 2000*, BHF Health Promotion Research Group, University of Oxford, Oxford

122 WHO (2002) *Food Safety and Foodborne Illness*, WHO Information Fact Sheet 237, World Health Organization, Geneva, January

123 Scholte, J. A. (2000) *Globalization: A Critical Introduction*, Harper Collins, London

124 Lang, T. (1999) 'Food and nutrition: The relationship between nutrition and public health', in Weil, O., McKee, M., Brodin, M. and Oberle, D. (eds) *Priorities for Pub-*

lic Health Action in the European Union, Societe Franchise de Sante Publique, Paris, Vandoeuvre-Les-Nancy & London

125 Acheson, D. (1999) *Independent Inquiry into Inequalities in Health: Report*, The Stationery Office, London

126 James, W. P. T., Nelson, M., Ralph, A. and Leather, S. (1997) 'Socioeconomic determinants of health: The contribution of nutrition to inequalities in health', *British Medical Journal*, vol 314, no 7093, pp1545–1549

127 Leather, S. (1996) *The Making of Modern Malnutrition*, Caroline Walker Trust, London

128 LIPT (1996) *Low Income, Food, Nutrition and Health: Strategies for Improvement*. A report by the Low Income Project Team for the Nutrition Task Force. Department of Health, London

129 Eisinger, P. K. (1998) *Towards an End to Hunger in America*, Brookings Institute Press, Washington, DC

130 Eisinger, P. K. (1998) *Towards an End to Hunger in America*, Brookings Institute Press, Washington, DC

21

Coming in to the Foodshed

Jack Kloppenburg, Jr, John Hendrickson and G. W. Stevenson

For virtually everyone in the North and for many in the South, to eat is to participate in a truly global food system. In any supermarket here in Madison, Wisconsin, we can find tomatoes from Mexico, grapes from Chile, lettuce from California, apples from New Zealand. And, in what we take to be an indicator of a developing slippage between the terms 'sustainable' and 'organic', we can even buy organic blackberries from Guatemala (which may be organically produced, but in all likelihood are not sustainably produced if sustainable is understood to encompass more than on-farm production practices and any reasonable element of social justice). We cannot, however, count on finding Wisconsin-grown tomatoes, grapes, lettuce, strawberries or apples in any supermarket in Madison, even when those crops are in season locally.

That food in the US travels an average of 1300 miles and changes hands half a dozen times before it is consumed (*The Packer*, 1992) is deeply problematic. What is eaten by the great majority of North Americans comes from a global everywhere, yet from nowhere they know in particular. The distance from which their food comes represents their separation from the knowledge of how and by whom what they consume is produced, processed and transported. If the production, processing and transport of what they eat is destructive of the land and of human community – as it very often is - how can they understand the implications of their own participation in the global food system when those processes are located elsewhere and so are obscured from them? How can they act responsibly and effectively for change if they do not understand how the food system works and their own role within it?

Recognizing the ecological and social destructiveness of the globally based food system, a variety of analysts have suggested an alternative founded on respect for the integrity of specific socio-geographic places (Herrin and and Gussow, 1989; Kneen, 1989; Berry, 1992; Crouch, 1993; Dahlberg, 1993; Friedmann, 1993; Gussow, 1993). Counterposed to the global food system in such analyses are self-reliant locally or regionally based food systems comprising diversified farms that use sustainable practices to supply fresher, more nutritious foodstuffs to small-scale processors and consumers to whom producers are linked by the bonds of community as well as economy. The landscape is understood as part of that community, and human activity is shaped to conform to knowledge and experience of what the natural characteristics of that place do or do not permit.

We find this vision of people living well and responsibly with one another and with the land on which they are placed to be deeply appealing. In our effort to work toward realization of that vision, we have found the notion of the foodshed to be particularly useful in helping us to analyse the existing food system, to imagine the shapes an alternative might take, and to guide our actions. It is our purpose in this essay to elaborate and extend that concept and to share out initial understandings of its utility.

The term 'foodshed' was coined as early as 1929 (Hedden, 1929), but we were introduced to it by an encounter with the article 'Urban foodsheds', written by Arthur Getz (1991). The idea of a foodshed immediately triggered a wide range of unexpected insights and evocative associations. The intrinsic appeal the term had and continues to have for us derives in part from its relation to the rich and well-established concept of the watershed. How better to grasp the shape and the unity of something as complex as a food system than to graphically imagine the flow of food into a particular place? Moreover, the replacement of 'water' with food' does something very important: it connects the cultural ('food') to the natural ('shed'). The term 'foodshed' thus becomes a unifying and organizing metaphor for conceptual development that starts from a premise of the unity of place and people, of nature and society.

The most attractive attribute of the idea of the foodshed is that it provides a bridge from thinking to doing, from theory to action. Thinking in terms of foodsheds implies development of what we might call foodshed analysis, the posing of particular kinds of questions and the gathering of particular types of information or data. And foodshed analysis ought in turn to foster change. Not only can its results be used to educate, but we believe that the foodshed – no less than Gary Snyder's watershed – is a place for organizing. In this unstable postmodern world, the foodshed can be one vehicle through which we reassemble our fragmented identities, reestablish community and become native not only to a place but to each other.

In expanding on these points we will depart from Getz's usage in one significant way. Getz defines the foodshed as 'the area that is defined by a structure of supply' and notes that 'our most rudimentary map of a foodshed might cover the globe' (Getz, 1991, p26). We want to establish an analytic and normative distinction between the global food system that exists now and the multiplicity of local foodsheds that we hope will characterize the future. Since we give the term 'foodshed' this normative meaning, 'global foodshed' is for us an oxymoron. Within the existing food system there already exist alternative and oppositionalist elements that could be the building blocks for developing foodsheds: food policy councils, community-supported agriculture, farmers' markets, sustainable farmers, alternative consumers. We will use the term 'foodshed' to refer to the elements and properties of that preferred, emergent alternative.

A foodshed in a moral economy

Where are we now? We are embedded in a global food system structured around a market economy that is geared to the proliferation of commodities and the destruction of the local. We are faced with transnational agribusinesses whose desire to extend and consolidate their global reach implies the homogenization of our food, our communities, and our landscapes. We live in a world in which we are ever more distant from one another and from the land, so we are increasingly less responsible to one another and to the land. Where do we go from here? How can we come home again?

There can be no definitive blueprint for the construction of some preferred future. Accordingly, we offer the foodshed not as a manifesto but as a conceptual vocabulary, not as a doctrine but as a set of principles. Below we set out five principles that seem particularly important to us. We do not claim that these are either exhaustive or particularly original. We have drawn inspiration and insight from a wide variety of people whom we consider to be engaged – whether they know it or not – in foodshed work. We invite others to join in that work.

Moral economy

A foodshed will be embedded in a moral economy that envelops and conditions market forces. The global food system operates according to allegedly 'natural' rules of efficiency, utility maximization, competitiveness and calculated self interest. The historical extension of market relations has deeply eroded the obligations of mutuality, reciprocity and equity that ought to characterize all elements of human interaction. Food production today is organized largely with the objective of producing a profit rather than with the purpose of feeding people. But human society has been and should remain more than a marketplace. E. P. Thompson (1966, p203) describes a 'moral economy' as exchange 'justified in relation to social or moral sanctions, as opposed to the operation of free market forces'. Wendell Berry (1993, p14) points to similar ethical precepts when he writes of the need for 'social and ecological standards' to guide us toward the aims of human freedom, pleasure, and longevity. The term 'moral economy' resonates for us and we use it here as a provisional shorthand phrase for the re-embedding of food production primarily within human needs rather than within the economist's narrow 'effective demand' (demand backed by ability to pay).

Adopting the perspective of the moral economy challenges us to view food as more than a commodity to be exchanged through a set of impersonal market relationships or a bundle of nutrients required to keep our bodies functioning. It permits us to see the centrality of food to human life as a powerful template around which to build non-market or extra-market relationships among persons, social groups and institutions who have been distanced from one another. The production and consumption of food could be the basis for the reinvigoration of familial, community and civic culture. We are all too well aware of the difficulty that will be involved in realizing this most fundamental principle of the foodshed. Nevertheless, we are encouraged by such innovations as Community-supported Agriculture (CSA) – 'partnerships of mutual commitment' between farmers and consumers (Van En and Roth, 1993). In CSA we have a concrete example of economic exchanges conditioned by pleasure, friendship, aesthetics, affection, loyalty, justice and reciprocity in addition to the factors of cost (not price) and quality.

The commensal community

CSA also serves as an illustration of our expectation that the moral economy of a foodshed will be shaped and expressed principally through communities. In *The Left Hand of Darkness* novelist, Ursula Le Guin (1969) imagines a society whose basic social unit is the Commensal Hearth. The word 'commensal' (from the Latin *mensa*, table) refers to those who eat together, and the word 'commensalism' is used in ecology to designate a relationship between two kinds of organisms in which one obtains food from the other without damaging it. We imagine foodsheds as commensal communities that encompass sustainable relationships both between people (those who eat together) and between people and the land (obtaining food without damage).

In human terms, building the commensal community means establishment or recovery of social linkages beyond atomistic market relations through the production, exchange, processing and consumption of food. Such social construction will occur among producers, between producers and consumers, and among consumers. Witness the recent proliferation of small-scale cooperative and collective production and marketing strategies implemented by farmers to meet growing consumer interest in organic, locally grown, non-industrial food. Other examples of such non-market cooperation from the upper Midwest include the mutual assistance commitments made within associations of small-scale producers of specialty cheeses, and the information and technology exchange that occurs through networks of farmers experimenting with the rotational grazing of dairy animals as an alternative to conventional, capital-intensive, confinement milk production systems (Hassanein and Kloppenburg, 1994). With respect to new relationships between producers and consumers, emerging cooperative linkages between fresh vegetable growers and neighborhood restaurants and consumer coops parallel the birth of CSA and the revitalization of farmers' markets (Hendrickson, 1994; Waters, 1990). Among consumers themselves, buying clubs, community gardens and changing patterns of food purchase reflect growing concern with the social, economic, ethical, environmental, health and cultural implications of how they eat.

While concrete precursors of what could conceivably become commensal communities are now visible, commitment to a moral economy requires that we work to make those communities as inclusive as possible. The sustainable agriculture movement has so far tended to be 'farm-centric' (Allen and Sach, 1991, p587) 'and has not yet seriously engaged issues of race, class, and gender even within – much less outside – rural areas. Hunger in the city is indeed an agricultural issue (Ashman et al, 1993; Clancy, 1993). The commensal community should confront and address the need not just for equitable access to food but also for broader participation in decision making by marginalized or disempowered groups. That progress is possible is evidenced by the activities of the Hartford Food System, which has made a priority of linking farmers directly to low-income consumers (Winne, 1994) and by initiatives to foster the acceptance of food stamps at farmers' markets. The 'food policy councils' now being created in a variety of US and Canadian cities are indicators of the plausibility of addressing foodshed issues by relating food affairs to other fundamental community dimensions such as economic development and nutrition and public health (Dahlberg, 1993; Toronto Food Policy Council, 1993).

Finally, the standards of a commensal community require respect and affection for the land and for other species. It is through food that humanity's most intimate and essential connections to the earth and to other creatures are expressed and consummated. In the commensal community, production, processing, distribution, consumption and waste disposal will be organized so as to protect and, where necessary, to regenerate the natural resource base. Responsible stewardship will involve sustainable cropping and humane livestock practices, reduced use of non-renewable energy sources, and a commitment to recycling and reuse.

Self-protection, secession, and succession

The dominant dynamics of the global food system actively erode both moral economy and community. We agree with those who believe that this destructiveness is an inherent property of that system, and that what is needed is fundamental transformation rather than simple reform (Allen and Sachs, 1991; Orr, 1992; Berry, 1993; Friedmann, 1993). Still, given the current dominance of the existing world food economy, people working toward foodshed objectives will need to carve out insulated spaces in which to maintain or create alternatives that will eventually bring substantive change. In opposition to the extension of the market system, there have always been examples of what Friedmann (1993, p218) calls 'movement(s) of self-protection'. From the Luddites of 19th century Britain to the Zapatistas of contemporary Chiapas, we have seen continuous refusal to submit without contest to the dictates of the globalizing food system. At the margins of consumer society and in the interstices between McDonald's and Monsanto and Philip Morris are all manner of alternative producers and eaters – Amish, vegetarians, rotational graziers, seed savers, food coop members, perennial polyculturists, bioregionalists, home gardeners, biodynamicists – who are producing and reproducing a rich set of alternative agrofood possibilities.

What these diverse people and groups share is that their activities and commitments involve various degrees of disengagement from the existing food system and especially from the narrow commodity and market relations on which it is based. We follow Berry (1993, pp17–18) and Orr (1992, p73) in our conviction that a fundamental principle of the foodshed is the need for 'secession'. This principle is based on a strategic preference for withdrawing from and/or creating alternatives to the dominant system rather than challenging it directly. Certainly, in many circumstances direct opposition to elements of the global food economy is appropriate and necessary (the situation in Chiapas, or the current manipulation of the Green Bay Cheese Exchange by food corporations such as Kraft and Pizza Hut). A primary strategy of the secession principle is 'slowly hollowing out' (Orr, 1992, p73) the structures of the global food system by reorganizing our own social and productive capacities. This is essentially what grazier groups are engaged in as they rediscover their own, indigenous capacity for producing the knowledge they need to be 'grass farmers', and as they withdraw from the agribusiness firms and agricultural scientists who have been doing their thinking for them (Hassanein and Kloppenburg, 1994).

A second and corollary point is that of 'succession', or the conscious and incremental transfer of resources and human commitments from old food-associated relationships and forms to new ones. Neither people nor institutions are generally willing or prepared to embrace radical change. The succession principle finds expression in a strategy of 'slowly moving over' from the food system to the foodshed. Food presents people with hundreds of small opportunities to take increasingly important steps away from the global market economy and toward the moral economy. An example is the consumer who decides not to purchase milk produced using recombinant bovine growth hormone (rBGH). While the motivation for that initial, simple step may be narrowly based on personal health considerations, the potential is there for making further connections. Once the link between rBGH and Monsanto is made, the consumer may become aware of the corporate/chemical/food link more generally, and begin moving a progressively higher percentage of the household food budget into purchases from alternative food sources. Similarly, restaurants or schools may be encouraged to purchase more of their food supplies from local producer cooperatives as these foodshed alternatives generate capacity.

Proximity (locality and regionality)

We see certain key spatial components to the secession and succession dynamics as characterizing the foodshed. If mitigation of the deleterious effects of distancing is the central challenge posed by the operation of the global food system, then greater attention to proximity – to that which is relatively near – should be an appropriate response. But apart from the principle of relative proximity, it is not clear precisely where the revised boundaries ought to be drawn. The limits of a foodshed will be a function of the shapes of multiple sets of boundaries; that is, of the aggregated boundaries of the climatic features, plant communities, soil types, ethnicities, cultural traditions, culinary patterns, and the like, of which foodsheds are composed. Hence we identify proximity rather than locality or regionality per se as a fundamental principle of the foodshed. Though their precise boundaries will rarely be sharply defined, we insist that foodsheds are socially, economically, ethically and physically embedded in particular places.

We do not, however, imagine foodsheds as isolated, parochial entities. While they might be – in Marge Piercy's (1976) term – as 'ownfed' as possible, we see them as self-reliant rather than self-sufficient. Self-reliance implies the reduction of dependence on other places but does not deny the desirability or necessity of external trade relationships (Friedmann, 1993, p228; Gussow, 1993, p14). For too long, however, trade in the global food economy has meant farmers selling low-value commodities to distant markets and processors and the subsequent reimportation of finished food products at high prices. In the foodshed, efforts would be made to increase the level of local and intra-regional food production, processing and distribution and so to retain economic value and jobs. Since economic concentration is a prime engine of distancing, secessionist and successionist alternatives ought to be built around small and midsized enterprises (dairies, cheese factories, smithies, greenhouses, canneries, restaurants, specialty markets) capable of responding affirmatively to the opportunities and responsibilities of the emergent commensal community.

The self-reliance associated with proximity is closely linked to both social and environmental sustainability. A community that depends on its human neighbours, neighbouring lands and native species to supply the majority of its needs must ensure that the social and natural resources it utilizes to fulfill those needs remain healthy. A consequence of proximate self-reliance is that social welfare, soil and water conservation, and energy efficiency become issues of immediate practical concern. For example, it is difficult for most city dwellers to be concerned about preserving farmland unless the destruction of farmland directly affects their food supply, or unless they know and care for the paving over of the land. Awareness of and affection for one's place can forestall the ethical distancing so characteristic of the global food system. In the foodshed, collective responsibility for stewardship of people and of the land becomes a necessity rather than an optional virtue.

Nature as measure

We understand the foodshed to be a socio-geographic space – human activity embedded in the natural integument of a particular place. That human activity is necessarily constrained in various ways by the characteristics of the place in question. Ignoring those natural constraints or overriding them with technology is one of the besetting sins of the global food system, the ecological destructiveness of which is now unambiguously apparent even to its apologists. In the foodshed, natural conditions would be taken not as an obstacle to be overcome but as a measure of limits to be respected.

While restraints on human activity will indeed often be required, to interpret natural parameters in terms of 'deficiency' rather than 'capacity' is to fail to transcend the conventional industrial mindset. Nature may be understood not just as a set of limits but as an exemplar of the possible, as an almanac of potential models for human conduct and action (Jackson, 1980; Orr, 1992, p33; Quinn, 1993). For example, from the perspective of the foodshed, one answer to Berry's (1987, p146) query, 'What will nature help us do here?' points toward the development of regional palates based on 'moving diets' of locally and seasonally available food. Who knows what lessons nature may offer us should we free ourselves to see its 'capacity'? These opportunities are by no means obvious. They must be discovered in intimate, extended conversation with the land. By acting with respect and affection for the natural world, we may begin to produce and eat *in* harmony with and within the rhythms and patterns of the places in which we live.

Ironically, much foodshed analysis will necessarily involve examination and explication of the structure and dynamics of the existing global food system. That food system exists and is a powerful and dominating structure indeed. Secession – even for so solitary a group as the Amish – can now be only partial and contingent. Emergent elements of what might become foodsheds are presently embedded in and often constrained by the rules, interests and operations of regional and global actors and institutions.

Aldo Leopold (1970, p137) suggested that we need to learn to 'think like a mountain'; that is, to think ecologically, to engage the hidden and unlooked-for connections among the elements of a system or between different levels of a system. Until and unless we know where we are in the larger social and political ecology of the global food system, we may not be able to move effectively toward realization of a foodshed locally. We do not necessarily have to accept the demands of the global food system, but we must understand and realistically address the constraints it imposes if we are to identify the space it permits for secessionist activities or simple self-protection.

Foodshed analysis will not eschew engagement with issues at the national or even the global level. It will ask that this extra-local investigation serve the objective of framing the prospects for successfully implementing concrete initiatives or changes within a particular socio-geographic place. Foodshed analysis will involve investigation of the existing food system in order to inform strategic decisions regarding opportunities for self-protection and secession. Such analysis should include the identification, celebration and study of existing and emergent alternatives to the food system. Ultimately foodshed work should seek to link such elements in a system of mutual support and integration, with the objective of fostering emergence of a truly alternative system: the foodshed. While as a general rule it is advisable to think and act as proximately as we can, we must recognize that the appropriate and necessary locus of both thought and action in the foodshed may sometimes be regional, national or even global.

Concretely, what would foodshed analysis entail? In simplest terms, it means answering Getz's basic question, 'Where is our food coming from and how is it getting to us?' For us, a substantial part of the appeal of the term 'foodshed' has to do with the graphic imagery it evokes: streams of foodstuffs running into a particular locality, their flow mediated by the features of both natural and social geography. Measuring the flow and direction of these tributaries and documenting the many quantitative and qualitative transformations that food undergoes as it moves through time and space toward consumption is the central methodological task of foodshed analysis.

What unit of analysis is appropriate for such study; what, after all, are the boundaries of a foodshed? What kinds of data or information ought to be collected? Answers to these questions will vary as a function of who is engaging in the analysis and what their objectives and resources are. The foodshed is not a determinate thing; foodshed analysis will be similarly variable. It may involve collection of data on local exports of corn or the capacity of the local landfill, on the distribution of edible plant species or the patterns of human hunger, on the organization of harvest festivals or the composition of the county board, on the content of school lunch menus or the forage preferences of diary cows.

Foodshed analysis will not be constructed to conform to some predetermined theoretical and methodological framework, but will be constituted by the concrete activities of those who seek to learn about the food system in order to change it. Many such projects have been completed or are under way at a variety of levels. The Cornucopia Project, organized by Rodale Press in Pennsylvania in the 1980s, chose states as its unit of analysis and emphasized collection of aggregate state-level data suited to the project's objective of raising the general public's awareness of the vulnerabilities of the national food system through state-specific reports and publicity (Rodale, 1982; Rural Wisconsin Cornucopia Task Force, 1982). Also at the state level, several studies by nutritionists have been undertaken in order to explore the parameters and implications for human health of sustainable, regional diets (Hamm, 1993; Herrin and Gussow, 1989).

Using cities as their socio-geographic framework, a variety of 'food policy councils' have been created to address issues of sustainability and equity in the food system (Hartford Food System, 1991; Dahlberg, 1993; Toronto Food Policy Council, 1993). The students and staff at several colleges have taken their own institutions as the basic unit of analysis and explored the rationale and mechanisms for getting commitments from their colleges to buy local food (Bakko and Woodwell, 1992; Valen, 1992). Local food projects at Hendrix College in Arkansas and Saint Olaf and Carleton colleges in Minnesota were successful in reorienting food purchasing patterns to more proximate sources. The degree of resolution characteristic of the lens of foodshed analysis can become very fine grained indeed. One of the most impressive and revealing analyses we have encountered is a self-study of a personal foodshed – 'from gut to ground' (Peterson, 1994) – that explores individual consumption and its implications for personal responsibility in the global food system.

An example of foodshed analysis that focuses on the urban poor is an initiative undertaken under the auspices of the Southern California Interfaith Hunger Coalition (IHC). The IHC's report, 'Seeds of Change: Strategies for Food Security for the Inner City', is an ambitious and finely realized effort to take an 'integrated, whole-systems approach' to assessing the need and prospects for reforming the existing food system in a specific and delimited place (Ashman et al, 1993). The IHC document is also of interest because the research and analysis for the report was undertaken largely by students and faculty from the University of California at Los Angeles. Much criticism has been directed toward universities (especially toward the land-grant colleges) for their subservience to industrial interests and their failure to orient knowledge production to local or regional needs.

'Seeds of Change' is striking evidence that academics can work effectively with advocacy groups oriented to transformation of the food system.

Although few of those whose efforts we have described think of what they do as 'foodshed analysis', we feel they are moving in directions similar to ours. To the extent that these diverse projects and undertakings are complementary, they constitute a rich set of conceptual and methodological resources for thinking about and assessing the nature and structure of the global food system in which we are now embedded, and for helping us to consider how and where we can realistically expect to make changes.

Radical reformism

It is apparent to increasing numbers of people that fundamental changes are needed in the global food system. Of course, we see that the question of food is simply a specific case of the general failure of late capitalism, or post-industrialism, or post-modernism, or whatever you wish to call this period of intense commodification and of accelerating distancing from one another and from the earth. We could equally well be calling for fundamental changes in the global health system, the global industrial system, the global political system, the global monetary system or the global labour system. Ultimately, what sustainability requires of us is change in global society as a whole. We need the recovery and reconstitution of community generally, not simply in relation to food. Although we may strive to think like mountains, we must act as human beings. To start the global task to which we are called, we need a specific place to begin, a specific place to stand, a specific place to initiate the small, reformist changes that we can only hope may some day become radically transformative.

We start with food. Given the centrality of food in our lives and its capacity to connect us materially and spiritually to one another and to the earth, we believe that it is an appropriate place to begin. We offer the term 'foodshed' to encompass the physical, biological, social and intellectual components of the multidimensional space in which we live and eat. We understand the foodshed as a framework for both thought and action. If our use of the term has any virtue, perhaps it is to help people see the relatedness of apparently disparate elements, and to perceive the complementarity of different but parallel initiatives for change. We also think it useful to make a semantic distinction between where we are now and where we wish to be in the future. Thinking and acting in terms of the foodshed is an indication of our commitment to work not simply to reform the food system but to transcend that system entirely. And while a system can be anywhere, the foodshed is a continuous reminder that we are standing in a specific place; not anywhere, but here.

We need to keep place firmly in our minds and beneath our feet as we talk and walk our way toward a transformed future. Because the path is long and because we must build it as we go – the foodshed offers a project, not a blueprint – our actions will be 'slow small adjustments in response to questions asked by a particular place' (Berry, 1990, p121). We share Orr's (1992, p1) hope for 'a rejuvenation of civic culture and the rise of an ecologically literate and ecologically competent citizenry who understand global issues, but who also know how to live well in their places'. If we are to become native to our places, the foodshed is one way of envisioning that beloved country.

References

Allen, P. L. and Sach, C. E. (1991) 'The social side of sustainability', *Science as Culture*, vol 2, no 13, pp569–590

Ashman, L., de la Vega, J., Dohan, M., Fisher, A., Hippler, R. and Romain, B. (1993) *Seeds of Change: Strategies for Food Security in the Inner City*, Interfaith Hunger Coalition, Los Angeles

Bakko, E. B. and Woodwell, J. C. (1992) 'The campus and the biosphere initiative at Carleton and Saint Olaf Colleges', in Egan, D. J. and Orr, D. W. (eds) *The Campus and Environmental Responsibility*, Jossey-Bass Publishers, San Franscisco

Berry, W. (1987) *Home Economics*, North Point Press, San Francisco

Berry, W. (1992) 'Conservation is good work', *Amicus Journal*, Winter, pp33–36

Berry, W. (1993) *Sex, Economy, Freedom and Community*, Pantheon Books, New York

Clancy, K. L. (1993) 'Sustainable agriculture and domestic hunger: Rethinking a link between production and consumption', in Allen, P. (ed) *Food for the Future*, John Wiley, New York

Crouch, M. (1993) 'Eating our teachers: Local food, local knowledge', *Raise the Stakes*, Winter, pp5–6

Dahlberg, K. (1993) 'Regenerative food systems: Broadening the scope and agenda of sustainability', in Allen, P. (ed) *Food for the Future*, John Wiley, New York

Friedmann, H. (1993) 'After Midas's feast: Alternative food regimes for the future', in Allen, P. (ed) *Food for the Future*, John Wiley, New York, pp213–233

Getz, A. (1991) 'Urban foodsheds'. *Permaculture Activist*, vol 24, pp26–27

Gussow, J. D. (1993) 'But what can I eat in March?' *Natural Farmer*, Spring, pp14–15

Hamm, M. W. (1993)'The potential for a localized food supply in New Jersey', presented at the conference on Environment, Culture and Food Equity, Pennsylvania State University, 3–6 June

Hartford Food System (1991) 'Solutions to hunger in Hartford: Rebuilding our local food system, 1991 action guide', Hartford Food System, Hartford

Hassanein, N. and Kloppenburg, J., Jr (1994) 'Where the grass grows again: Knowledge exchange in the sustainable agriculture movement', unpublished

Hedden, W. P. (1929) *How Great Cities Are Fed*, D. C. Heath, Boston

Hendrickson, J. (1994) 'Community supported agriculture', Direct Marketing, no 41, University of Wisconsin, Madison – Extension

Herrin, M. and Gussow, J. D. (1989) 'Designing a sustainable regional diet', *Journal of Nutrition Education*, December, pp270–275

Jackson, W. (1980) *New Roots for Agriculture*, Friends of the Earth, San Francisco

Kneen, B. (1989) *From Land to Mouth: Understanding the Food System*, NC Press, Toronto

Le Guin, U. K. (1969) *The Left Hand of Darkness*, Ace Books, New York

Leopold, A. (1949, reprint 1970) *A Sand County Almanac*, Ballantine Books, New York

Packer, The (1992) 'From grower to consumer: An elaborate odyssey', *The Packer*, 13 June, p11

Peterson, R. (1994) 'From gut to ground: A personal case study of a foodshed', unpublished

Piercy, M. (1976) *Woman on the Edge of Time*, Fawcett-Crest, New York

Quinn, D. (1993) *Ishmael*, Bantam Books, New York

Rodale, R. (1982) *The Cornucopia Papers*, Rodale Press, Emmaus

Rural Wisconsin Cornucopia Task Force (1982) *The Wisconsin Cornucopia Project: Toward a Sustainable Food and Agriculture System*, Rural Wisconsin Cornucopia Task Force, Madison

Thompson, E. P. (1966) The *Making of the English Working Class*, Vintage Books, New York

Toronto Food Policy Council (1993) *Developing a Food System Which is Just and Environmentally Sustainable*, Toronto

Valen, G. L. (1992) 'Hendrix College local food project', in Egan, D. J. and Orr, D. W. (eds) *The Campus and Environmental Responsibility*, Jossey-Bass Publishers, San Franscisco

Van En, R. and Roth, C. (1993) 'Community supported agriculture', University of Massachusetts Cooperative Extension System, Amherst

Waters, A. (1990) 'The farm–restaurant connection', in Clark, R. (ed) *Our Sustainable Table*, North Point Press, San Francisco

Winne, M. (1994) 'Community food planning: An idea whose time has come', *Seedling*, vol 1–4, p8

Part 5

Perspectives from Developing Countries

Introduction to Part 5: Perspectives from Developing Countries

Jules Pretty

The final section of this *Reader in Sustainable Agriculture* focuses on six perspectives from developing countries. These begin with an overview of the rethinking opportunities for agriculture (by Erick Fernandes, Alice Pell and Norman Uphoff), and are followed with assessments of change towards sustainability in central America (by Roland Bunch and Gabino Lopez), of agroecological transformations in Brazil's southern state of Santa Caterina (by Sergio Pinheiro), of Cuba's experience with sustainable agriculture (by Peter Rosset and Martin Bourque), of the benefits of agroforestry in east and southern Africa (by Pedro Sanchez), and finally on how agricultural sustainability in more than 200 projects across 52 countries has had substantial positive effects on yield increases (by Jules Pretty, James Morison and Rachel Hine).

In the first article, Erick Fernandes and colleagues summarize the types of transitions effected by the Green Revolution, and then set out a vision for agriculture centred on field-culture, with sensitivities towards patterns in space and time. Monocultures are often erroneously seen as real agriculture, yet it is polycultures that have long offered rural people opportunities to maintain on-farm diversity of products and their functions. Multifunctional systems with many components are more resilient and meet many needs compared with mono-functional systems. This chapter sets out four ideas that need revising: that pest control always needs pesticides, that soil fertility constraints always need chemical fertilizers, that solving water problems needs new irrigation, and that raising productivity only needs genetic and breeding approaches. There are many productive opportunities that can arise by adopting more biological and people-centred approaches to agricultural development and its sustainability.

The second article, by Bunch and Lopez, describes the remarkable work of various NGOs in the central American countries of Guatemala and Honduras during the 1980s and early 1990s. It is an evaluation of the long-term effectiveness of soil recuperation programmes that focused primarily on the use of green manures and cover crops in maize-based systems. This impact study investigated what farmers were still doing some years after the end of three programmes in Cantarranas, Guinope and San Martin Jilotepeque. These programmes had already documented substantial increases in productivity on small farms, and this study found that there was evidence of considerable spread and adaptation of technologies after termination of official projects. Farmers were voluntarily taking up these new technologies, adapting them to their own conditions, and then spreading them to others. This question of persistence of technological innovations is a critical element of any debates about what constitutes agricultural sustainability.

In the third article, Sergio Pinheiro describes the experience of sustainable and agroecological development of rural areas of the state of Santa Caterina in Brazil. One driver of the process has been the Ecological Farmers Association of the Hillsides of Santa Caterina State (Agreco), who have sought to help transform production systems towards organic models through cooperation, solidarity and team work. This collective model for both production and marketing is essential if small farmers are to make transitions in a competitive local as well as global food system. Those farmers involved have increased the diversity of their farm produce, as well as increased demand for labour in rural areas. These farmers' organizations and networks are also adding value to farm produce through local agroindustry, this is selling food with a story rather than simply as a commodity.

The fourth article by Peter Rosset and Martin Bourque is the introductory chapter to the book entitled *Sustainable Agriculture and Resistance*, which tells the story of Cuba's recent transition to agricultural sustainability following the cessation of external support with the collapse of the Soviet bloc at the beginning of the 1990s. This is, in itself, a unique set of circumstances, yet the transformation of agricultural practices and institutions over a short period of time has been remarkable. At the turn of the century, Cuba was the only developing country with an explicit national policy for sustainable agriculture. To the end of the 1980s, Cuba's agricultural sector was heavily subsidized by the Soviet bloc. It imported more than half of all calories consumed, and 80–95 per cent of wheat, beans, fertilizer, pesticides and animal feed. It received three times the world price for its sugar. But in 1990, trade with the Soviet bloc collapsed, leading to severe shortages in all imports, and restricting farmers' access to petroleum, fertilizers and pesticides. The government's response was to declare an 'Alternative Model' as the official policy – an agriculture that focused on technologies that substituted local knowledge, skills and resources for the imported inputs. It also emphasized the diversification of agriculture, oxen to replace tractors, integrated pest management to replace pesticides, and the promotion of better cooperation among farmers both within and between communities. It has taken time to succeed. Calorific availability was 2600 kilocalories (kcal) per day in 1990, fell disastrously to between 1000–1500 soon after the transition, leading to severe hunger, but subsequently rose to 2700kcal per day by the end of the 1990. Two important strands to sustainable agriculture in Cuba have emerged. First, intensive organic gardens have been developed in urban areas – self-provisioning gardens in schools and workplaces (*autoconsumos*), raised container-bed gardens (*organoponicos*) and intensive community gardens (*huertos intensivos*). There are now more than 7000 urban gardens, and productivity has grown from 1.5kg m^{-2} to nearly 20kg m^{-2}. Second, sustainable agriculture is encouraged in rural areas, where the impact of the new policy has already been substantial.

The fifth article, by Pedro Sanchez, describes the benefits of agroforestry in farming systems in Kenya and Zambia. Agroforestry is the practice of integration of trees and other perennials into farming systems, and is, of course, an ancient practice. During the 1980s, agroforestry was widely promoted for conservation purposes, but has had somewhat limited impact on people's livelihoods. This paper shows, however, how agroforestry can be fully integrated into farm systems of poor households, leading both to soil fertility replenishment and substantially increased yields. Some of the practices can appear counter-intuitive to farmers at first – such as leaving land fallow for perennial growth over a season or more – yet the evidence is unequivocal. Get the mix of crops and perennials right, and farm families and their environments can benefit substantially.

For the sixth paper, Jules Pretty and colleagues conducted a large survey of sustainable agriculture improvements in developing countries. The aim was to audit progress towards agricultural sustainability, and assess the extent to which such initiatives, if spread on a much larger scale, could feed a growing world population that is already substantially food insecure. They analysed more than 200 projects in 52 countries, including 45 in Latin America, 63 in Asia and 100 in Africa, and calculated that almost 9 million farmers were using sustainable agriculture practices on about 29 million hectares, more than 98 per cent of which emerged during the 1990s. These methods were working particularly well for small farmers, as about half of those surveyed are in projects with a mean area per farmer of less than one hectare, and 90 per cent in areas with less than two hectares each.

Improvements in food production were found to be occurring through one or more of four different mechanisms. The first involves the intensification of a single component of farm system, with little change to the rest of the farm, such as home garden intensification with vegetables and/or tree crops, vegetables on rice bunds, and introduction of fish ponds or a dairy cow. The second involves the addition of a new productive element to a farm system, such as fish or shrimps in paddy rice, or agroforestry, which provides a boost to total farm food production and/or income, but which do not necessarily affect cereal productivity. The third involves better use of nature to increase total farm production, especially water (by water harvesting and irrigation scheduling), and land (by reclamation of degraded land), so leading to additional new dryland crops and/or increased supply of additional water for irrigated crops, and thus so increasing cropping intensity. The fourth involves improvements in per hectare yields of staples through the introduction of new regenerative elements into farm systems, such as legumes and integrated pest management, and new and locally appropriate crop varieties and animal breeds. The study found that sustainable agriculture has led to an average 93 per cent increase in per hectare food production. The relative yield increases are greater at lower yields, indicating greater benefits for poor farmers, and for those missed by the recent decades of modern agricultural development.

22

Rethinking Agriculture for New Opportunities

Erick Fernandes, Alice Pell and Norman Uphoff

Over the last 30 years, the creation and exploitation of new genetic potentials of cereal crops, leading to what is called the Green Revolution, has saved hundreds of millions of people around the world from extreme hunger and malnutrition, and tens of millions from starvation. However, these technologies for improving crop yields have not been maintaining their momentum. The rate of yield increase for cereals worldwide – around 2.4 per cent in the 1970s and 2 per cent in the 1980s – was only about 1 per cent in the 1990s. Although the global food production system has performed well in recent decades, will further support of conventional agricultural research and extension programmes increase yields sufficiently to meet anticipated demand?

The next doubling of food production will have to be accomplished with less land per capita and with less water than is available now (Postel, 1996). The gains needed in the productive use of land and water are so great that both genetic improvements and changes in management will be required. The world needs continuing advances on the genetic front. However, food production is more often limited by environmental conditions and resource constraints than by genetic potential. Preoccupation with the methods that brought us the Green Revolution can divert attention from opportunities that can increase food supply without adversely affecting the environment, which are considered in this book.

Given appropriate research, policies, institutions and support, food production could be doubled with the existing genetic bases. Many of the needed advances in food production could be achieved by developing agricultural systems that capitalize more systematically on biological and agroecological dynamics rather than by relying so much on agrochemicals, mechanical and petrochemical energy and genetic modification.[1] This will require, however, some rethinking of what constitutes agriculture.

Although it has been argued that agricultural output will decline if 'modern' agriculture is not promoted to the maximum (Avery, 1995), 'low-tech' methods can be very productive with now-better-understood scientific bases. Where economically justifiable, these methods use available resources more efficiently than do high-input approaches. Farming systems such as those for rice in Madagascar, maize and beans in Central America and potatoes and barley in the Andes demonstrate that output can be raised substantially,

Note: Reprinted from *Agroecological Innovations* by Uphoff, N. (ed), copyright © (2002) Earthscan, London

sometimes several-fold, with limited dependence on external resources. These crops are staples that are essential for meeting world food needs.

The potential of non-mainstream methods cannot be known until agroecological approaches are taken more seriously and evaluated systematically. Gains made through genetic improvement and use of external capital and chemical inputs over the last four decades have been substantial, and the first Green Revolution, despite the shortcomings some critics have pointed to, was one of the major accomplishments of the century.[2] But what will agricultural science do for an encore? While biotechnology holds out many promises, most of its benefits continue to be anticipated more than realized. Access to and widespread distribution of biotechnology's prospective benefits remain uncertain. The widely publicized 'golden rice' is still years from production in farmers' fields.

The challenge facing agriculture worldwide involves more than just achieving higher production, justifiable as that goal has been for previous scientific innovation when serious food deficits were an ominous possibility. Valid ecological and social considerations now make it imperative that further advances be environmentally friendly as well as economically sustainable and socially equitable. Also, more than increased food supply is needed; we should aim to ensure balanced and adequate supplies of nutrients that people can afford. In particular, adverse environmental and health externalities that result from modern agricultural methods – soil erosion, chemical hazards, soil and water pollution – are things that nobody would like to see increased, let alone doubled, as we seek to double the production of food.

Should resources for agricultural research be devoted, for example, to developing genetically engineered rice with high levels of vitamin A, assuming that cereal grain monoculture will continue to predominate? Or should we strive to incorporate nitrogen-fixing and nutrient-rich legumes and livestock into farming systems to better meet people's nutritional requirements with diversified diets – while simultaneously maintaining soil fertility? Such questions need to be addressed.

The next Green Revolution will depend at least in part on enlarging upon and diversifying the ideas that have guided past development efforts. The paradigms that presently organize and direct agricultural research and extension have been helpful for planning activities and producing theoretical explanations. But they have also created certain blind spots. The task of meeting world food needs will be more difficult if our vision of what is possible is limited by constraining conceptions of how best to raise agricultural output in effective, efficient and sustainable ways.

Agriculture as field-culture: An etymological perspective

The very concept of agriculture as it has been understood and practised in the West has been shaped by its semantic origins, coming from the Latin word *ager*, 'field'. Agriculture is mostly understood as the growing of plants in fields. (Similarly, in South Asia, most words for agriculture derive from the Sanskrit word for plough, *krsi*, so that agriculture in that region is characterized as 'plough work'.) Such a conceptualization, however tacit, makes the raising of livestock, fish, trees and other activities less central to the agricultural enterprise, except where cattle or oxen are necessary for ploughing, or where monocrop tree plantations substitute for fields. The full range and richness of the agricultural enterprise has not been well captured in the word that we use to refer to it.

Etymologically, it is not clear where livestock, fish, insects, microbes and trees fit in. Few sustainable farming systems exist that do not include several of these groups in addition to plants. But most often, those who work on other flora or on fauna have been accorded marginal status within agricultural ministries, or been assigned to separate ministries, leaving crop and soil specialists in charge of the agricultural sector.

Fishery departments are invariably marginal if located within an agricultural ministry, even though *aquaculture* integrated within farming systems has great potential. Indeed, until 'agroforestry' was discovered (King, 1968; Bene et al, 1977) and the International Centre for Research in Agroforestry (ICRAF) was established, there was little concern with trees as part of agriculture, except in large-scale plantations where tree crops were commercially profitable. Otherwise, trees got respect and attention only if looked after by a separate ministry that was more concerned with forests or plantations than with farms.

Although *agroforestry* may sound like a kind of forest management, it is a comprehensive landuse management strategy that includes a range of woody perennials (particularly trees but also shrubs) in spatial and temporal associations with non-woody perennials, grasses and annual crops, together with a variety of animals, including cattle, sheep, goats, pigs, chickens, guinea pigs, fish and even bees (Lundgren and Raintree, 1982). While some agroforestry practices are extensive – for example, most agrosilvopastoral systems – these practices generally contribute to intensified production that is agroecologically sound and maintains soil fertility (Fernandes and Matos, 1995). Fortunately, the integration of perennial plants into otherwise annual farming systems is increasingly recognized as a mainstream opportunity to increase per-hectare output in future decades.

A bias in favour of fields means that *horticulture* gets somewhat marginalized in most institutions dealing with agriculture, including universities. Gardens and orchards, being smaller, have lower status than fields, even if they produce several times more value per unit of land when intensively managed. Horticulture is devalued in part also because its produce is mostly perishable and hard to denominate. Heads of cabbage and baskets of apples are hard to compare with bags of rice or tonnes of wheat, their nutritional value notwithstanding. Historically, governments have gained more wealth and security from grains because these could be stored (or seized) more easily than fruits and vegetables.

Farming systems of most rural households around the world depend crucially upon *livestock and poultry*, large and/or small, together with home gardens and orchards and often with fish ponds and hedgerows. Efforts to improve single components of farming systems are likely to produce limited results unless the interdependence of land use, labour supply and seasonal activities for all of these farm enterprises is acknowledged.[3] In many areas of Asia, acceptance of the short-stalked, high-yielding cereal varieties that made the Green Revolution was low, for example, because the quantity and quality of the fodder produced by the new varieties was insufficient to meet livestock requirements. The goal of plant breeders had been to increase grain yield without considering forage needs. Farmers were willing to accept lower yields of grain in order to be able to feed their animals, which provided them with the manure they needed to maintain soil fertility and the traction required for tilling their land.

An argument sometimes made against livestock production is that animals are inherently wasteful; more calories can be produced per hectare from plants than from animals. If animals are fed on forages and byproducts, however, rather than competing with humans for edible grain, such 'wastefulness' can be beneficial. In extensive and semi-extensive systems, animals that range freely during the day harvest plant nutrients from non-arable areas; at night when they are penned, most of these nutrients are deposited in their enclosure, later to be distributed as manure onto cropland. In parts of West Africa, pastoralists often negotiate grazing contracts with crop-growing neighbours. Pastoralists are encouraged to graze their cattle on fields with crop residues because the cattle deposit manure: their owners may even receive additional compensation for this service. If animals were in fact highly efficient in their conversion of harvested nutrients, there would be less transfer of nutrients from rangelands to croplands.

When green and animal manures are judiciously used in combination, nutrient availability can be nicely synchronized to meet plant demands. Manure is an important product of livestock raising. In sub-Saharan Africa, 25 per cent of agricultural domestic product comes from livestock even without considering manure or traction; when these are considered, this figure rises to 35 per cent (Winrock International, 1992). The quality and quantity of manure produced depends on what the animal consumes; in Java where 'cut and carry' tree-based fodder systems are common, animals are given extra feed to improve the quality of their manure (Tanner et al, 1995). Thus, animal production can be beneficial in ecological as well as human nutritional terms.

An additional consideration obscured by a preoccupation with fields is that *common property resources* for grazing and for forest products are an essential part of many households' economic operations (Berkes, 1989; Jodha, 1992).

Common lands often are the sites from which grazing livestock harvest nutrients that are brought back to the farm at night. As these areas are not fields, however, and do not belong to any specific user, evaluating their contributions to production is admittedly difficult. This is not, however, sufficient reason to overlook their role and potential, leaving their productivity to languish.[4] Privatization of these commons, often advised, removes the flexibility people need to withstand drought in dry regions. Farming systems improvement should encompass all the area and resources available to farmers and pastoralists.

Developing an adequate knowledge base for more productive and sustainable agriculture should start with explicit acknowledgment that agriculture involves much more than fields and field crops. To be sure, fields are commonly the main component of most farm production strategies. *Staple foods* are, after all, what their name implies – essential for food security. The world in general needs more, rather than less, of them, especially for the 800 million people who are currently undernourished. But other sources of calories are also important – potatoes, cassava, yams, sorghum, millet, sweet potatoes, taro, fish, meat, milk and so on – and these have been given much less support than rice, wheat and maize.[5] Calories, while necessary for survival, are not sufficient for human health. To achieve balanced diets, including essential micronutrients, the whole complex of flora and fauna that rural households manage to achieve food security and maintain their living standards should be better understood and utilized.

Not only should fixation on individual crops be avoided, but a broader understanding of the biophysical unit for agriculture is needed. A narrow focus on fields is giving way to a broader focus on *landscapes* and/or *watersheds*, within which fields function as interdependent units, especially as we gain a better agroecological understanding of agriculture (Conway, 1987; Carrol et al, 1990; Altieri, 1995).

Assumptions associated with field-centred agriculture

Several limitations arise from this long-standing concept of agriculture. In different ways, each works against strategies for intensified and sustainable agricultural development that use the full set of local resources most productively.

The time dimension of agriculture: A cyclical view

In lore and literature, agriculture is described and celebrated as 'the cycle of the seasons'. How is agriculture practised with its field-based definition? By ploughing, planting, weeding, protecting and finally harvesting. Farmers then wait until the next growing season to plough, plant, weed, protect and harvest again, and wait once more for the next planting time. Planting defines agriculture in our minds as does the activity of harvesting. Yet if one looks beyond this standardized seasonal conception of agriculture, one finds trees that keep their leaves year-round, sheep that lamb twice a year, and microbes that continuously decompose soil organic matter with generation intervals measured in hours or minutes. These different time frames all affect agricultural performance.

Fixation on an annual cycle of agriculture has arisen from its practice in temperate climates, where most modern scientific advances have been made. There, summer and winter seasons are the central fact of agricultural life. The year-round agriculture of tropical zones seems somehow irregular, almost unnatural, since it lacks periodic cultivation. This view is reflected in reports from early colonial administrators in tropical countries who regarded indigenous populations as 'lazy' because they did not work hard to produce their sustenance. There was no annual cycle of ploughing, planting and so on, which counterparts in colder climates had to maintain. People who harvested what they had not planted, or had not planted recently, were not regarded as 'real agriculturalists' by people from temperate zones.

There is seasonality in tropical regions, to be sure. The contrast between wet and dry seasons can be as stark as that between summer and winter. But with agriculture seen primarily as a matter of *cultivation*, annual crops get more attention and status than perennials. The latter have very important roles to play, however, particularly if one is concerned with the sustainability of agriculture. Their growth usually does not disturb or tax the soil as much, or as often, as does annual cropping. The latter invests in myriad biological 'factories' that produce food or fibre and then demolishes them at the end of the season. On the other hand, trees, vines or crops that rattoon keep all or most of that biological factory intact from year to year.

Since, usually, very little biomass is discarded in the farming systems operated by poorer farm families – it is used for fodder, fuel, mulch or other purposes – our point here is directed to research and extension priorities rather than to farmers. The latter have long known that combining a variety of perennials with annuals, animals and horticultural crops creates opportunities for more total output from given areas of land during the year, and with less pressure on soil resources; energy and nutrient flows are more efficient, and adverse pest and environmental impacts can be reduced by growing perennials rather than annuals.[6] Especially if the sustainability of agricultural production is an objective, giving perennials a larger role in agriculture makes sense.

Within agriculture understood in annualist terms, fallows are periods of rest and recuperation for the soil, a kind of gap in the cropping calendar. Many farmers, however, have thought of fallows differently, managing them so that they are more productive than land that is simply left alone. 'Managed fallows' are not an oxymoron but rather a source of supplementary income, providing fodder, fruit or other benefits while enriching the soil when leguminous species or plants otherwise considered to be weeds are allowed or encouraged to grow.[7] Cropping cycles are best looked at in terms of how soil fertility can be continuously enhanced while utilizing a wide variety of plant and animal species – a strategy described as 'permaculture' by Mollison (1990) – looking beyond crops that are planted periodically.

Spatial dimensions of agriculture: Thinking in terms of soil volume instead of surface

Agriculture has been defined and limited by a mental construction of agricultural space in much the same way that it has been stereotyped in terms of annual cycles. While farmers have long appreciated that agriculture is an enterprise best conducted in three dimensions, most agronomic and economic assessments consider agriculture essentially in *two dimensions*, as an enterprise carried out on a plane. The practice of agriculture is epitomized by ploughing, which breaks the surface of the soil in order to plant seeds and grow crops. This strategy suffices so long as the soil is deep, fertile and well supplied with water. But agriculture can be made more productive by conceiving and treating soil in *three-dimensional* terms, as volume, doing more than just breaking its surface and working it two-dimensionally.

Indeed, working the soil is a better term for agriculture than ploughing it, since working encompasses many functions.[8] This concept includes incorporating organic matter of various sorts into the soil and altering soil topography to capture and hold water, or to drain it. Getting crop residues and animal manures into the soil can promote greater synchrony between nutrient release from those residues and crop nutrient demand; soil organic matter promotes better water infiltration and retention at the same time that it creates better habitats for soil microflora and for micro- and macrofauna. In many traditional farming systems around the world, one finds soil being mounded into raised beds and even raised fields; terraces are constructed to retain and improve the soil and to make watering it easier, and drains are often installed. Soil-working activities are intended not just to exploit the soil's fertility but to improve it.

Alternately, in some farming systems one finds no ploughing, just the planting of seeds in undisturbed soil. This might be considered one-dimensional agriculture with activities concentrated on points rather than a surface, leaving the volume of soil beneath intact to nurture macro- and microbiological communities. To be sure, two-dimensional thinking accomplishes some important activities such as weed control and breaking the soil crust, but disturbances of the soil contribute to major erosional losses. Weeds can be controlled by other means than ploughing, and 'no-till agriculture' is now widely accepted as a modern practice, as noted below.

In the coming decades, efforts to raise yields per hectare should not take the quality and durability of soil for granted, as the health and fertility of the soil are critical for productive and sustainable agriculture. Soil should be understood and managed in terms of its *volume* rather than its *surface*. Raising output sustainably will require more than working chemical fertilizers into the top horizon. Thinking of soil three-dimensionally should be part of any strategy for sustainable agricultural intensification.

Monoculture as 'real' agriculture

The standard view of agriculture as limited in time and space favours monocropping for achieving control and efficiency in production. Applying inputs is made easier with monoculture, whether calculating fertilizer applications or using mechanical power for weeding. But the conclusion that this is always the most productive way to use land is mistaken. This production method can raise the economic returns to labour or to capital, but it does not necessarily increase the returns to land. The latter resource will become ever more important in coming decades as the availability of arable land per capita declines.

Polyculture systems employing a combination or even a multitude of plants commonly have higher total yields per hectare, absorbing and generally requiring higher inputs of labour and nutrients. Where labour is relatively abundant and land is relatively scarce, this can be an efficient and economic system of resource use. The advantage of monocropping is that it makes mechanization, substituting capital for labour, more effective.[9] Only where mechanical power can bring into cultivation land that manual power cannot is greater physical production likely to result from mechanization. This generally makes agriculture more extensive than intensive.

Even when population is high in relation to arable area, it can be difficult to attract or retain labour to work in farm operations. Much of the impetus for farm mechanization has come from labour scarcities in the more economically advanced countries. When tractors and other machines have been introduced into developing countries with the mistaken idea that this will raise production, they have done more to displace labour than to make land more productive. Tractorization can raise profits for those who have greater access to land and capital, but it seldom leads to higher output per unit of land than using hand labour and animal traction, other things being equal.[10] In contrast to tractors, animals used for traction reproduce themselves, pay returns on the farmer's investment, and provide food, fuel and fertilizer at the same time. Since capital is so often subsidized by government policies, one should not consider the private profitability of using tractors and other capital inputs as a sole or sufficient justification for their use without analysing the full range of social costs and benefits.[11]

Because polyculture is less amenable to mechanization, it requires an adequate and reasonably skilled supply of labour. Many of the practices we discuss here are relatively labour-demanding, using human energy and skill instead of capital and chemicals to get more production from limited land resources. To the extent that investments of labour are made more productive by agroecological innovations, they can be better remunerated and lead to improvements in the agricultural sector and the rest of the economy.

It is widely believed, with more emotion than calculation, that clean-ploughed fields, sown uniformly in a single crop, planted neatly in rows with all extraneous plants removed, is the best kind of agriculture. Mulch makes fields look messy, and crop mixtures look chaotic rather than productive. But this assessment is more a matter of aesthetics than of science. Yields, yield stability and nutritional quality per unit of land from polyculture, although harder to measure, are usually greater than with monoculture.[12] Furthermore, keeping soils covered protects them against erosion.

Polycropping supported by a strategy of managing and recycling organic inputs offers many advantages and can raise yields with equivalent inputs. When maize and soybeans are intercropped, for example, there is about a 15 per cent gain in production that cannot be explained simply by the inputs applied, an increase reflecting synergy within the crops' growing environments (Vandermeer, 1989). Plant-animal intercropping yields comparable benefits. There are many situations, determined more by economic than by agronomic considerations, where monoculture will be a preferable strategy. But its superiority should not be assumed without proof, as happens now.

Mechanical conceptions of agriculture

Monocropping implicitly regards agriculture as a mechanical process, with inputs being converted into outputs by some fixed formula, whereas polycropping recognizes the inherently biological nature of agriculture. The relation posited between inputs and outputs is different for mechanical and biological paradigms. In the first, the ratio of outputs to inputs is predictable and proportional, fixed and usually linear. In the realm of nature, on the other hand, relationships are less predictable and seldom proportional. Large investments of inputs can come to nought, while under favourable conditions and with good management, modest inputs have many-times-larger effects.

Until something like the perpetual motion machine is invented, such disproportionality is not possible with mechanical phenomena, which depend on continuous inputs for their operation. Biological processes, on the other hand, can be self-sustaining and can adapt and evolve unassisted. Moreover, biological inputs can reproduce themselves. How one regards and utilizes *inputs* thus differs in subtle but important ways according to whether they are understood within a mechanical framework or in a biological context.

One area where 'modern' agriculture has rediscovered the advantages of biology is with so-called minimum tillage or no-till systems, now given the positive appellation 'conservation tillage' (Avery and Avery, 1996). Twenty years ago this was considered atavistic agriculture, harking back to the dibble stick in a modern era when heavy tractors and field machinery should be used to plough, plant, weed and harvest 'clean' fields. Yet no-till agriculture has now become state-of-the-art in many areas of the United States. Mechanical corn harvesters are designed to chop up plant stalks, leaves, husks and cobs to return this biomass to the land in biodegradable form to preserve soil fertility. In addition to recycling nutrients, conservation tillage protects the soil's surface and reduces wind and water erosion. The main limitation with little or no tillage is that weeds can become more of a problem unless farmers can afford chemical herbicides or use hand labour. (This new/old technology has become popular with businesses that sell herbicides to control weeds when there is no ploughing.)

Innovative practices like the use of mulches, cover crops and green and animal manures, which were until recently largely ignored in 'modern' agriculture, can solve the problem of weeds. These techniques capitalize on the large dividends that nutrient recycling can pay because of the multiplicative dynamics of biological processes. Whereas mechanical advantage is a well-accepted principle in physics and engineering, agricultural science should capitalize on the analogous and even more powerful principle of *biological advantage*.

Four equations in need of revision

Efforts to raise agricultural productivity have been guided for many decades by four presumptions. These have produced some impressive results, so our objection is not that they are wrong. Rather, they have become too dominant in our thinking, with too hegemonic an influence on policy and practice. It has been taken for granted that they represent superior ways to boost production. This thinking can be stated in four tacit equations that have shaped contemporary agricultural research, extension and investment.

1   Control of pests and diseases = application of pesticides or other agrochemicals.
2   Overcoming soil fertility constraints = application of chemical fertilizers.
3   Solving water problems = construction of irrigation systems.
4   Raising productivity beyond these three methods = genetic modification.

Equating certain kinds of solutions with broad categories of problems limits the search for other methods to solve those problems, even when alternative practices might have a lower cost and be more beneficial in environmental and social terms. More progress in agriculture will be made if the above propositions are broadened. Fortunately, there is a good precedent in the way that the first equation has been substantially modified over the past 15 years.

Crops and animals can be protected by non-chemical means

The modern-input paradigm for raising production has been most directly challenged with regard to pest and disease control through what is called integrated pest management (IPM). Adverse effects on human health as well as on the environment caused some scientists to explore ways to produce crops and animals with little and even no use of chemicals. Biological controls as well as alternative crop management practices have often turned out to be more cost-effective, and sometimes simply more effective. The chemical-based strategy of 'zero tolerance' for pests and diseases, rather than being a solution, exacerbates the problem, killing beneficial insects that are predators of crop pests. The widespread use of agrochemicals, particularly broad-spectrum ones, has had the consequence of making pest attacks worse.[13] Routine use of antibiotics to treat diseases and promote the growth of livestock has, unfortunately, increased the antibiotic resistance of pathogens that can infect humans and/or animals.

An IPM strategy does not preclude the use of chemicals. But the first lines of defence against pests and diseases are biological, trying to utilize the defensive and recuperative powers of plants and animals as well as the activity of beneficial and predator insects to farmers' advantage.[14] The Indonesian IPM programme, for example, taught farmers that spiders, previously viewed with antagonism, should be protected and preserved. Demonstrations showed that rice beyond a certain stage can sustain extensive leaf damage from insects, as much as 25 per cent, without depressing effects on yield, and even possibly some gain. When sheep in Australia and South Africa were fed leguminous forages containing tannins as part of their diets, their internal parasite loads were reduced, reducing expenditures on antihelmintic medicines and providing an alternative treatment when antihelmintic resistance is a problem. The presumptions of modern agricultural science regarding chemical means for pest and disease control have been broadly challenged, with such means being increasingly reduced and avoided where possible.

Soil fertility can be enhanced, often more effectively, by non-chemical means

The most broadly successful component of modern agriculture has been the introduction and use of inorganic fertilizers to supply soil nutrients, particularly nitrogen, phosphorous and potassium, where these were lacking. But this success has led many policy makers and some scientists to equate soil fertility improvement with the application of fertilizers when, in fact, fertility depends on many additional factors. Indeed, the misuse or overuse of chemical fertilizer results in adverse effects on yield by negatively affecting the physical and biological properties of soil. The advantage of inorganic fertilizers is that they are easier to apply, often cheap (if subsidized) and have more predictable nutrient content. Also, organic nutrients are sometimes simply not available in sufficient supply.

When inorganic fertilizers are added to soils that possess good physical structure, with adequate soil organic matter and sufficient cation-exchange capacity, they can produce impressive improvements in yield. Where soils are acidic (low pH) and the nutrients needed for plants are in short supply, the application of appropriate amounts of lime (calcium carbonate) along with inorganic fertilizers can result in spectacular crop yield increases and can greatly improve farmer income. But in many circumstances, especially in the tropics, soils are not so well structured or well endowed. Then, inorganic fertilizers, especially if used in conjunction with tractors that compact the soil, can lead to changes in soil physics and biology that are counterproductive and diminish, sometimes sharply, the returns from adding chemical nutrients.

We have suggested to dozens of soil scientists in the US and overseas that probably 60 to 70 per cent of soil research over the past 50 years worldwide has focused on soil chemistry and about 20 to 30 per cent on soil physics. This means that less than 10 per cent of soil research has been devoted to improving our understanding of its biology. This estimate has not been challenged by agronomists to date. Why such preoccupation with soil chemistry? It is the easiest kind of soil deficiency to study, giving quick, precise and replicable results, which point to simple remedies. The results of soil chemistry analyses are easy to interpret; by adding certain amounts and combinations of fertilizer nutrients, one can expect predictable increments to production. Moreover, such research gets funding easily, given the interests of fertilizer producers in such knowledge.

Yet even brief consideration of these three domains affecting soil fertility suggests that the amount of effort going into each, even if not necessarily equal, should be closer to parity. Any national research programme that deliberately allocated its scientific resources in the above disproportions would be considered misguided. Microbial activity is essential for nutrient availability and uptake. When one walks on ground that has been converted by leguminous species, compost, mulch or manures from something resembling concrete into absorbent, friable soil underfoot with good tilth, the contribution of soil microbiology is self-evident. But studying biological processes is more difficult than assessing differences in soil structure, and many times more difficult than measuring the chemical composition of soil samples.

Similarly, plant scientists with whom we have spoken have agreed that 90 per cent or more of their research effort over the past 50 years has been devoted to those parts of plants that are above ground, and less than 10 per cent to what is below ground. Indeed, plant scientists usually suggest that less than 5 per cent of their research has investigated subsurface processes and dynamics. Yet any assessment of how plants grow and thrive suggests that a more balanced distribution of effort is desirable, with much more attention paid to the growth and functions of roots than in the past. However, just as it has been easier to study the chemistry of soil, it has been easier to analyse leaves and stalks than to probe the underground mechanisms of roots for uptake and transport of nutrients and water. Changing the soil's temperature by just a few degrees can alter significantly the microbial populations underground, for example, which makes such research difficult to replicate and validate.

Modern agricultural research's focus on soil chemistry and above-ground portions of plants has led to solutions that favour chemical and mechanical means. The belief that chemical fertilizers are the best way to deal with soil fertility limitations has arisen from – and has reinforced – the image of agriculture as a kind of industrial enterprise, where producing desired outputs is mostly a matter of investing certain kinds and amounts of inputs. Consequently, viewing agriculture more as a biological than as a mechanical process attaches greater value to the use of organic inputs. In recent years there has been a major increase in the application of biologically based technologies, such as vermiculture (raising worms) to enhance soil fertility and ameliorate the negative effects of industrial and agricultural wastes on soil (Appelhof et al, 1996; Acharya, 1997).

As in most things, combinations of factors are more likely to approach the optimum than one factor by itself. It is well known that for plants to utilize chemical fertilizer effectively, the soil in their root zone must have substantial capacity to retain and exchange nutrient cations, and that exchange capacity is considerably enhanced as soil organic matter content increases. Research shows the benefits of utilizing organic means to maintain soil fertility and also of adding some inorganic nutrients in combination with organic inputs to get the best results.[15]

Adding appropriate amounts and combinations of chemical nutrients can increase both plant productivity and the amount of crop residues (shoots and roots) that become available to increase and maintain soil organic matter. Augmenting organic matter is especially necessary in tropical soils, which, due to climatic and edaphic conditions, are more likely to need maintenance and restoration of organic material and nutrients. The bottom line is that chemical fertilizers by themselves are no substitute for incorporation of soil organic matter. Ideally both will be used in synergistic ways.[16]

Irrigation is not the only way to deal with water limitations

A mechanistic conception of agriculture reinforces the millennia-old fixation on irrigation as the best if not the only means of providing water for plants in water-scarce environments. In many places, given hydrological cycles and opportunities, irrigation is certainly necessary for the practice of agriculture. But its success over several thousand years has led people to look to this technology as the universal solution to water scarcity problems. When crops need water, the first thought is how to provide irrigation from surface or groundwater sources.

But there are other ways to meet crop requirements besides capturing water in a reservoir, by river diversion or by pumping it from some body of water above or below ground, and then conveying it through canals and other structures to deliver it to particular fields, in amounts and at times when it is needed.[17] In much the same way that assuming soil fertility problems are best solved by fertilizer applications, seeing water shortages as best handled by irrigation has made water harvesting and conservation almost lost arts. When farmers in semi-arid Burkina Faso, assisted by Oxfam, demonstrated that they could grow much better millet crops simply by placing rows of stones across their fields, to slow water runoff and store it in the soil, this was seen as a remarkable technology (Harrison, 1987, pp165–170). Numerous case studies with similar results have been documented in Reij et al (1996). Such practices should become part of the repertoire of soil and water management practices that farmers can adopt to utilize available rainfall most advantageously. Using mulch to capture water and slow evaporation is another simple method.

Measures to conserve and utilize water, like planting crops in certain rotations or seeding a new crop in a standing one to capitalize on residual moisture, should not be seen as something novel but rather as something normal, making the best use of water in combination with soil. Methods including collecting and storing water in small catchment dams, large clay jars or simply in porous soils should be experimented with to determine what designs can provide enough water to crops and animals (and for human uses) to justify the expenditure of labour and capital and sometimes land. Small catchment ponds are becoming more attractive and feasible options, providing water supplies in situ. We should also understand better how land preparation practices affect water retention and utilization.[18]

Irrigation will surely remain a major means for solving water problems, and we should be learning how to use scarce irrigation water more efficiently and effectively through means of social organization (Uphoff, 1986a, b; 1996). But irrigation is not the only means to ensure that growing plants and animals have the water they need. Water scarcity will surely increase for agriculture around the world, so all possible means to acquire and conserve water need to be considered.

Genetic manipulation is not always necessary to raise production significantly

The modern approach to agricultural improvement has stressed better plant and animal breeding, especially since the advent and success of the Green Revolution. Without denying the value of such efforts, or that there will be some future benefits from biotechnology, we think more attention should be paid to cultural practices, to soil preparation and management, to use of organic inputs, to more productive cropping patterns and systems, and to species that have previously been overlooked or underutilized.

A good example is the System of Rice intensification (SRI) developed in Madagascar which can boost yields from any variety of rice by 100 to 200 per cent or more by changing management practices and without requiring any use of purchased inputs. There are other examples of major yield increase potentials with staple crops. In the 1970s, a programme in Guatemala was able to help farmers raise their maize and bean yields from 400–600kg/ha to about 2400kg in just seven years, at a cost of about US$50 per household. Farmers who had become acquainted with experimentation and evaluation methods proceeded to double yields once more on their own after external assistance was withdrawn. Very poor farmers working with an NGO in the high Andean regions have found that they could double or triple their yields of potatoes and barley by using lupine, a leguminous plant, as a green manure to add nitrogen to the very poor mountain soils and increase soil organic matter. This method, like SRI in Madagascar, works with whatever varieties farmers are already planting and uses organic rather than chemical inputs from outside the community. Leguminous fallows can raise maize yields in southern Africa by two to four times. The Mukibat technique, named after the farmer who devised it in Indonesia almost 50 years ago, can increase the yield of cassava by five times or more. It involves grafting cassava tubers onto the root of a wild rubber tree of the same genus as cassava, which gives the growing tubers more access to sunlight and nutrients (Foresta et al, 1994). That this technology has aroused, so little scientific attention, and was not reported in the literature until more than 20 years after it was devised (Bruijn and Dharmaputra, 1974), may reflect the indifference among most researchers towards cassava, a low-status staple crop on which hundreds of millions of people depend for much of their sustenance. Or perhaps it reflects a lack of interest in innovations that do not come from the scientific community.

Smallholding farmers around the world at present are probably exploiting less than 50 per cent of the existing genetic potential of various crops due to less than optimal management. In many cases this is because the returns to labour are not high enough to justify intensification, but often it is a matter of not knowing how to capitalize on synergies that could raise these returns. Reducing the yield variability of traditional varieties and taking fuller advantage of their genetic potential through nutrient cycling and better soil and water management within complex farming systems could, we think, be a cost-effective strategy that complements longer-run and higher-cost biotechnological efforts being undertaken to produce new and better varieties. Increased production of other food sources, including fish culture, small animals and various indigenous plants, can augment in non-competing ways whatever nutrients are provided by staples.

Even if these alternative methods by themselves cannot achieve a doubling of world food production, they could contribute substantially to this, making up the difference that is unlikely to be produced by more modern means that are heavily dependent on inputs of energy, chemicals and water. Capitalizing on 'non-modern' opportunities will require reorientation of socio-economic as well as biophysical thinking. It necessitates looking beyond the farm and its fields, and beyond particular crop cultivars, animal species and cultivation practices, to institutions and policies.[19]

Utilizing these productive opportunities

Doing 'more of the same' in either the so-called modern or traditional sectors of agriculture is not likely to be sufficient for meeting food needs in the decades ahead. Researchers, extensionists and policy makers who wish to assist households around the world to become more food-secure, healthy and well-off need to consider how to make broadly based improvements in output through evolving systems that are more intensive and more complex. These will resemble but improve upon present practices that are not fully or sustainably utilizing soil, biological and other resources.

Traditional farmers are for the most part quite resource constrained. The technologies offered by extension services were usually developed for larger, simpler production systems that are not appropriate for the kinds of systems that the majority of farmers in the world are managing. There are wide variations in productivity within and across farming communities, with some producers tapping production potentials better than others. We look towards 'hybrid' strategies to raise production, combining the best of farmers' current practices with insights derivable from modern science to tap the power of plant and animal germplasm nurtured under optimal conditions.

There is no reason to believe that the elements of 'modern' agriculture are wrong, but neither is there a warrant to consider them (yet) complete. They offer many advantages of productivity and profit for large numbers of agricultural producers – but not for all of them, and maybe not even for a majority of farming households around the world today. Our analysis here calls into question the presumption, whether it is argued or assumed, that mainstream approaches are the best or the only way to advance agriculture in the future. For the sake of productivity and sustainability, it will be advisable to 'backcross' some of the modern varieties of agriculture, which are most suitable for advantaged producers and regions, with often more traditional methods so as to develop a more robust 'hybrid' agriculture, one that can better meet the world's needs for food, health, employment and security in this century.

Notes

1 This is not a statement in opposition to research on genetic modification, a controversial subject these days. Transgenic research has some potentially valuable, legitimate and safe uses and we would not want to see it curtailed – though more oversight and regulation and a different international property rights regime would make this enterprise more defensible and beneficial. Improvements in pest- and drought-resistance, for example, if achieved through advanced technology, could be great boons, particularly for the poor. Our focus on opportunities to raise production through different, more intensive management practices aims at a diversified strategy of agricultural development, one which will include work on genetic improvements.

2 'Had the cereal yields of 1961 still prevailed in 1992, China would have needed to increase its cultivated cereal area more than three-fold and India about two-fold, to equal their 1992 harvests' (Borlaug and Dowswell, 1994).

3   One of the pre-eminent agricultural development projects in the 1960s and 1970s, Plan Puebla in Mexico, was set up to benefit rural smallholder households by increasing their production of maize under rainfed conditions. Maize was considered their main crop. Yet a survey in the Puebla area showed that animal production provided 28 per cent of households' income, more than the 21 per cent that came from maize and almost as much as from the sale of all crops, 30 per cent. In addition, 40 per cent of household income came from off-farm employment (Diaz Cisneros et al, 1997, p123). The project made little progress with small farmers until it sought to improve production of beans along with maize, as these crops when grown together produced more than maize grown by itself and also contributed more to family nutrition. Farmers' cooperation also increased when other lines of production were assisted by the project. A more recent survey of 206 households selected randomly in four villages in the northern Philippines found that livestock contributed almost as much to household incomes (90 per cent as much) as did their rice production (Lund and Fafchamps, 1997).

4   In a watershed development programme in the Indian state of Rajasthan, where a participatory approach to technology development was taken that aimed to capitalize on local knowledge, fodder production on rainfed common lands was increased eight- to ten-fold with corresponding improvements in soil conservation (Krishna 1997, pp261–262). While such areas usually face serious physical constraints on increased production because they have been so neglected by researchers and extension personnel, they often offer substantial opportunities, previously ignored, for raising output.

5   This is discussed by Chambers (1997, especially p47). While rice, wheat and maize have received the lion's share of research funding, at least four of the international agricultural research centres in the CGIAR system have some of these other staple crops as a central part of their mandates. There are also centres now working on animals, agroforestry and aquaculture, though the centre on horticulture has yet to become part of the system (due to political reasons). The centres responsible for working on rice (International Rice Research Institute – IRRI) and wheat and maize (International Centre for the Improvement of Wheat and Maize – CIMMYT) are increasingly undertaking research that relates these staples to the growing of other crops.

6   As with most generalizations, this has some exceptions. Some perennial crops make heavy demands on soil nutrients, and others such as pineapples can require heavy agrochemical applications. On the general value of perennials in cropping systems, see Piper (1994) and Piper and Kulakow (1994).

7   Managed fallows have been largely ignored in the existing agricultural literature. To remedy this lack, a Southeast Asian regional workshop on intensification of farming systems was held in Bogor, Indonesia, in June 1997, with over 80 papers prepared for this collaborative effort of ICRAF, Cornell International Institute for Food, Agriculture and Development, the International Development Research Centre of Canada, and the Ford Foundation. Documentation of these resource management systems, mostly developed by farmers, is published in Cairns (2000).

8   The German and Dutch words for agriculture, *Landbau* and *Landbouw*, are more congenial to a three-dimensional conception of agriculture as they mean land-building.

9 'Mechanization' as used here refers to tractorization. Other forms of mechanization such as water pumps can be very valuable for increasing production, but they are not necessarily linked to monocropping in the way that tractorization is.

10 Those who can afford tractors usually own the best-quality land, making their practice of agriculture appear better.

11 When the labour power available for agricultural production is a constraint in some countries, this often reflects the fact that the low prices paid for agricultural commodities are keeping rural wage rates correspondingly low, influenced by urban-biased national policies and/or agricultural production subsidies in industrialized countries. National policies in developing countries have generally favoured urban consumers over rural producers, leading to low prices for food. Low food prices also reflect the extent of poverty, which depresses the purchasing power of the poor who have need for more food but do not have the means (effective demand) with which to acquire it. In such situations, low wages and low labour productivity for agriculture do not reflect either a true equilibrium or an efficient use of resources in terms of meeting human needs.

12 See Steiner (1982). That monocrop yields, being single, are easier to measure has contributed to the popularity of monocropping as a subject for agricultural research and extension. More effort is required to assess polycropping precisely. The UN Food and Agriculture Organization (FAO)'s world census of agriculture in the 1980s specifically ignored all crop mixtures, deciding to record crops only as monocultures (Chambers, 1997, p95).

13 This has been seen and documented most dramatically in Indonesia, where an IPM programme started with FAO assistance showed that rice yields would not decline, and in some instances increased, when use of chemicals was drastically cut back (more than 50 per cent), and in some cases terminated where cultural practices were changed. The key was giving farmers effective hands-on training in agroecosystem management, so that they began to diagnose problems themselves and experiment with solutions, developing alternatives to chemical dependence (Oka, 1997). Widespread use of chemicals had increased the problem of pest attacks on rice, inducing build-up of pesticide resistance in pest populations at the same time that it reduced the population of spiders and other 'beneficials' that prey on pests.

14 Recent research on rice IPM has found that maintaining the populations of 'neutral' insects in rice paddies, insects that are neither pests nor beneficials, is important. Their presence can sustain the populations of beneficials when pests have been eliminated, keeping these populations vigorous and available to deal with any new increases in pest populations. Keeping sufficient organic matter in the soil to support populations of neutrals is becoming part of an IPM strategy (Peter Kenmore, during Bellagio conference, pers comm).

15 See Fernandes et al (1997). On infertile acid soils, farmers often need to use certain chemical nutrients such as phosphorous and calcium to prime biological processes such as nutrient recycling and nitrogen fixation. Research in Costa Rica found that when cultivating beans, mulches of organic matter prevent phosphorous fertilizer from becoming bound to aluminium and other ions in the acid soil, making it more available for plant nutrition. Phosphorous applied in conjunction with organic

material produced as good or better yields as when three times as much phosphorous was applied directly to the soil (Schlather, 1998).

16 There is research indicating that the application of inorganic nitrogen fertilizer suppresses potentials for biological nitrogen fixation by reducing micro-organisms' production of the enzyme nitrogenase which enables soil microbes to transform nitrogen from the atmosphere into forms usable by the roots (Van Berkum and Sloger, 1983). This suggests that naturally-occurring nitrogen can be made unavailable by the application of nitrogen fertilizers, but it does not negate the point that organic and inorganic sources of nutrients are best managed in a complementary manner. It is worth contemplating the fact that since 1950, applications of nitrogen fertilizer have increased about 20-fold (Smil, 2000, p109), while crop yields have gone up at most three-fold. While nitrogen is often a limiting factor for plant growth, if it were of overwhelming importance for plant production, we should see more proportional increases in yield, rather than such sharply diminishing returns.

17 'The importance of water-control techniques in contrast with irrigation is consistently underestimated in the literature. There is a wide range of these techniques, including those that just hold water in the sandier soils [by increasing soil organic matter] as well as a series of measures to reduce runoff where crusting is the problem. These are not just indigenous techniques. The most important ones in the next decade have large potential yield effects (when combined with inorganic fertilizers) and need to be undertaken during the crop season, generally with animal traction, and not just as emergency measures on the most degraded or most easily degraded regions (hillsides)' (Sanders, 1997, p19). On this point generally, see FAO (1994).

18 In the rice-wheat rotation systems widely used in the Indo-Gangetic Plains of South Asia, certain kinds of ploughing techniques, adjusted by depth and timing, can retain enough water from the rice season for the following wheat season, so that the amount of water needed for the latter crop is reduced (Craig Meisner, CIMMYT/CIIFAD, pers comm). Seeding wheat in the standing rice crop towards the end of its growing season enables the wheat crop to benefit from residual soil moisture, reducing the need for irrigation (Peter Hobbs, CIMMYT, pers comm).

These low-till methods are being promoted by CIMMYT and IRRI because they can save water, raise yields, lower production costs, reduce weeds and herbicide use, plus reduce greenhouse gas emissions ('New Movement Among Farmers to Give Up the Plow Takes Root', press release from Future Harvest, The Hague, 2 October 2001, http://futureharvest.org/new/lowtill.shtml).

19 Most of the ideas in this chapter have been prompted from the co-authors' interactions with colleagues at Cornell University and in developing countries where CIIFAD has been engaged in collaborative, interdisciplinary programmes since 1990 to further the prospects for sustainable agricultural and rural development (Uphoff, 1996a). It is hard to know where ideas come from, and to give full or proportional credit where it is due. We take responsibility for presenting these ideas for critical consideration by researchers and practitioners, not claiming personal credit for all of them, and acknowledging our indebtedness to colleagues at Cornell and elsewhere for the stimulation and challenge they have contributed to this thinking. Critical review by Rainer Asse and Christopher Barrett of the whole manuscript was particularly helpful.

References

Acharya, M. S. (1997) 'Integrated vermiculture for rural development', *International Journal of Rural Studies*, vol 4, no 1, pp8–10

Altieri, M. A. (1995) *Agroecology: The Science of Sustainable Agriculture*, 2nd edition), Westview Press, Boulder, Colorado

Appelhof, M., Webster, K. and Buckerfield, J. (1996) 'Vermicomposting in Australia and New Zealand', *BioCycle*, vol 37, no 6, pp63–66

Avery, D. T. (1995) *Saving the Planet with Pesticides and Plastic: The Environmental Triumph of High-Yield Farming*, Hudson Institute, Indianapolis, Indiana

Avery, D. T and Avery, A. (1996) 'Farming to sustain the environment'. Hudson Briefing Paper no 190, May, Hudson Institute, Indianapolis, Indiana

Bene, J. G., Beall, H. W. and Cote, A. (1977) *Trees, Food and People*, International Development Research Centre, Ottawa

Berkes, F. (ed) (1989) *Common Property Resources: Ecology and Community-based Sustainable Development*, Belhaven Press, London

Borlaug, N. and Dowswell, C. R. (1994) 'Feeding a human population that increasingly crowds a fragile planet', in International Society of Soil Science, Supplement to Transactions of the 15th World Congress of Soil Science, Acapulco, Mexico, Chapingo, Mexico

Bruijn, G. H. de and Dharmaputra, T. S. (1974) 'The Mukibat system: A high-yielding method of cassava production in Indonesia', *Netherlands Journal of Agricultural Science*, vol 22, no 1, pp89–100

Cairns, M. F. (ed) (2000) *Voices from the Forest: Farmer Solutions Towards Improved Fallow Husbandry in Southeast Asia*, Bogor, International Centre for Research in Agroforestry, Southeast Asian Regional Research Programme. Proceedings of a Regional Conference on Indigenous Strategies for Intensification of Shifting Cultivation in Southeast Asia, Bogor, Indonesia, 23–27 June 1997

Carrol, C. R., Vandermeer, J. H. and Rosset, P. M. (1990) *Agroecology*, McGraw-Hill, New York

Chambers, R. (1997) *Whose Reality Counts? Putting the First Last*, Intermediate Technology Publications, London

Conway, G. R. (1987) 'The properties of agroecosystems', *Agricultural Systems*, vol 24, no 1, pp95–117

Diaz Cisneros, H. et al (1997) 'Plan Puebla: An agricultural development program for low-income farmers in Mexico' in Krishna et al (eds) *Reasons for Hope: Instructive Experiences in Rural Development*, Kumarian Press, West Hartford, Connecticut, pp120–136

FAO (1994) *Water Harvesting for Improved Agricultural Production*, Food and Agriculture Organization, Rome

Fernandes, E. C. M. and Matos, J. C. (1995) 'Agroforestry strategies for alleviating soil chemical constraints to food and fiber production in the Brazilian Amazon' in Seidl, P. R. et al (eds) *Chemistry of the Amazon: Biodiversity, Natural Products and Environmental Issues*, American Chemical Society, Washington, DC, pp34–50

Fernandes, E. C. M., Motavalli, P., Castilla, C. and Mukurumbira, L. (1997) 'Management control of soil organic matter dynamics in tropical land-use systems', *Geoderma*, vol 79, no 1, pp49–67

Foresta, H. de, Basri, A. and Wiyono (1994) 'A very intimate agroforestry association: Cassava and improved homegardens – The Mukibat technique', *Agroforestry Today*, vol 6, no 1, pp12–14

Harrison, P. (1987) *The Greening of Africa*, Penguin Books, New York

Jodha, N. S. (1992) *Common Property Resources: A Missing Dimension of Development Strategies*, World Bank, Washington, DC

King, F. H. (1911) *Farmers over Forty Centuries, or Permanent Agriculture in China, Korea and Japan*, Rodale Press, Emmaus, Pennsylvania (republished 1973)

Krishna, A. (1997) 'Participatory watershed development and soil conservation in Rajasthan, India' in Krishna et al (eds) *Reasons for Hope: Instructive Experiences in Rural Development*, Kumarian Press, West Hartford, Connecticut, pp255–272

Lund, S. and Fafchamps, M. (1997) 'Risk-sharing networks in rural Philippines', unpublished paper, Stanford University, Department of Economics, Stanford

Lundgren, B. O. and Raintree, J. B. (1982) 'Sustained agroforestry' in Nestel, B. (ed) *Agricultural Research for Development: Potentials and Challenges in Asia*, International Service for National Agricultural Research, The Hague, pp37–49

Mollison, B. (1990) *Permaculture: A Practical Guide for a Sustainable Future*, Island Press, Washington, DC

Oka, I. N. (1997) 'Integrated crop pest management with farmer participation in Indonesia' in Krishna et al (eds) *Reasons for Hope: Instructive Experiences in Rural Development*, Kumarian Press, West Hartford, Connecticut, pp184–199

Piper, J. K. (1994) 'Neighborhood effects on growth, seed yield, and weed biomass for three perennial grains in polyculture', *Journal of Sustainable Agriculture*, vol 4, no 2, pp11–31

Piper, J. K. and Kulakow, P. A. (1994) 'Seed yield and biomass allocation in sorghum bicolor and F1 and backcross generations of *S. bicolor* × *S. halepense* hybrids', *Canadian Journal of Botany*, vol 72, no 4, pp468–474

Postel, S. (1996) 'Dividing the water: food security, ecosystem health, and the new politics of scarcity', Worldwatch Paper no 132, Worldwatch Institute, Washington, DC

Reij, C., Scoones, I. and Toulmin, C. et al (eds) (1996) *Sustaining the Soil: Indigenous Soil and Water Conservation in Africa*, Earthscan, London

Sanders, J. H. (1997) 'Developing technology for agriculture in Sub-Saharan Africa: Evolution of ideas, some critical questions, and future research', Discussion Paper, International Food Policy Research Institute, Washington, DC

Schlather, K. (1998) 'The dynamics and cycling of phosphorous in mulched and unmulched bean production systems indigenous to the humid tropics of Central America', unpublished PhD thesis, Cornell University, Ithaca, New York

Smil, V. (2000) *Feeding the World: A Challenge for the Twenty-First Century*, MIT Press, Cambridge, Massachusetts

Steiner, K. G. (1982) 'Intercropping in tropical smallholder agriculture with special reference to West Africa', GTZ Publication no 137, Gesellschaft für Technische Zusammenarbeit, Eschborn

Tanner, J., Holden, S. J., Winugroho, M., Owen, E. and Gill, M. (1995) 'Feeding livestock for compost production: A strategy for sustainable upland agriculture on Java' in Powell, J. M., Fernández-Rivera, S., Williams, T. O. and Renard, C. (eds) *Livestock and Sustainable Nutrient Cycling in Mixed Farming Systems of Sub-Saharan Africa*, International Livestock Centre for Africa, Addis Ababa, pp115–128

Uphoff, N. (1986) *Improving International Irrigation Management with Farmer Participation: Getting the Process Right*, Westview Press, Boulder, Colorado

Uphoff, N. (1996a) *Learning from Gal Oya: Possibilities for Participatory Development and Post-Newtonian Social Science*, Intermediate Technology Publications, London

Uphoff, N/ (1996b) 'Collaborations as an alternative to projects: Cornell experience with university-NGO-government networking', *Agriculture and Human Values*, vol 13, no 2, pp42–51

Van Berkum, P. and Sloger, C. (1983) 'Interaction of combined nitrogen with the expression of root-associated nitrogenase activity in grasses and with the development of N2 fixation in soy bean (*Glycine max* L Mon)', *Plant Physiology*, vol 72, pp741–745

Vandermeer, J. (1989) *The Ecology of Intercropping*, Cambridge University Press, Cambridge, UK

Winrock International (1992) *Assessment of Animal Agriculture in Sub-Saharan Africa*, Winrock International Institute for Agricultural Development, Morrilton, Arkansas

23

Soil Recuperation in Central America: How Innovation was Sustained after Project Intervention

Roland Bunch and Gabinò López

Introduction

Much has been said and written about the sustainability of agricultural development. But there have been few studies on programme impact after the outside intervention has ended. This chapter describes a study of three agricultural development efforts in Guatemala and Honduras and assesses impacts up to 15 years after the termination of outside intervention. The study was carried out by the Honduran organization (Associación de Consejeros una Agricultura Sostenible, Ecológica y Humana, COSECHA).

The study assessed the impact that soil recuperation interventions have had over many years. The results show considerable increases in productivity after intervention, and indicate that while specific technologies do not generally have long-term sustainability, the process of agricultural innovation does. The study points to a need for future agricultural development programmes to design their work in such a way that villagers are given strong motivation to innovate.

The three areas studied

This chapter concentrates on the following three areas of Guatemala and Honduras.

The Cantarranas area

Between 1987 and 1993, the Cantarranas Integrated Agricultural Development Program, financed by Catholic Relief Services and managed by World Neighbors, worked in some 35 villages around the central Honduran town of Cantarranas (Bunch, 1990). Using in-row tillage and intercropped green manures as its cutting edge, it expanded into a general programme of agricultural development and preventive health.

Note: Reprinted from *Fertile Ground: The Impact of Participatory Watershed Management,* by Hinchcliff, F., Thompson, J., Pretty, J., Guijt, I. and Shah, P. (eds), copyright © (1999) with permission from Technology Publishers, London

Cantarranas lies at about 300m in elevation in a narrow valley about 40km long, between two parallel mountain ranges that rise to over 1800m. The programme worked almost entirely with small farmers with 2–5 hectare landholdings. These hillsides vary in slope, with an average of about 30 per cent. The forests have been seriously degraded. The climate of the Cantarranas area varies from hot and semi-arid, with frequent and severe droughts in the bottom of the valley, to a cool climate, with sufficient rainfall for six months each year.

The Guinope area

Between 1981 and 1989, a similar World Neighbors programme worked in 41 villages, most of which are included in the townships of Guinope, San Lucas, and San Antonio de Flores in southeastern Honduras (Bunch, 1988). This programme also worked heavily in soil recuperation, basic grains and diversification, as well as preventive health. The programme's lead technologies were drainage ditches (at 0.5 per cent slope) with live barriers and the use of chicken manure.

The Guinope area contains the same variations in altitude and rainfall as Cantarranas, but with less severe slopes. Nevertheless, an impenetrable subsoil underlies the 15cm to 50cm deep topsoil. When this thin layer of topsoil has eroded away, agriculture becomes impossible. Before 1981, emigration from the Guinope area was heavy; some residents referred to it as a 'dying town'.

The San Martin Jilotepeque area

The San Martin Integrated Development Programme was financed by Oxfam-UK and carried out by World Neighbors between the years 1972 and 1979 (Bunch, 1977). It was a highly integrated programme, working in everything from agriculture and health to road construction, functional literacy and cooperative organization. The Programme used contour ditches and a side-dressing of nitrogen on maize as the initial technologies to motivate people.

The San Martin township lies just 50km west of Guatemala City. The southern half of the township, where the Programme worked in some 45 villages, varies in altitude from about 800m to 2000m, and has enough rainfall for a good maize crop in most years. The mainly Cakchiquel Indian population is extremely land poor, owning an average of less than 0.5ha of seriously depleted land per family.

Methodology for the impact study

The methodology used in the study varied from area to area. The most explicitly participatory work occurred in Guinope and Cantarranas. In these cases, team members knew the areas well, having worked in the programmes being studied. The four COSECHA personnel involved in the San Martin study were all Cakchiquel Indian farmers originally from the area, who had been trained in the programme and gradually progressed from programme participant to extensionist to programme director. The methodology consisted of a combination of:

- observation of the plots in the study villages, including visual productivity estimates and the use of a checklist of questions about easily observed factors in the fields (the existence of contour live barriers, contour ditches and fruit trees);
- individual open-ended interviews;
- open-ended informal conversations held with people known to the study team;
- participatory rural appraisal (PRA) methods with groups of villagers which included men, women and children – these included mapping exercises, priority exercises and participatory economic analyses of specific crops;
- a review of programme documents, including evaluations made of programme impact.

COSECHA personnel made a list of all the 121 villages, and divided these into three roughly equal categories:

1   those in which they judged the impact to have been best;
2   those in which the impact was moderate;
3   those in which the impact was relatively poor.

A composite list was then made, averaging the ratings in the three separate lists. One village was then selected from the best category, two from the middle category, and one from the poorer impact category. These villages were selected so as to provide an even geographic spread within the programme area, afford fairly good access during the wet season (when the study was carried out), and avoid contamination of the findings by subsequent work of other development agencies.

Findings of the impact study

Changes in technologies used

Different technologies had been given different emphases by each programme, and in different villages within each programme. Nevertheless, the figures displayed in the tables below give some idea of the relative sustainability of the technologies. For each technology, the number of farmers using the technology is shown at the time of programme initiation, programme termination, and at the time of the study (1994).

The villages studied in Guinope and Cantarranas are presented in decreasing order of previously judged quality of impact. That is, the first village listed is that chosen from the one-third of the villages with best impact, the next two from the group of average impact villages, and the fourth from the group of least impact. These results show that the overall level of continuing innovation, despite programme termination, is remarkable.

Changes in productivity

In Central America, maize is the basic staple. As it is very sensitive to soil fertility, maize productivity is a good indication of overall soil fertility, and of productivity in general.

Table 23.1 *Changes in adoption of resource-conserving technologies in three programmes in Central America during and after projects (no of farmers with technologies)*

	At initiation	At termination[a]	In 1994
Contour grass barriers			
San Martin	1	100	203
Guinope	0	44	33
Cantarranas	0	48	44
Contour or drainage ditches			
San Martin	1	136	162
Guinope	0	56	43
Cantarranas	0	39	34
Green manures			
San Martin	0	21	38
Guinope	0	0	2
Cantarranas	0	14	12
Crop rotation			
San Martin	0	6	10
Guinope	12	46	125
Cantarranas	0	97	119
No longer burn fields or forests			
San Martin	0	0	0
Guinope	2	83	117
Cantarranas	0	77	108
Organic matter as fertilizer			
San Martin	10	42	124
Guinope	4	100	213
Cantarranas	30	53	60

Note: a Project termination dates were: San Martin 1979; Guinope 1989; Cantarranas 1991.

Table 23.2 shows that major increases in productivity have been achieved after the programmes ended.

All averages of harvest data are rounded off to the nearest 100kg/ha in the tables. Most of the harvest data come from farmers, based on the number of bags they carry home during the harvest of the mature grain. These calculations, therefore, exclude grain lost to thievery, eating ears harvested early, and occasional ears given to labourers as partial recompense for their work.

The figures are especially dramatic in San Martin, where the time since programme termination is longest. Thus these figures are probably the most important in the entire

Table 23.2 *Productivity of maize (in kg/ha) at project initiation, termination and at the time of the study*

	Initiation	Termination	1994
San Martin			
San Antonio Correjo	400	2400	4800
Las Venturas	400	2800	5200
Xesuj	300	2000	3200
Pacoj	500	2800	4800
Guinope			
Pacayas	600	3200	4200
Manzaragua	600	2000	2000
Lavenderos	600	2000	2000
Cantarranas			
Guacamayas	800	nd	none
Joyos del Caballo	400	2000	2200
Guanacaste	800	1900	1900

Note: The lack of maize production in Guacamayas is due to the fact that farmers make more money from vegetables, and prefer to buy maize. Thus, the lack of maize is precisely because of the dramatic increases in yields and value of production achieved in this village since programme termination.

study, clearly demonstrating that productivity has continued to improve after the programmes' termination. Although the increase in yields is not as dramatic as during the programmes' existence, these figures leave no doubt that even though some practices have been abandoned, the productivity of the better villages (and all the villages in San Martin) has continued to rise after the programmes ended.

Table 23.3 shows the productivity of beans in the villages where they are grown.

Further impacts

Other positive impacts in the project areas have also occurred:

- increased wage rates;
- increased land values;
- decreasing or reversed emigration from the project areas (Table 23.4);
- decreased resource degradation;
- increased numbers of trees planted;
- almost total elimination of the use of herbicides through hand weeding or green manures;
- significant reduction in the use of chemical fertilizers with a variety of organic fertilizers now used;

Table 23.3 *Productivity of edible beans (Phaseolus vulgaris) in kg/ha at project initiation, termination and at the time of the study*

	Initiation	*Termination*	*1994*
San Martin			
San Antonio Correjo	200	1200	1800
Las Venturas	100	200	800
Xesuj	200	1200	2000
Guinope			
Lavenderos	100	800	800
Cantarranas			
Guacamayas	none	900	1500
Joyos del Caballo	500	1500	1500
Guanacaste	500	900	900

Table 23.4 *Impact of programmes on migration*

Migration (no of households)	*Initiation*	*Termination*	*1994*
San Martin:			
San Antonio Correjo	65	nd	4
Las Venturas	85	nd	4
Guinope: 3 villages	38	0	(2)[a]
Cantarranas: 3 villages	nd	10	(6)[a]

Note: a These represent negative outmigration, or families moving back to rural areas.

- increased crop diversity and practice of intercropping;
- an increase in local savings, leading to a decreasing dependence on formal credit and increased investment in education, land improvement and purchasing animals;
- marked improvements in diets, including the consumption of more vegetables, native herbs, milk and cheese;
- improved resilience and resistance to drought and climatic variability;
- increased involvement in local groups, such as producers' associations, agricultural study groups, community improvement committees, or groups formed by whole villages to protect communal forests from loggers or corrupt municipal officials.

Less success has been achieved in reducing dependence on the use of insecticides and fungicides. Central America in general has been slow to find feasible alternatives for these chemicals. The programmes taught people the dangers of pesticides, the importance of integrated pest management, and encouraged them to try any others they could. But few effective alternatives have been found. Finding alternatives to insecticides is a major area for additional work.

The most disappointing finding is the lack of spontaneous technological spread *between* villages. Whereas spontaneous spread within villages does occur, spread between villages is negligible. This might be attributable to the Guatemalan counter-insurgency campaign of the 1980s, which purposely turned villages against each other, and the Honduran villagers aversion to walking from one village to another (there being no major village markets in the programme areas). Nevertheless, there is no evidence that the technology would have spread in the absence of these factors.

Local innovation

Local innovation is critical to villagers' becoming the 'subjects' of their *own* development. Within the study sites, the amount of continuing innovation has been remarkable. In San Martin, over 30 innovations have been adopted successfully since programme termination. These include the introduction of new crops (cauliflower, broccoli and herbs), adoption of new green manures (velvetbean and *Tephrosia*), planting improved pastures (such as Kikuyu grass), and building stables for animals.

Probably the most important is that each village has developed at least one whole new system of production. In one village, a whole system of intensified cattle raising has been developed, in which improved pastures are planted to supplement the Napier grass barriers, legumes are being tried to increase protein, animals are stabled, pastures rotated, and cheese making increases the value of the milk before it is marketed. In other villages, much land has gone into coffee or fruit production. In Las Venturas, a system of sustainable forest management has become a major economic factor, where villagers are planting out seedlings to fill clearings, and are cutting a certain number of the largest trees each year.

In Honduras, innovations have occurred in virtually all the villages.

- In Guacamayas, new crops include avocados, lemons, potatoes, tomatoes, green beans, and cauliflower; farmers are also experimenting with new green manures *(Mucuna pruriens* and *Phaseolus coccineus)*, organic vegetable production, and have developed a simple way of processing coffee pulp.
- In Guanacaste, people now grow fruit and cassava on a commercial scale for the first time, mainly because of their observed resistance to drought.
- In Lavanderos, chilli peppers, cabbage, carrots, beets and strawberries are all common crops which did not exist in the village at programme termination.
- In Pacayas alone, people counted 16 innovations adopted since programme termination. These consist of four new crops (chilli peppers, beets, onions and carrots); two green manures; two new species of short grass for use as contour barriers in vegetable fields; a zero-cash-cost chicken pen made entirely of king-grass planted on a rectangle marigolds used to control nematodes; the feeding of both lablab bean *(Dolichos lablab)* and velvetbean to cattle and chickens; numerous cases of nutrient recycling from fish ponds and animals to vegetables and field crops and back; the use of human waste through composting latrines; home-made sprinklers; and the use of Napier grass on cliff edges to stop further caving in.

Discussion

The persistence of specific technologies

The technologies that have proven sustainable over a 15-year period in San Martin Jilootepeque without significant abandonment are contour grass barriers fertilization with organic matter and crop rotation. In central Honduras, however, the grass barriers were losing popularity after only five years, and the local green manure systems have not fared well – largely because they still need some improving. The cessation of agricultural burning has also continued to spread, and should continue to do so.

However, fertilization with organic matter is really not a single technology, but rather a range of quite varied technologies. Farmers in San Martin are using methods of organic fertilization that are quite different from those originally introduced. The ending of burning is not really a technology either, but rather the absence of a previous technology, which has been superseded by the use of alternative sources of organic nutrients. Therefore, only one technology – crop rotation – has really survived in its original form for at least 15 years.

Might this lack of technological sustainability be caused by poor selection of the technologies? It would appear not. These same technologies work well elsewhere. It is much more likely that most of the technologies have fallen by the wayside because changing circumstances, such as emerging markets, disease and insect pests, land tenure, soil fertility, labour availability and costs, and the adoption of new technologies, have reduced or eliminated their usefulness. Wheat growing, a major programme technology in San Martin (adopted by over 600 farmers), was lost completely because cauliflower and broccoli – also cash crops – paid much better. Broccoli and cauliflower, in turn, disappeared when farmers nearer the processing plant took over San Martin's market.

Similarly, when grass barriers that trap eroding soil build up a natural terrace (some four to six years after adoption), farmers stop cleaning out their contour ditches. As many villagers have explained, 'If my ditches never fill up with water any more, why should I keep cleaning them out?' Both in-row tillage and cover crops can make contour barriers irrelevant. This study has led us to believe that the half-life of well-chosen technologies for farmers is probably about six years.

The sustainability of the development process

The results clearly indicate that even though the vast majority of specific technologies disappeared, farmers' productivity continued to climb. In some of the best villages, yields are continuing to increase at rates comparable with those achieved during the programmes' presence. Thus, the sustainability of specific technologies may well be largely irrelevant. Much more relevant to farmers' wellbeing and productivity is the sustainability of the development process. That this process can lead to significant increases in people's wellbeing and can be carried on by the villagers themselves, is probably the most important single finding of this study.

Some might wonder if this continuing improvement was not the result of other outside programmes' work in the area. Yet no other programmes worked for any significant period of time on soil conservation or crop production in these villages during the ensuing years, except in minor ways in the cases of Pacayas and Pacoj. Another possibility would be that agricultural productivity was increasing during these years in any case, and the villages studied merely shared in a more general improvement. However, in villages near the studied villages, yields presently average less than 1.6ha (compared to Las Venturas' 5.2 t ha^{-1}), and most of the last two decades relatively small increase in yields is directly attributable to heavier use of expensive chemical fertilizers.

What has happened is that the process of agricultural innovation was greatly accelerated by the programmes to the point that it is capable of improving yields over the medium and long term. This increase in the intensity of the innovation process requires that villagers:

- learn the rudiments of simple scientific experimentation;
- learn a minimum of very basic theoretical ideas about soils and agriculture, in order to orient their experiments in useful directions;
- learn to share with each other the results of their experiments;
- become motivated to do all of the above sustainably.

The key to designing a sustainable soil conservation or agricultural programme does not consist, therefore, of choosing a group of technologies that will be sustainable. Rather, the key is choosing a very few technologies that will motivate farmers to become involved in a process of innovation, to search for new ideas, experiment with them, adopt those that prove useful and share the experimental results with others.

One of the striking features of soil conservation technologies is that they rarely accomplish any of the above. We have heard farmers say dozens of times, 'But I can't eat a grass barrier'. For farmers to accept soil conservation technologies and become involved in a sustainable process, the technologies must be combined with a technology that enhances yields. It is the increase in yields that convinces the farmers of the value of soil conservation.

Through such a process, subsidies in the form of food-for-work or direct financial incentives become irrelevant. If the yields have increased or costs decreased, artificial incentives are not needed. If the yields have not increased, no artificial incentive will make the technology's adoption sustainable.

Recommendations for sustainable technologies

Combine soil conservation or recuperation technologies with technologies that raise yields or reduce costs

Farmers adopt technologies that bring rapid recognizable success. But soil conservation technologies cannot bring rapid increases in yield. For villagers to become motivated to maintain and/or improve soil conservation technologies, they must achieve such increases while adopting soil conservation measures.

Use intercropped green manures (or other green manures that can be produced on land with no opportunity cost) wherever possible

Most small farmers already use virtually all the organic matter they produce, and can seldom afford to buy more. Therefore, we must seek to increase the high-nitrogen bio-mass that can be produced on farm. One of the best-known ways to do this is to produce green manures or cover crops. These can usually be produced without spending cash and in ways that involve no opportunity cost – they can be intercropped, on wasteland, under trees, during periods of frost, or during the dry season. The best of these are multi-purpose crops – they can fix nitrogen, prevent soil erosion, increase soil fertility dramatically, control weeds, control nematodes, and/or provide both highly nutritious animal fodder and human food (Bunch, 1993; Flores, 1991–96).

Use simple, low-cost and appropriate technologies

These should give positive ecological impact, rapid recognizable results, and the possibility that the technology can serve as a basis for many other technological innovations.

Maintain flexibility in technological recommendations

Giving exact specifications and making only one recommendation for solving a problem reduces the space for villagers to experiment and make the technology their own.

Initiate the process with the smallest number of technologies consistent with achieving significant success

Since the objective of the initial technologies is not to introduce permanent innovations, but rather to get people involved a process of self-generated innovation, there is no reason to introduce a large number of technologies, or even a 'technology package'. In time, as in the villages studied, the programme and the farmers themselves will expand the array of technological innovations being used. This is the only way to sustain a positive impact on agricultural productivity.

References

Admassie, Y. (1992) *The Catchment Approach to Soil Conservation in Kenya. Regional Soil Conservation Unit Report No 6.* RSCU, Swedish International Development Authority, Nairobi

Anderson, D. (1984) 'Depression, dust bowl, demography, and drought: The colonial state and soil conservation in East Africa during the 1930s, *African Affairs*, pp321–343

Bunch, R. (1977) 'Better use of land in the highlands of Guatemala', in Stamp, E. (ed) *Growing Out of Poverty*, Oxford University Press, Oxford

Bunch, R. (1988) 'Guinope integrated Development Programme, Honduras', in Conroy, C. and Litvinoff, M. (eds) *The Greening of Aid: Sustainable Livelihoods in Practice*, Earthscan, London

Bunch, R. (1990) 'Low input soil restoration in Honduras: The Cantarranas farmer-to-farmer extension programme', *Gatekeeper*, vol 23, International Institute for Environment and Development, London

Bunch, R. (1993) *The Use of Green Manures by Villager Farmers, What We Have Learned to Date*, Technical Report no 3, Second Edition, CIDICCO, Tegucigalpa

Beinart, W. (1984) 'Soil erosion, conservationism and ideas about development: A southern African exploration, 1900–1960', *Journal of Southern African Studies II*, pp52–83

Bennett, H. H. (1939) *Soil Conservation*, McGraw-Hill, New York, CIIR in association with James Currey, London

Eckbom, A. (1992) *Economic Impact Assessment of Implementation Strategies for Soil Conservation. A comparative Analysis of the On-farm and Catchment Approach in Trans Nzoia, Kenya*, Unit for Environmental Economics, Department of Economics, Göthenberg University, Sweden

Figueiredo, P. (1986) *The Yield of Crops on Terraced and Non-terraced Land. A Field Survey in Kenya*, Swedish University of Agricultural Sciences, Uppsala

Flores, M. (1991–1996) Series of *Crop Cover Newsletters*, CIDICCO, Tegucigalpa

Grönvall, M. (1987) *A Study of Land Use and Soil Conservation on a Farm in Mukurweini Division, Central Kenya*. Swedish University of Agricultural Sciences, Uppsala

Holmgren, E. and Johansson, G. (1988) *Comparison Between Terraced and Non-terraced Land in Machakos District, Kenya*, Swedish University of Agricultural Sciences, Uppsala

Howell, J. (ed) (1988) *Training and Visit Extension in Practice*, ODI, London

Hunegnaw, T. (1987) *Technical Evaluation of Soil Conservation Measures in Embu District*, Report of a minor field study, IRDC, Swedish University of Agricultural Sciences, Uppsala

Kiara, J. K., Segerros, M., Pretty, J. N. and McCracken, J. (1990) *Rapid Catchment Analysis in Murang'a District, Kenya*, Soil and Water Conservation Branch, Ministry of Agriculture, Nairobi

Lindgren, B.-M. (1988) *Economic Evaluation of a Soil Conservation Project in Machakos District, Kenya*, Swedish University of Agricultural Sciences, Uppsala

Lundgren, L. (1993) *Twenty Years of Soil Conservation. SIDA Report No 9*, Regional Soil Conservation Unit, Nairobi

MOA (1981) *Soil Conservation in Kenya. Especially in Small-scale Farming in High Potential Areas Using Labour Intensive Methods*, 7th edition, SWCB, Ministry of Agriculture, Nairobi

MOA/MALDM (passim) *Reports of Catchment Approach Planning and Rapid Catchment Analyses*, SWCB, Ministry of Agriculture, Livestock Development and Marketing, Nairobi, 1988–1993

Moris, J. (1990) *Extension Alternatives in Tropical Africa*, ODI, London

Mullen, J. (1989) 'Training and visit system in Somali: Contradictions and anomalies', *Journal of International Development*, vol 1, pp145–167

Pretty, J. N. (1990) Rapid *Catchment Analysis for Extension Agents*, Notes on the 1990 Kericho Workshop for the Ministry of Agriculture, Kenya, IIED, London

Pretty, J. N., Kiara, J. K. and Thompson, J. (eds) (1993) *The Impact of the Catchment Approach to Soil and Water Conservation: A Study of Six Catchments in Western, Rift Valley and Central Provinces, Kenya*, SWCB, Ministry of Agriculture, Livestock Development and Marketing, Kenya

Raikes, P. (1988) *Modernising Hunger: Famine, Food Surplus and Farm Policy in the EEC and Africa*, Catholic Institute for International Relations in Collaboration with James Currey, London, and Heinemann, Portsmouth, New Hampshire

24

Sustainable Rural Life and Agroecology, Santa Catarina State, Brazil

Sergio Pinheiro

Summary and introduction

This case study analyses and summarizes a sustainable development experience based on agroecology developed by the Ecological Farmers Association of the Hillsides of Santa Catarina State (Agreco), an NGO in southern Brazil. One of the main characteristics of this ecological system is that it considers not only the organization, management and control of the bio-physical systems (called 'production systems' or 'hard-systems'), but also aims to understand and develop the interactions which characterize more abstract and complex systems (called 'soft-systems'), particularly the human relations and the sustainable development of the territory in which the ecosystem operates.

This experience was initiated by people who were born in the territory but, like most Brazilians, migrated to cities. Many of these 'new urban people' stayed in contact with relatives and friends who remained behind and began to take advantage of a development opportunity that started with the sale of organic products produced in the territory. They realised that agroecology and organic agriculture could be the basis for a philosophical and methodological development strategy for the small family farmers living at the hillsides of Santa Catarina State.

Consequently, Agreco was formed, initially involving an organized network of several small agroindustries which were created to process and add value to the primary organic products and to assist with marketing. Other development actions and projects soon complemented the initiative:

- a rural tourism project, consolidating relationships between rural and urban people and creating new income opportunities;
- a credit cooperative, offering an alternative to the official financial system to which small family farmers often do not have access.

Note: Reprinted from *Organic Agriculture, Environment and Food Security* by El-Hage, Scialabba, N. and Hattan, C. (eds), Environment and Natural Resources Servise Series no 4, copyright © (2002) with permission from the Food and Agriculture Organization of the United Nations, Rome

In addition, two important Forums were organized: the Solidarity Economy Forum, bringing together urban consumers and rural producers, and the Hillsides Development Forum, involving stakeholders interested in collective action for sustainable development in the region.

History and characteristics of the territory

The hillside region of Santa Catarina State in southern Brazil is a very beautiful territory (approximately 2000km^2) in which many strategic rivers have their sources and develop for Santa Catarina. They form one of the largest catchments of the State, which include some of the most populated cities of Santa Catarina, such as *Florianópolis* (State Capital), *São José*, *Palhoça* and *Biguaçu*. Mineral and thermal waters emerge from the subsoil. The region is characterized by hilly topography with altitudes varying from 400 to 1800m above sea level, offering a variety of climatic conditions and vegetation types. From a sustainable development perspective, these characteristics offer opportunities for the creation of natural parks, intensification of agroecotourism and other environmental management projects.

Most villages share similar problems and characteristics: a small population, a traditional rural economy and a location far away from the main roads, tourist and urban consumer centres. Today, agriculture is the main economic activity of approximately 80 per cent of the families, particularly small family farmers.

During the first colonization period at the end of the 19th century and beginning of the 20th century, a 'traditional' type of agriculture was practised, characterized by a diversification of crops and livestock used mainly for subsistence. From the early 1960s, agriculture experienced its first transformation process with the partial modernization of the tobacco crop. Tobacco became the main cash crop for most small family farmers in the region, who began to buy modern inputs (chemical fertilizers, pesticides and fungicides) and sell the harvest to the tobacco companies. By the early 1990s, as in most regions of Brazil and other developing countries, this partial modernization process (which occurred particularly in crops like tobacco, soybeans, irrigated rice and apples) was resulting in negative social and environmental impacts: rural exodus, poverty, urban violence, environmental degradation, health problems and other socio-environmental effects reaching levels never before observed in the region.

In attempt to obtain another source of income and as a result of the tobacco crisis (mainly due to negative social and environmental impacts, together with low prices and low farmers' incomes), many farmers launched into intensive pig and poultry production systems designed and controlled by large agroindustries. Others increased native wood extraction, charcoal production and animal hunting for survival. These activities increased environmental degradation such as water pollution, soil erosion and devastation of native vegetation.

As a consequence, many people born in the region migrated to the cities; however, they maintained connections with the territory and with relatives and friends who had remained behind. Members of the Florentino Schmidt family, for example, decided to market produce from the region. In 1982, motivated by negative effects in both environmental and human health (eg water course pollution, soil contamination, intoxication and diseases due to agrochemicals), the Schmidt family decided to stop tobacco production and return to diversified production of vegetables, eggs, cheese and other products. They sold their products directly to consumers in urban centres, providing deliveries to households, universities and other places. Later, other neighbouring families joined this process, increasing the job and income opportunities in the area.

The foundation of Agreco and the ecological development option

Since May 1991, with the first *Gemüse Fest* (Vegetable Festival), the relationship between rural people who went to study and work in the cities and families who stayed behind has been strengthened. The *Gemüse Fest* has since become an annual event that mobilizes many other families in the region, facilitating meetings between farmers and urban-based friends or relatives. From these meetings partnerships have evolved with the aim of reinforcing economic opportunities and alternatives, and in 1996 the Ecological Farmers Association in the Hillsides of Santa Catarina State (Agreco) was founded.

Agreco is a civil non-profit organization (an NGO which includes mainly local people but also some 'foreigners') based in Santa Rosa de Lima. The organization aims, through agroecology and organic agriculture, cooperation, solidarity and team work, to contribute to the transformation of the production systems of its associated family farmers (from chemical to organic systems), to add value to production through processing and marketing, to consolidate relations between rural and urban people and to create new income opportunities through rural ecological tourism, technical and administrative assistance, farm management advice, and access to financial resources as well as facilitating and motivating the organization of family farmers groups and small agroindustrial units.

Following the formation of Agreco, and in response to the negative environmental impacts, Agreco defined in its Association Internal Norms and Policies document (Regimento Interno da Associação, 1997) that all associates must develop agroecological systems and promote sustainable management on their farms. As a result, agroecological systems and agritourism, among other activities promoted by Agreco, have contributed to the decrease in intensive tobacco, pig and poultry systems as well as wood extraction, charcoal production and animal hunting while Agreco's associates have increased their activities towards the protection of native animals and vegetation.

In addition to these objectives, two important Forums were organized: the Solidarity Economy Forum, approximating urban consumers and rural producers, and the Hillsides Development Forum, which involves both urban and rural stakeholders interested in collective action for sustainable development in the region.

Sustainable development based on agroecology

Agreco's ideology for sustainable development based on agroecological systems and solidarity is placed in the context of the social-environmental crisis experienced by modern agriculture. Agreco's approach aims to develop agricultural policies and systems moved by cooperation and solidarity instead of competition and individualism. It also seeks to associate traditional agricultural practices still applied by family farmers in the region with sustainable development principles and ecological knowledge accumulated by science over recent decades. To this end, Agreco has established partnerships with the Federal University of Santa Catarina (UFSC), with the Scientific and Technological Council of Brazil, with the Ministry of Rural Development, with the Government of Santa Catarina and with other stakeholders and several Municipal Governments in the region.

In 1997, the first organic production system of legumes, honey, grains and fruits involved 20 families and about 50 people associated with Agreco. Later, other organic production systems were organized, expanding Agreco's operational territory, including more families and diversification of production. By 1998, Agreco included about 200 associates and involved 50 families. Today there are now about 500 associates, involving over 200 families.

This increase in associates was stimulated by the organization of a network of 53 small agroindustries financed by the National Programme to Empower Small Family Farmers (Pronaf), through loans particularly designed to meet the requirements and capacities of small family farmers. The network was constructed with a view to processing and adding value to the various primary organic products, as well as to facilitating marketing and creating jobs for family farmers. There are now 27 agroindustries processing horticultural products, jam, tinned food, sugarcane brandy, honey, milk, eggs, free-range chicken and bread providing 206 jobs to 120 families. A further 299 jobs from another 120 families have been provided in the primary organic production. Most primary production units are in the transformation process from conventional to organic systems while all agroindustries have been operating in the organic system since the beginning. In August 2001, the First Organic Products Festival of the region was held.

The network of agroindustries is coordinated and assisted by a central support unit providing management and technical advice to the individual agroindustries, particularly in the planning of the marketing and production process. It is a kind of 'virtual network' (a 'soft-system') which functions in a very limited physical space consisting of four rooms: secretary, computers, technicians and library, very different from the large traditional cooperatives. All marketing activities are carried out directly by the central support unit. After the production plan for each agroindustry is established, the agroindustries interact with each other in order to exchange some primary products and family labour. They also 'talk' to each other about the organization and participate in Agreco's meetings and events.

The network of agroindustries has been very important in reversing the economic stagnation of the area and in the consequent provision of new job alternatives for family farmers. It has stimulated associates to work in groups, to add value and improve the quality of their products, processing them according to market demand and to facilitate the approximation between producers and consumers.

Organic products from the region are now being sold in more than ten supermarket chains all over Santa Catarina State, the volume of sales increasing every year. Products are marketed at local fairs as well as through direct delivery of small baskets to urban consumers in Florianopolis, the State Capital. Schools in Santa Catarina have also begun to serve organic products from Agreco.

The sustainable rural life project

Drawing on Agreco's experience, in January 2001, the Sustainable Rural Life Project was implemented with technical and financial support from the National Service for Support of Small and Medium Business (Sebrae Nacional, group of private enterprises with informal links to the Government). This project aims to consolidate the new sustainable development process in the hillsides of Santa Catarina offering theoretical and methodological references by which similar initiatives in other regions of Santa Catarina and Brazil can be orientated. The focus is the development of a learning support process aiding the training of Agreco's associates as well as other interested people.

This project is subdivided into six subprojects (or thematic topics): (1) interest-raising; (2) learning (including training); (3) organic production and processing; (4) market and marketing; (5) implementation of quality control, management and quality certification processes; and (6) agritourism, communication and culture. The towns of Santa Rosa de Lima and Anitápolis are the main participants, but 15 others are also included. The first year of activities was based on awareness raising; consolidating the Agreco label (through self-certification) as a synonym of quality and ecological production; and in diversification and expansion of the marketing process through the definition of a policy stating the basic principles and rules for the conversion from conventional to organic systems.

The overall aim is to increase the Agreco system through cooperation with other farmers' organizations and to create a network for marketing organic products throughout Brazil. In the areas of farm management, control quality and certification, the characteristics and cost of each product have been established and are monitored in order to support the price policy. The price policy establishes the final price of each product and the income of each person involved in the production chain according to the respective costs at each phase. Two parameters are used for setting the price range, firstly production costs and secondly the market price. The aim for 2002 is to sell half of Agreco's production in the institutional market (mainly schools) and through consumer organizations but also to consolidate the price policy based on production costs. One of the main characteristics of Agreco's system is that its planning process starts at the end of the 'multiple production chains'. In other words, it starts at the market. The institutional market offers not only a possibility to expand production but also a way to bring producers and consumers closer and contribute to increasing the number of informed consumers and producers. In this process, schools offer not only a potential market but also a learning opportunity for children and their parents, who will hopefully become well-informed consumers.

Agritourism as a means to create new income opportunities and bridge the rural–urban gap

Since 1997, Agreco has been encouraging family farmers to start agritourism activities on their farms following the framework of a programme included in its operational work plan and motivated by the increasing number of visitors to the area. Besides the natural beauty of the area, the organic production, processing and marketing system developed by Agreco has attracted the interest of many, not only but including technicians and other farmers.

In 1999, an agritourism association *Acolhida na Colônia* was created, involving 50 organic farming members of Agreco. The Association follows the principles and name of the French association 'Accueil Paysan'[1] which has supported this activity. Financial support for Agreco's agritourism activities has been provided by Pronaf and by the Brazilian tourism enterprise, Embratur (a governmental organization).

One of the chief benefits of the agritourism association, *Acolhida na Colônia,* is the creation of new income and employment opportunities for many family farmers in the region, particularly for woman and young farmers, who are usually the first to migrate to cities. Another important benefit is the consolidation of relations between rural and urban people, with the latter are increasingly visiting the region to know the producers and where and how they live, but also to ensure that the products they are buying and eating are effectively organic.

The credit cooperative: Access to finance for family farmers

The transition from conventional to organic agriculture as well as the processing and marketing of products require substantial financial means. Yet, small family farmers in the region are suffering the negative consequences of the economic processes that have occurred in Brazil and most developing countries over the last few decades, such as increasing poverty and rural exodus, rising urban violence and the concentration of incomes in ever fewer hands.

One of the greatest difficulties for small farmers is gaining access to the official financial system, as they cannot offer the minimum guarantees or the usual bank requirements. Many family farmers do not even own the land they farm. To overcome this problem, Agreco joined up with other local organizations to create a credit cooperative called *CrediColônia* (or *Credi*, as it is known), in order to finance the production, processing and marketing of organic products. Presently, all Agreco's financial resources go through the credit cooperative as do most of its associates. The credit cooperative operates like any traditional bank except that it does not require collateral or other financial guarantees from small family farmers. It focuses its investments on developing agroecological activities and systems in the region.

In addition to facilitating family farmers' access to credit, the *CrediColônia* cooperative is being developed as a strategic institution in the mobilization of local savings as well as financial support from other regions or countries, while managing and applying financial resources for the promotion of regional sustainable development. *Credi* hopes soon to expand its activities to the whole of Agreco's territory.

The hillsides development forum: From individualism and competition to collective action and solidarity

As in most other regions of Brazil, the majority of infrastructure problems (such as energy supply, roads construction and maintenance, communication, education and health services) are common to all small villages in the hillsides of Santa Catarina. As in most other Brazilian regions, these problems remain without solution for years due to their physical scale and isolation. Solving these types of problem usually requires collective action planned at a regional level.

This situation motivated Agreco to join forces with stakeholders interested in collective action for the sustainable development of the region. As a result, the Hillsides Development Forum was created with the aim of escaping from individualism and competition and promoting collective action and solidarity among the diverse institutions and stakeholders in the region for the solution of common problems. Based on the principles of participation, solidarity and agroecology, the Forum engages partnerships at local, regional and national levels.

The Hillsides Development Forum involves participants from 15 cities, all characterized by essentially agricultural economies practised by small family farmers and suffering a severe rural exodus. The Forum had its first open meeting in May 2001 at Santa Rosa de Lima and since then has alternated its monthly meetings among the cities of the territory with the organizational support from important institutions like the Federal University of Santa Catarina (UFSC).

In these monthly meetings (open to all interested people) diverse stakeholders are invited to discuss actions and policies (manly infrastructure problems such as energy supply, road construction and maintenance, communication, education and health services). At each meeting a different stakeholder organizes and facilitates the discussion. The meetings have occurred in a very relaxed and friendly environment, facilitating the solidarity among the participants and definition of collective actions. The main priority actions established in 2001 involved improvements in the communication and energy systems together with improvements in roads, health and education services (particularly student lunch meals and transportation). These activities have been financed by diverse stakeholders, according to the theme, for example, road construction and maintenance and educational activities have been financed by both the state and/or municipal governments through their specific departments while other activities have been financed by governmental, non-governmental and private organizations.

The solidarity economy forum: Expanding the system's boundaries and approximating urban consumers and rural producers

Agreco's experience has interested several similar organizations located in both rural and urban regions. As a result a Solidarity Economy Forum was created in 2001 involving 22 farmers' and urban consumers' organizations from the most populated coastal cities of Santa Catarina.

The Solidarity Economy Forum expands the limits of Agreco's sustainable rural development project and links urban consumers and rural producers. The main actions involve the organization of organic lunches for schools and twice weekly deliveries of organic baskets in urban centres (about 150–200 baskets each time). This process facilitates participants' understanding towards issues of food security, water contamination with agrochemicals, increasing poverty, environmental degradation and rural exodus and helps demonstrate that increasing urban violence and revenue concentration are interrelated phenomena which are caused by both urban and rural human activities.

These are viewed as common problems which interest and impact upon both urban and rural citizens. As a consequence, the Solidarity Economy Forum's activities generate potential for increased collaboration and solidarity between urban and rural people for the promotion of sustainable life in both rural and urban regions.

Lessons learnt, results and outlook

During its five years of existence, Agreco has promoted positive impacts in both environmental conservation and in job and income generation as well as in improving the socio-economic outlook and the enthusiasm of the family farmers who live in or have commitments in the territory. Other positive impacts have been observed in terms of the methodologies related to the joint management of the network of small agroindustries (each run by a group of families). However, the project has faced some difficulties. Only about half of the 53 initially planned small agroindustries have been set up. The main reason behind this is that the principle source of financial support for these agroindustries is Pronaf, which uses the traditional official banking system, the requirements for which many small family farmers cannot fulfil.

The 27 agroindustries involve 120 families (out of 211 originally expected in the 53 agroindustries) and generate 505 new part-time and full-time jobs. These are important results for a territory characterized by small villages but represent just a small indication of success given the employment and income requirements of the whole country. A very positive aspect which can be observed is the change of attitude of most farmer families and local and regional leaders towards the regional development. From accommodation and resignation to economic stagnation and rural exodus, these actors started observing new economic alternatives as well as new jobs, revenue and development opportunities and income generated from these activities has usually been sufficient to cover family needs, with other non-farm activities only being occasionally necessary.

In this context, the option for collective action and organic agriculture systems as the main sustainable development strategies deserve to be highlighted and Agreco is presently expanding its operational activities over the whole region. Despite some initial inertia to change, and thanks to the example of the original success of the project, the willingness and enthusiasm of associates to work in groups, and trust amongst associates and other interested actors and stakeholders has increased.

The expansion of activities over the hillside region generated operational difficulties but also helped to promote collective action, particularly in cultural exchange and agri-tourism activities as well as in primary organic production, processing and marketing. New markets are now being investigated and created, including the expansion of the student organic meals and delivery of organic baskets in urban centres, together with sales through local fairs and regional supermarket chains. The potential for commercialization on the international market is also being investigated and discussed together with the possibilities for certifying products for the overseas market with internationally accepted certification bodies which operate in Brazil (such as Ecocert).

However, Agreco's policy is not to expand throughout Santa Catarina State or other Brazilian regions. The idea is to serve as an example and interact with other farmer and consumer associations, each one with its own particular characteristics but operating as part of a larger organized network linked by the same interest in agroecological principles and sustainable development.

Agreco is an ecological farmers' association that has attracted the attention of institutions and people from all over the country and from overseas. It has become a reference of success for similar initiatives aimed at constructing sustainable development based on agroecological systems and collective action. During its five years of operation, the project has received support from many governmental and non-governmental organizations, but a key to its success is doubtless the enthusiastic participation of local and regional leaders who have promoted the interaction among institutions and stakeholders. The methodologies and principles used to develop Agreco's system have been systematized in order to serve as a reference for the implementation of similar experiences in other regions or countries. Methodologies have been based on the principles of agroecology, sustainability and solidarity as well as on the concept that human beings must be the main actors of their own history.

In synthesis, Agreco's experience has validated organic production and ecological principles, together with an organized net of small businesses as a basis for a sustainable development project. It has also empowered both the family farmer and the urban consumer who together constructed the sustainable life project.

Note

1   The main principles of the Accueil Paysan association are: tourism must complement (and not substitute) the usual farm activities and farmers' families should keep living in their farms. Accommodation must use existing farm buildings after improvement if necessary. Farmers' families must wish to receive visitors and exchange life experiences with them. Farmers' families must also be enthusiastic about improving

the quality of their products and services while taking proper care of the environment and offering reasonable prices to visitors.

25

Lessons of Cuban Resistance

Peter Rosset and Martin Bourque

At the Institute for Food and Development Policy (Food First), we have spent 25 years studying hunger around the world, and its relationship to agriculture and rural development. Over the years we have seen many countries enter into food crises. The proximate causes have been many, ranging from wars to droughts or floods, though invariably the ultimate causes have been more tightly linked to inequality or lack of social justice in some form, be it in *access* to land, jobs, government assistance, or the structure of the world economy. Such crises have often led to famines, which have only been resolved by massive international intervention and food aid, usually leaving the afflicted region or country less able to feed itself in the future and more dependent on food imports from the West than ever before (Lappé and Collins, 1977; Lappé et al, 1998).

The experience of Cuban resistance to external shocks during the 1990s stands in sharp contrast to this panorama. When collapsing trade relations plunged this island nation into a food crisis, foreign assistance and food aid were scarcely available, thanks to the tightening of the US trade embargo. Cuba was forced to turn inward, toward its own natural and human resources, and tap both old and new ways to boost production of basic foods without relying on imports. It wasn't easy, but in many ways the Cuban people and government were uniquely prepared to resist – the well-educated and energetic populace put their dynamism and ingenuity to the task, and the government its commitment to food for all and its support for domestic science and technology. Cubans and their government pulled through the crisis, and their tale offers powerful lessons about resistance, self-reliance, and alternative policies and production methods that could well serve other countries facing their own rural and food crises (Rosset and Altieri, 1994; Rosset, 1998).

Our global food system is in the midst of a multifaceted crisis, with ecological, economic and social dimensions. To overcome that crisis, political and social changes are needed to allow the widespread development of alternatives. While there are many examples of farmer-driven and community based alternative agricultural development models throughout the world that demonstrate that the alternatives work and are economically viable, Cuba offers one of the few examples where fundamental policy shifts and serious governmental resources have supported this movement. It is therefore important all people interested in developing food systems that are socially just, environmentally sustain-

Note: Reprinted from *Sustainable Agriculture and Resistance*, by Funes, F., Garcia, L., Bourque, M., Perez, N. and Rosset, P. (eds), copyright © (2002) with permission from Food First Books, Oakland, CA

able, and economically viable, pay close attention to current policy and technological developments in Cuba.

The world food system, based largely on the conventional 'Green Revolution' model, is productive – there should be no doubt about that – as per capita food produced in the world has increased by 15 per cent over the past 35 years. But that production is in ever fewer hands, and costs ever more in economic and ecological terms. In spite of these per capita increases, and an overall per capita surplus of calories, protein and fats, there are at least 800 million people in the world who do not reap adequate benefits from this production. And it is getting worse. In the last 20 years the number of hungry people in the world – excluding China – has risen by 60 million (Lappé et al, 1998).

Ecological impacts of industrial-style farming include adverse effects on groundwater due to abuse of irrigation and to pesticide and fertilizer runoff, on biodiversity through the spread of large scale monocultures and the elimination of traditional crop varieties, and on the very capacity of agroecosystems to be productive into the future (Rosset and Altieri, 1997; Rosset, 1997a).

Economically, production costs rise as farmers are forced to use ever more expensive machines and farm chemicals, while crop prices continue a several-decade-long downward trend, causing a cost-price squeeze which led to the loss of untold tens of millions of farmers worldwide through bankruptcies. Socially, we have the concentration of farmland in fewer and fewer hands as low crop prices make farming on a small scale unprofitable (despite higher per acre total productivity of small farms), and agribusiness corporations extend their control over more and more basic commodities (Rosset and Altieri, 1997).

Clearly the dominant corporate food system is not capable of adequately addressing the needs of people or of the environment. Yet there are substantial obstacles to the widespread adoption of alternatives. The greatest obstacles are presented by political-corporate power and vested interests, yet at times the psychological barrier to believing that the alternatives can work seems almost as difficult to overcome. The oft-repeated challenge is: 'Could organic farming (or agroecology, local production, small farms, farming without pesticides) ever really feed the entire population of a country?' (Rosset, 1999). Recent Cuban history – the overcoming of a food crisis through self-reliance, smaller farms and agroecological technology – shows us that the alternatives can indeed feed a nation, and thus provides a crucial case study for the ongoing debate.

Around the globe, farmers, activists, and researchers are working to create a new model for agriculture that responds to the multiple facets of the crisis. The goals of this model are to be environmentally sound, economically viable, socially just, and culturally appropriate. The Cuban experience presented in this volume offers many new ideas to this movement.

A brief history

When trade relations with the Soviet Bloc crumbled in late 1989 and 1990, and the US tightened the trade embargo, Cuba was plunged into economic crisis. In 1991 the government declared the 'Special Period in Peacetime', which basically put the country on a wartime economy-style austerity program. An immediate 53 per cent reduction in oil imports not only affected fuel availability for the economy, but also reduced to zero the foreign exchange that Cuba had formerly obtained via the re-export of petroleum. Imports of wheat and other grains for human consumption dropped by more than 50 per cent, while other foodstuffs declined even more. Cuban agriculture was faced with an initial drop of about 70 per cent in the availability of fertilizers and pesticides, and more than 50 per cent in fuel and other energy sources produced by petroleum (Rosset and Benjamin, 1994).

Suddenly, a country with an agricultural sector technologically similar to California's found itself almost without chemical inputs, with sharply reduced access to fuel and irrigation, and with a collapse in food imports. In the early 1990s average daily caloric and protein intake by the Cuban population may have been as much as 30 per cent below levels in the 1980s.

Fortunately, Cuba was not totally unprepared to face the critical situation that arose after 1989. It had, over the years, emphasized the development of human resources, and therefore had a cadre of scientists and researchers who could come forward with innovative ideas to confront the crisis. While Cuba has only two per cent of the population of Latin America, it has almost 11 per cent of the scientists (Rosset and Benjamin, 1994).

Alternative technologies

In response to this crisis Cubans and their government rushed to develop and implement alternatives. Because of the drastically reduced availability of chemical inputs, the state hurried to replace them with locally produced, and in most cases biological, substitutes. This has meant biopesticides (microbial products) and natural enemies to combat insect pests, resistant plant varieties, crop rotations and microbial antagonists to combat plant pathogens, and better rotations and cover cropping to suppress weeds. Scarce synthetic fertilizers were supplemented by biofertilizers, earthworms, compost, other organic fertilizers, animal and green manures, and the integration of grazing animals.

In place of tractors, for which fuel, tyres and spare parts were often unavailable, there was a sweeping return to animal traction (Rosset and Benjamin, 1994).

When the crisis began, yields fell drastically throughout the country. But production levels for domestically consumed food crops began to rise shortly thereafter, especially on Agricultural Production Cooperatives (APCs) and on the farms of individual small holders or *campesinos*. It really was not all that difficult for the small farm sector to effectively produce with fewer inputs. After all, todays small farmers are the descendants of generations of small farmers, with long family and community traditions of low-input production. They basically did two things: remembered the old techniques – like intercropping and manuring – that their parents and grandparents had used before the advent of modern chemicals, and simultaneously incorporated new biopesticides and biofertilizers into their production practices (Rosset, 1997b, 1997c).

The state sector, on the other hand, faced the incompatibility of large monocultural tracts with low-input technology. Scale effects are very different for conventional chemical management and for low external input alternatives. Under conventional systems, a single technician can manage several thousand hectares on a 'recipe' basis by simply writing out instructions for a particular fertilizer formula or pesticide to be applied with machinery on the entire area. Not so for agroecological farming. Whoever manages the farm must be intimately familiar with the ecological heterogeneity of each individual patch of soil. The farmer must know, for example, where organic matter needs to be added, and where pest and natural enemy refuges and entry points are (Altieri, 1996). This partially explains the difficulty of the state sector to raise yields with alternative inputs. A partial response was obtained with a programme that began before the Special Period, called *Vinculando el Hombre con la Tierra*, which sought to more closely link state farm workers to particular pieces of land, but it wasn't enough (Enriquez, 1994).

In September 1993 Cuba began radically reorganizing the state sector in order to create the small-scale management units that seemed most effective in the Special Period. The government issued a decree terminating the existence of the majority of state farms, turning them into Basic Units of Cooperative Production (BUCPs), a form of worker-owned enterprise or cooperative. Much of the 80 per cent of all farmland that was once held by the state, including sugarcane plantations, was essentially turned over to the workers. The UBPCs allow collectives of workers to lease state farmlands rent free, in perpetuity. Property rights remain in the hands of the state, and the UBPCs must still meet production quotas for their key crops, but the collectives are owners of what they produce. What food crops they produce in excess of their quotas could be freely sold at newly opened farmers' markets. This last reform, made in 1994, offered a price incentive to farmers to make effective use of the new technologies (Rosset, 1997a).

The pace of consolidation of the UBPCs has varied greatly in their first years of life. With a variety of internal management schemes, in almost all cases the effective size of the management unit has been drastically reduced. It is clear that the process of turning farm workers into farmers will take some time – it simply cannot be accomplished overnight – and many UBPCs are struggling, while others are very successful. On the average, small farmers and CPAs probably still obtain higher levels of productivity than do most UBPCs, and do so in ways that are more ecologically sound.

Food shortage overcome

By the latter part of the 1990s the acute food shortage was a thing of the past, though sporadic shortages of specific items remained a problem, and food costs for the population had increased significantly. The shortage was largely overcome through domestic production increases which came primarily from small farms, and in the case of eggs and pork, from booming backyard production (Rosset, 1998). The proliferation of urban farmers who produce fresh produce has also been extremely important to the Cuban food supply (Murphy, 1999; GNAU, 2000). The earlier food shortages and the resultant increase in food prices suddenly turned urban agriculture into a very profitable activity for Cubans, and once the government threw its full support behind a nascent urban gardening movement, it exploded to near epic proportions. Formerly vacant lots and backyards in all Cuban cities now sport food crops and farm animals, and fresh produce is sold from stands throughout urban areas at prices substantially below those prevailing in the farmers' markets. There can be no doubt that urban farming, relying almost exclusively on organic techniques, has played a key role in assuring the food security of Cuban families over the past two to three years.

An alternative paradigm?

To what extent can we see the outlines of an alternative food system paradigm in this Cuban experience? Or is Cuba just such a unique case in every way that we cannot generalize its experiences into lessons for other countries? The first thing to point out is that contemporary Cuba turned conventional wisdom completely on its head. We are told that small countries cannot feed themselves; that they need imports to cover the deficiency of their local agriculture. Yet Cuba has taken enormous strides toward self-reliance since it lost its key trade relations. We hear that a country can't feed its people without synthetic farm chemicals, yet Cuba is virtually doing so. We are told that we need the efficiency of large-scale corporate or state farms in order to produce enough food, yet we find small farmers and gardeners in the vanguard of Cuba's recovery from a food crisis. In fact, in the absence of subsidized machines and imported chemicals, small farms are more efficient than very large production units. We hear time and again that international food aid is the answer to food shortages – yet Cuba has found an alternative in local production.

Abstracting from that experience, the elements of an alternative paradigm might therefore be:

- *Agroecological technology instead of chemicals*: Cuba has used intercropping, locally produced biopesticides, compost, and other alternatives to synthetic pesticides and fertilizers.
- *Fair prices for farmers*: Cuban farmers stepped up production in response to higher crop prices. Farmers everywhere lack incentive to produce when prices are kept artificially low, as they often are. Yet when given an incentive, they produce, regardless of the conditions under which that production must take place.
- *Redistribution of land*: Small farmers and gardeners have been the most productive of Cuban producers under low-input conditions. Indeed, smaller farms worldwide produce much more per unit area than do large farms (Rosset, 1999). In Cuba redistribution was relatively easy to accomplish because the major part of the land reform had already occurred, in the sense that there were no landlords to resist further change.
- *Greater emphasis on local production*: People should not have to depend on the vagaries of prices in the world economy, long distance transportation, and superpower 'goodwill' for their next meal. Locally and regionally produced food offers greater security, as well as synergistic linkages to promote local economic development. Furthermore such production is more ecologically sound, as the energy spent on international transport is wasteful and environmentally unsustainable. By promoting urban farming, cities and their surrounding areas can be made virtually self-sufficient in perishable foods, be beautified, and have greater employment opportunities. Cuba gives us a hint of the underexploited potential of urban farming.

The Cuban experience illustrates that we can feed a nation's population well with an alternative model based on appropriate ecological technology, and in doing so we can become more self-reliant in food production. Farmers must receive higher returns for their produce, and when they do they will be encouraged to produce. Expensive chemical inputs – most of which are unnecessary – can be largely dispensed with. The important lessons from Cuba that we can apply elsewhere, then, are agroecology, fair prices, land reform, and local production, including urban agriculture.

References

Altieri, M. (1996) *Agroecology: The Science of Sustainable Agriculture*, Westview Press, Boulder, Colorado

Enriquez, L. (1994) *The Question of Food Security in Cuban Socialism*, Institute of International and Area Studies, University of California at Berkeley, Berkeley

GNAU (2000) *Manual Tecnico de Organoponicosy Huertos Intensivos*, ACTAF, INIFAT, MINAG, Havana

Lappé, E. M., and Collins, J. (1977) *Food First: Beyond the Myth of Scarcity*, Ballantine Books, New York

Lappé, E. M., Collins, J. and Rosset, P. M. with Esparza, L. (1998) *World Hunger: Twelve Myths*, 2nd edition, Grove Press, New York

Murphy, C. (1999) *Cultivating Havana: Urban Agriculture and Food Security in the Years of Crisis. Food First Development Report No. 12*, Institute for Food and Development Policy, Oakland

Rosset, P. M. (1997a) 'La crisis de la agricultura convencional, la sustitución de insumos, y el enfoque agroecológico. (Chile)', *Agroecologia y Desarrollo*. vol 11/12, pp2–12

Rosset, P. M. (1997b) 'Alternative agriculture and crisis in Cuba', *Technology and Society*, vol 16, no 2, pp19–25

Rosset, P. M. (1997c) 'Cuba: Ethics, biological control, and crisis', *Agriculture and Human Values*, vol 14, pp291–302

Rosset, P. M. (1998) 'Alternative agriculture works: The case of Cuba', *Monthly Review*, vol 50, p3

Rosset, P. M. (1999) *The Multiple Functions and Benefits of Small Farm Agriculture in the Context of Global Trade Negotiations. Food First Policy Brief No. 4*, Institute for Food and Development Policy, Oakland

Rosset, P. M., and Altieri, M. (1994) 'Agricultura en Cuba: Una experiencia nacional en conversion organica. (Chile)', *Agroecologia y Desarrollo*, vol 7/8

Rosset, P. M. and Altieri, M. (1997) 'Agroecology versus input substitution: A fundamental contradiction of sustainable agriculture', *Society & Natural Resources*, vol 10, no 3, pp283–295.

Rosset, P. M. and Benjamin, M. (eds) (1994) *The Greening of the Revolution: Cubas Experiment with Organic Agriculture*, Ocean Press, Australia

26

Benefits from Agroforestry in Africa, with Examples from Kenya and Zambia

Pedro A. Sanchez

Agroforestry – integrating trees and other perennials into farming systems for the benefit of farm families and the environment – is an ancient practice that began moving from the realm of indigenous knowledge into agricultural research only about 25 years ago (Bene et al, 1977). During the 1980s, agroforestry was promoted widely as a sustainability-enhancing practice with great potential to increase crop yields and conserve soil and recycle nutrients, while producing fuelwood, fodder, fruit and timber (Steppler and Nair, 1987; Nair, 1989). At that time, agroforestry was considered almost a panacea for solving landuse problems in the tropics. Many development projects pushed agroforestry technologies that were without foundations in solid research. During the past decade, however, agroforestry studies have become more empirical, based on process-oriented research (Sanchez, 1995; Young, 1997; Buck et al, 1999).

Agroforestry is now recognized as an applied science based on principles of natural resource management (NRM) (TAC, 1999; Izac and Sanchez, 2001). The application of such principles includes the following practices:

- participatory, multidisciplinary and analytical approaches;
- technical and policy research;
- working at and across different spatial and temporal scales;
- beneficiaries identified at the community, national and global levels;
- working along the whole research-development continuum;
- working in partnership with governmental and non-governmental organizations (NGOs);
- moving rapidly into on-farm research with a decreasing degree of researcher control;
- assessing impacts in economic, social and environmental terms;
- being a credible partner in development.

Agroforestry is in fact a very widespread practice, found from the Arctic to the southern temperate regions, but most extensive in the tropics. Approximately a fifth of the world's population (1.2 billion people) depend directly on agroforestry products and services in rural and urban areas of developing countries (Leakey and Sanchez, 1997).

Note: Reprinted from *Agroecological Innovations* by Uphoff, N. (ed), copyright © (2002) Earthscan, London

Agroforestry products include fuelwood, livestock fodder, food, fruits, poles, timber and medicines. Agroforestry services include erosion control, soil fertility replenishment, improved nutrient and hydrological cycles, boundary delineation, poverty reduction and enhanced food security, household nutrition, watershed stability, biodiversity, microclimate enhancement and carbon sequestration. Many agroforestry systems are superior to other landuse systems at global, regional, watershed and farm scales because they optimize trade-offs among increased food production, poverty alleviation and environmental conservation (Izac and Sanchez, 2001). Being complementary to rather than competitive with arable or pastoral practices makes agroforestry an important part of strategies to produce sufficient food in the decades ahead in ways that meet both human and environmental needs.

While the original impetus for agroforestry was very practical and empirical, it is supported increasingly by scientific foundations that permit its extension and extrapolation across the tropics. The complex agroforests of Indonesia are one example (Michon, 1997). Research based on the principles of competition for light, water and nutrients (Ong and Huxley, 1996) or the complexity of interacting socio-economic and biophysical factors (Sanchez, 1995) has led to new agroforestry components that increase the sustainability and profitability of existing farming systems.

Much of the information for determining the biophysical performance, profitability and acceptability of agroforestry comes from on-farm trials (Franzel et al, 1998). At the same time, farmers have become increasingly interested in on-station work and have often suggested technologies that address constraints and identify opportunities showing promise upon further investigation. Farmers' informal and formal visits to research station work has led to keen interest for collaboration between researchers and farmers in testing and assessing agroforestry technologies on-farm.

Three broadly defined partnerships for on-farm trials have been adopted by the International Centre for Research in Agroforestry (ICRAF): those that are researcher-designed and -managed (type 1), researcher-designed and farmer-managed (type 2) and farmer-designed and -managed (type 3) (ICRAF, 1995; Franzel et al, 1998). The suitability of these different kinds of trials depends on the objectives; no type of trial is intrinsically 'better' than another type. Which type should be preferred depends on the objectives of the participants (facilitators and farmers) and on the particular circumstances.

The collaborating farmers for these on-farm trials are selected through locally based institutions such as extension services or farmer groups. Farmers are asked to volunteer and are selected to represent the range of different farmers in the study area, for example, large to small sizes of holdings, male and female, different wealth categories. Farmers choose among several alternative technologies to test, and except for receiving planting material and information, they are not provided with any further incentives for participating in trials with researchers.

Researchers are involved mainly in technical backstopping for farmers in trials of types 2 and 3, and they help lay out type 2 trials. There is great variation in the number of trials and the number of farmers involved, but in most cases small numbers of farmers start type 2 trials, on average about ten; after modification during the first year, the number is expanded up to about 50. The number of farmers involved in type 3 trials can be much larger. In Eastern Zambia in 1997, for example, extension services and NGOs were helping 2800 farmers test improved tree fallows in type 3 trials (Franzel et al, 1998).

Two agroforestry components that have been extensively researched and adopted are the domestication of indigenous trees (Leakey et al, 1996; Leakey and Tomich, 1999) and soil fertility replenishment (Buresh et al, 1997). In the process, some previously promoted practices such as alley cropping that have not met science-based tests are no longer advocated on a large scale. There are, however, some indigenous alley cropping systems in Indonesia that are many decades old and popular with farmers where they fit certain ecosystem niches (Piggin, 2000).

Some agroforestry innovations that can be applied to meet the particular agricultural challenges in Africa are discussed here – how to assure food security, reduce poverty and enhance ecosystem resilience at the scale of thousands of smallholder farmers. There are many examples of successful agroforestry innovations in other parts of the world that could be cited (Buck et al, 1999).

Redressing soil fertility problems

When smallholding farmers throughout the sub-humid and semi-arid tropics of sub-Saharan Africa, hereafter referred to as Africa, are involved in diagnosis and design exercises, they invariably identify soil fertility depletion as the fundamental reason for declining food security in this region. Scientists concur. No matter how effectively other constraints are remedied, per capita food production in Africa will continue to decrease unless soil fertility depletion is effectively addressed (Sanchez and Leakey, 1997; Sanchez, Shepherd et al, 1997; Sanchez, Buresh and Leakey, 1997; Pieri, 1998).

During the 1960s, the fundamental cause of declining per capita food production in Asia was the lack of rice and wheat varieties that could respond efficiently to increases in nutrient availability. Food security was only effectively addressed with the advent of improved germplasm in this region for higher-yielding varieties. Then other key aspects of agricultural development that had been previously less important – enabling government policies, irrigation, seed production, fertilizer use, pest management, research and extension services – came into play in support of the spread of new varieties.

The need for soil fertility replenishment in Africa is now analogous to the need for Green Revolution germplasm in Asia three decades ago.[1] A full description of the magnitude of nutrient depletion, its underlying socio-economic causes, the consequences of such depletion and various strategies for tackling this constraint are described elsewhere (Buresh et al, 1997; Sanchez and Leakey, 1997; Sanchez, Shepherd et al, 1997; Sanchez et al, 2001). Fortunately, strategies for soil fertility enhancement and agroforestry can be combined based on much research (Sanchez and Leakey, 1997).

Nitrogen and phosphorous are the most severely depleted nutrients in smallholder African farms. Although such constraints can be alleviated with imported mineral fertilizers, economic, infrastructural and policy constraints make the use of mineral fertilizers extremely limited in such farms. However, Africa has ample nitrogen and phosphorus resources – nitrogen in the air and phosphorus in many rock phosphate deposits. The challenge is to get these natural resources to where they are needed and in plant-available forms. For nitrogen, this can be achieved through biological nitrogen fixation by leguminous woody species utilized in fallows. For phosphorus, there can be beneficial direct application of reactive, indigenous rock phosphate combined with biomass transfers of non-leguminous shrubs.

Two-year leguminous fallows, leaving land uncultivated but with selected leguminous species growing on it, can accumulate 200kg of nitrogen per hectare in plant leaves and roots. Incorporating these into the soil, with subsequent mineralization, provides sufficient nitrogen for two or three crops. Resulting maize yields can be two to four times higher (Kwesiga and Coe, 1994; Kwesiga et al, 1997; Kwesiga et al, 1999).

The greatest impact of this work so far has been in Southern Africa, where about 10,000 farmers are now using *Sesbania sesban*, *Tephrosia vogelii*, *Gliricidia septum* and *Cajanus cajan* in two-year fallows followed by maize rotations for two to three years (Rao et al, 1998). The species used in such fallows produce nutrients that would cost US$240/ha for an equivalent amount of mineral fertilizer, well beyond the reach of farmers in this region who make less than US$1 per day.

The provision of nutrients through such plant and soil management methods, which require hardly any cash, repays the labour invested very well. The results of such practices were summarized by one farmer, Sinoya Chumbe, who lives in Kampheta village near Chipata, Zambia, when he stated: 'Agroforestry has restored my dignity. My family is no longer hungry; I can even help my neighbours now' (interview, 12 April 1999).

In many high-potential areas of East Africa, smallholder farms are depleted of both nitrogen and phosphorous, necessitating the combined use of organic and mineral sources of nutrients (Palm et al, 1997). Short-term improved fallows of up to 16 months' duration using *Tephrosia vogelii*, *Crotalaria grahamiana* and *Sesbania sesban* are an effective and profitable way of adding about 100kg of nitrogen per hectare and of recycling other nutrients in the nitrogen-depleted soils of Western Kenya. With fast-growing trees, fallows as short as six months have tripled maize yields in villages where farmers are now practising a fallow-crop rotation every year in a bimodal rainfall environment (Niang et al, 1998; Rao et al, 1998).

In phosphorous-deficient soils, Minjingu rock phosphate from Northern Tanzania has proven to be as effective as imported triple superphosphate, as well as more profitable for small farmers (Sanchez and Leakey, 1997; Niang et al, 1998; Sanchez et al, 2001). Basal applications of 125–250kg of phosphorous per hectare as a capital investment are beginning to be used by farmers with an expected residual effect of five years. In addition, biomass transfers from hedges of wild sunflower, tithonia (*Tithonia diver si folia*). have shown tremendous effects on yields of maize and high-value crops such as vegetables in Western Kenya (Gachengo et al, 1998; Jama et al, 2000).

Tithonia biomass has high concentrations of nitrogen, phosphorous and potassium and decomposes very rapidly in the soil (Palm et al, 1997). Given the large additions of soluble carbon and nutrients to the soil when tithonia leaves decompose, it appears that these processes enhance phosphorous cycling and therefore the conversion of mineral forms of phosphorous into organic ones (Nziguheba et al, 1998).

Combining the application of tithonia biomass with phosphorous fertilizer has been shown to be particularly effective (Rao et al, 1998). Tithonia grows abundantly along roadsides and in farm hedges at intermediate elevations throughout sub-humid Africa, making it an easily available natural resource that can be utilized to replenish soil fertility.

About 4000 farmers are currently trying these techniques in Western Kenya. Most of the dissemination work has been done at the village level as a pilot development project (Niang et al, 1998). An assistant chief of Barsauri sublocation in the Siaya District of Nyanza Province, Hosea Omollo, summarized the results of this agroforestry technology as follows: 'For the first time there have been no hunger periods in this village. Only two ears of maize have been reported stolen this year' (interview, 7 July 1998).

Many farmers who have adopted tithonia biomass transfers to their fields have shifted now from maize to high-value vegetables, which can be readily sold in nearby towns, effectively bringing them into the cash economy. One farmer, Charles Ngolo of Ebuyango in Vihiga District, Western Province, reported to ICRAF that his annual cash income had increased from US$100 to US$1000 through the sale of kale, locally known as *sukuma wiki (Brassica oleracea cv. Acepbala)*. He commented: 'My wife and I are living the tithonia life. I built a new house with a tin roof, and we are going to be able to send our children to school' (interview, 4 June 1997).

These people now farming on replenished soils have achieved food security, and Mr Ngolo is an example of a smallholder who is beginning to work his way out of poverty. Economic analysis has shown high net present values for these technologies (Sanchez and Leakey, 1997). The potentials of agroforestry practices can no longer be discounted as hypothetical. The question now is how to scale-up their delivery, from thousands to millions of farmers, a major challenge facing national governments and international agencies (ICRAF, 1998). Enabling policies at the national, district and community levels are beginning to emerge in support of technological advances. These include increasing the availability of phosphorous fertilizer and high quality seeds, providing microcredit and levying fines on farmers who let their cattle eat their neighbours' sesbania fallows (Sanchez and Leakey, 1997; Sanchez, Shepherd et al, 1997 ICRAF, 1998). The Kenyan government has established and funded a pilot project on soil fertility recapitalization and replenishment for Western Kenya. Further, the government as a member of the Consultative Group on International Agricultural Research (CGIAR) is providing financial support for ICRAF to conduct strategic research to underpin replenishment efforts. Technological and policy research are both needed in agroforestry. Their joint impact can enable soil fertility replenishment through biological means to make a major contribution to food security in Africa.

Opportunities

It would be unwarranted to generalize that all agroforestry interventions will have similar degrees of success. But there are surely many more innovations still to be identified and evaluated. Agroforestry is not the best landuse option for all tropical areas, and some practices have met with widespread failure when they were not based on solid technical and policy research. Science-based agroforestry pursued in cooperation with farmers can, on the other hand, assuredly produce economically, socially and environmentally sound results. These examples and many others that are emerging and spreading throughout the world, where trees and other perennials are integrated with other farming components and practices, can raise productivity and security for several billion people who will benefit from combining ancient practices and modern science.

Note

1   Two of the 'fathers' of the Green Revolution agree with this analogy: Norman Borlaug (Borlaug and Dowswell, 1994) and M. S. Swaminathan (July 1998, pers
comm).

References

Bene, J. G., Beall, H. W. and Cote, A. (1977) *Trees, Food and People*, International Development
Research Center, Ottawa

Borlaug, N. and Dowswell, C. R. (1994) 'Feeding a human population that increasingly crowds
a fragile planet', in International Society of Soil Science, Supplement to Transactions of the
15th World Congress of Soil Science, Acapulco, Mexico, Chapingo, Mexico

Buck, L. E., Lassoie, J. P. and Fernandes, E. C. M. (eds) (1999) *Agroforestry in Sustainable Agricultural Systems*, CRC Press, Boca Raton, Florida

Buresh, R. J., Sanchez, P. A. and Calhoun, F. (eds) (1997) *Replenishing Soil Fertility in Africa*,
SSSA Special Publication no 51, Soil Science Society of America, Madison, Wiskonsin

Franzel, S., Coe, R., Cooper, P. J., Place, F. and Scherr, S. J. (1998) 'Assessing the adoption potential of agroforestry practices: ICRAF's experiences in sub-Saharan Africa', paper presented at
international symposium of the Association for Farming Systems Research-Extension, Preto-
ria, South Africa, 29 November–4 December

Gachengo, C. N., Palm, C. A., Jama, B. and Othieno, C. (1998) 'Tithonia and senna green manures
and inorganic fertilizers as phosphorous sources for maize in western Kenya', *Agroforestry
Systems*, vol 44, pp21–36

ICRAF (1995) *Annual Report for 1994*, International Centre for Research in Agroforestry, Nairobi

ICRAF (1998) *Annual Report for 1997*, International Centre for Research in Agroforestry, Nairobi

Izac, A.-M. N. and Sanchez, P. A. (2001) 'Towards a natural resource management research paradigm: An example of agroforestry research', *Agricultural Systems*, vol 69, nos 1–2, pp5–25

Jama, B., Palm, C. A., Buresh, R. J., Niang, A. I., Gachengo, C., Nziguheba, G. and Amadalo,
B. (2000) '*Tithonia diversifolia* as a green manure for soil fertility improvement in western
Kenya: A review', *Agroforestry Systems*, vol 49, pp201–221

Kwesiga, F. R. and Coe, R. (1994) 'The effect of short rotation *Sesbania sesban* planted fallows on
maize yields', *Forest Ecology and Management*, vol 64, pp199–208

Kwesiga, F. R., Franzel, S., Place, F., Phiri, D. and Simwanza, C. P. (1999) '*Sesbania sesban*
improved fallows in Eastern Zambia: Their inception, development and farmer enthusiasm',
Agroforestry Systems, vol 47, pp49–66

Kwesiga, F. R., Phiri, D. and Raunio, A.-L. (1997) 'Improved fallows with sesbania in eastern
Zambia', in *Summary Proceedings of a Consultative Workshop*, ICRAF, Chipata, Zambia/Nai-
robi

Leakey, R. R. B. and Sanchez, P. A. (1997) 'How many people use agroforestry products?' *Agroforestry Today*, vol 9, no 3, pp4–5

Leakey, R. R. B., Temu, A. B., Melnyk, M. and Vantomme, P. (eds) (1996) *Domestication and
Commercialization of Non-timber Forest Products in Agroforestry Systems: Non-Wood Forest
Products*, Food and Agriculture Organization, Rome

Leakey, R. R. B. and Tomich, T. P. (1999) 'Domestication of tropical trees: From biology to economics and policy', in Buck, L. E., Lassoie, J. P. and Fernandes, E. C. M. (eds) *Agroforestry in Sustainable Agricultural Systems*, CRC Press, Boca Raton, Florida, pp319–335

Michon, G. (1997) 'Indigenous gardens: Re-inventing the forest', in Whitten, T. and Whitten, J. (eds) *The Indonesian Heritage, Vol I: Plants*, Grolier, Singapore, pp88–89

Nair, P. K. R. (ed) (1989) *Agroforestry Systems in the Tropics*, Kluwer Academic Publishers, Dordrecht, the Netherlands

Niang, A., de Wolf, J., Nyasimi, M., Hansen, T., Romelsee, R. and Mdewa, K. (1998) 'Soil fertility replenishment and recapitalization project in western Kenya', Progress Report February 1997–July 1998, Pilot Project Report no 9, Regional Agroforestry Research Centre, Maseno, Kenya

Nziguheba, G., Palm, C. A., Buresh, R. J. and Smithson, P. J. (1998) 'Soil phosphorus fractions and adsorption as affected by organic and inorganic sources', *Plant and Soil*, vol 198, pp59–168

Ong, C. and Huxley, P. A. (eds) (1996) *Tree–Crop Interactions: A Physiological Approach*, CAB International, Wallingford, UK

Palm, C. A., Myers, R. J. K. and Nandwa, S. M. (1997) 'Combined use of organic and inorganic nutrient sources for soil fertility maintenance and replenishment', in Buresh, R. J., Sanchez, P. A. and Calhoun, F. (eds) *Replenishing Soil Fertility in Africa*, Soil Science Society of America Special Publication 51, Soil Science Society of America, Madison, Wisconsin, pp193–217

Pieri, C. M. G. (1998) 'Soil fertility improvement: Key connection between sustainable land management and rural well being', 16th World Congress of Soil Science (CD-ROM), Montpellier, France

Piggin, C. (2000) 'The role of Leucaena in swidden cropping and livestock production in Nusa Tenggara Timur, Indonesia' in Cairns, M. F. (ed) *Voices from the Forest: Farmer Solutions Towards Improved Fallow Husbandry in Southeast Asia*, proceedings of a regional conference on Indigenous Strategies for Intensification of Shifting Cultivation in Southeast Asia, held in Bogor, Indonesia, 23–27 June 1997; International Centre for Research in Agroforestry, Bogor, Southeast Asian Regional Research Programme, pp278–290

Rao, M. R., Niang, A., Kwesiga, F., Duguma, B., Franzel, S., Jama, B. and Buresh, R. J. (1998) 'Soil fertility replenishment in sub-Saharan Africa: New techniques and the spread of their use on farms', *Agroforestry Today*, vol 10, no 2, pp3–8

Sanchez, P. A. (1995) 'Science in agroforestry', *Agroforestry Systems*, vol 30, pp5–55

Sanchez, P. A. and Leakey, R. R. B. (1997) 'Land-use transformation in Africa: Three determinants for balancing food security with natural resource utilization', *European Journal of Agronomy*, vol 7, pp15–23

Sanchez, P. A., Jama, B. and Niang, A. I. (2001) 'Soil fertility, small-farm intensification and the environment in Africa', in Barrett, C. and Lee, D. (eds) *Tradeoffs or Synergies? Agricultural Intensification, Economic Development and the Environment in Developing Countries*, CAB International, Wallingford, UK

Sanchez, P. A., Buresh, R. J. and Leakey, R. R. B. (1997) 'Trees, soils and food security', *Philosophical Transactions of the Royal Society of London, Series B*, no 353, pp949–961

Sanchez, P. A., Shepherd, K. D., Soule, M. I., Place, F. M., Buresh, R. J., Izac, A.-M. N., Mokwunye, A. U., Kwesiga, F. R., Ndiritu, C. G. and Woomer, P. L. (1997) 'Soil fertility replenishment in Africa: An investment in natural resource capital', in Buresh, R. J., Sanchez, P. A. and Calhoun, F. (eds) *Replenishing Soil Fertility in Africa*, SSSA Special Publication no 51, Soil Science Society of America, Madison, Wisconsin, p46

Steppler, H. A. and Nair, P. K. R. (eds) (1987) *Agroforestry: A Decade of Development*, ICRAF, Nairobi

TAC (1999) *Second External Review of ICRAF*, Technical Advisory Committee of the Consultative Group on International Agricultural Research, FAO, Rome

Young, A. (1997) *Agroforestry for Soil Management*, 2nd edition, CAB International, Wallingford, UK

27

Reducing Food Poverty by Increasing Agricultural Sustainability in Developing Countries

Jules Pretty, James Morison and Rachel Hine

Introduction

Over the past 40 years, per capita world food production has grown by 25 per cent, with average cereal yields rising from 1.2 tonnes per hectare (t ha^{-1}) to 2.52 t ha^{-1} in developing countries (1.71 t ha^{-1} on rainfed lands and 3.82 t ha^{-1} on irrigated lands), and annual cereal production up from 420 to 1176 million tonnes (FAO, 2000). These global increases have helped to raise average per capita consumption of food by 17 per cent over 30 years to 2760 kilocalories (kcal) day^{-1}, a period during which world population grew from 3.69 to 6.0 billion. Despite such advances in productivity, the world still faces a persistent food security challenge. There are an estimated 790 million people lacking adequate access to food, of whom 31 per cent are in east and southeast Asia, 31 per cent in south Asia, 25 per cent in sub-Saharan Africa, 8 per cent in Latin America and the Caribbean, and 5 per cent in North Africa and Near East (Pinstrup-Andersen et al, 1999). A total of 33 countries still have an average per capita food consumption of less than 2200 kcal day^{-1} (FAO, 2000).

An adequate and appropriate food supply is a necessary condition for eliminating hunger. But increased food supply does not automatically mean increased food security for all. A growing world population for at least another half century, combined with changing diets arising from increasing urbanization and consumption of meat products, will bring greater pressures on the existing food system (Popkin, 1998; Delgado et al, 1999; Pinstrup-Andersen et al, 1999; UN, 1999; ACC/SCN, 2000; Smil, 2000). If food poverty is to be reduced, then it is important to ask who produces the food, who has access to the technology and knowledge to produce it, and who has the purchasing power to acquire it? Modern agricultural methods have been shown to be able to increase food production, yet food poverty persists. Poor and hungry people need low-cost and

Note: Reprinted from *Agriculture, Ecosystems and Environment*, vol 95, Pretty, J., Morison, J. and Hine, R. 'Reducing food poverty by increasing agricultural sustainability in developing countries', pp217–234, copyright © (2003), with permission from Elsevier, London

readily-available technologies and practices to increase food production. A further challenge is that this needs to happen without further damage to an environment increasingly harmed by existing agricultural practices (Pretty et al, 2000; Wood et al, 2000; McNeely and Scherr, 2001).

Key questions for research on agricultural sustainability

There are three strategic options for agricultural development if food supply is to be increased:

1   expand the area of agriculture, by converting new lands to agriculture, but resulting in losses of ecosystem services from forests, grasslands and other areas of important biodiversity;
2   increase per hectare production in agricultural exporting countries (mostly industrialized), but meaning that food still has to be transferred or sold to those who need it, whose very poverty excludes these possibilities;
3   increase total farm productivity in developing countries which most need the food, but which have not seen substantial increases in agricultural productivity in the past.

In this research, we explore the capacity to which more sustainable technologies and practices can address the third option. We draw tentative conclusions about the value of such approaches to agricultural development. This is not to say that industrialized agriculture cannot successfully increase food production. Manifestly, any farmer or agricultural system with unlimited access to sufficient inputs, knowledge and skills can produce large amounts of food. But most farmers in developing countries are not in such a position, and the poorest generally lack the financial assets to purchase costly inputs and technologies. The central questions, therefore, focus on:

*   to what extent can farmers increase food production by using low-cost and locally available technologies and inputs?
*   what impacts do such methods have on environmental goods and services and the livelihoods of people who rely on them?

The success of industrialized agriculture in recent decades has often masked significant environmental and health externalities (actions that affect the welfare of or opportunities available to an individual or group without direct payment or compensation). Environmental and health problems associated with industrialized agriculture have been well documented (Balfour, 1943; Carson, 1963; Conway and Pretty, 1991; EEA, 1998; Wood et al, 2000), but it is only recently that the scale of the costs has come to be appreciated through studies in China, Germany, the UK, the Philippines and the US (Steiner et al, 1995; Pimentel et al, 1995; Pingali and Roger, 1995; Waibel and Fleischer, 1998; Norse et al, 2000; Pretty et al, 2000, 2001).

What do we understand by agricultural sustainability? Systems high in sustainability are making the best use of nature's goods and services whilst not damaging these assets

(Altieri, 1995; Pretty, 1995, 1998; Thrupp, 1996; Conway, 1997; Hinchliffe et al, 1999; NRC, 2000; Li Wenhua, 2001; McNeely and Scherr, 2001; Uphoff, 2002). The aims are to: (1) integrate natural processes such as nutrient cycling, nitrogen fixation, soil regeneration and natural enemies of pests into food production processes; (2) minimize the use of non-renewable inputs that damage the environment or harm the health of farmers and consumers; (3) make productive use of the knowledge and skills of farmers, so improving their self-reliance and substituting human capital for costly inputs; and (4) make productive use of people's capacities to work together to solve common agricultural and natural resource problems, such as pest, watershed, irrigation, forest and credit management.

Agricultural systems emphasizing these principles are also multifunctional within landscapes and economies. They jointly produce food and other goods for farm families and markets, but also contribute to a range of valued public goods, such as clean water, wildlife, carbon sequestration in soils, flood protection, groundwater recharge, and landscape amenity value. As a more sustainable agriculture seeks to make the best use of nature's goods and services, so technologies and practices must be locally adapted. They are most likely to emerge from new configurations of social capital, comprising relations of trust embodied in new social organizations, and new horizontal and vertical partnerships between institutions, and human capital comprising leadership, ingenuity, management skills and capacity to innovate. Agricultural systems with high levels of social and human assets are more able to innovate in the face of uncertainty (Uphoff, 1999; Pretty and Ward, 2001).

Research methods

The aim of the research was to audit recent progress in agricultural sustainability in developing countries. We accessed an international network of key professionals in the field of agricultural sustainability and food security, and asked them both to suggest projects and initiatives, and to pass on details of this research project to other relevant people or institutions. We also accessed other datasets (FAO, 1999). We asked for nominations for three types of initiatives: (1) research projects with active farmer involvement, but which may not yet have spread; (2) community-based projects with proven impacts; and (3) regional projects/initiatives that have spread to many communities. We use the term 'project/initiative' here, as these have emerged from many types of institutional context – some are international development projects, some are activities within government programmes, some are non-government organization or private company led, and some are promoted entirely by farmers' organizations themselves.

We developed a four-page questionnaire as the survey instrument, with a short descriptive rationale on agricultural sustainability and the aims of this research project. The questionnaire survey instrument was based on an assets-based model of agricultural systems, and was developed to understand both the role of these assets as inputs to agriculture and the consequences of agriculture upon them (Conway, 1997; Pretty, 2000; Pretty and Hine, 2000). The questionnaire addressed key impacts on total food production and on natural, social and human capital, the project/initiative structure and institutions, details of the context and reasons for success, and spread and scaling-up (institutional, technical and policy constraints). The questionnaire was sent out in English, French and Spanish to all potential projects by email and conventional post in early 1999. Field operatives from nodal organizations were contacted with specific requests and questionnaires. Each project was contacted with a personalized covering letter and questionnaire, and the resultant high response rate (some 60 per cent of those contacted replied with some information) appears to be a consequence of this personal contact. Some 200 reminders and questionnaires were sent out by email in autumn 1999 to attempt to access those who had not yet answered. Follow-up contacts were made to many of these during the course of the year 2000. In a number of instances, we received secondary data on the project rather than a completed questionnaire. We collated returned questionnaires and secondary material, and added these to country files. All datasets were re-examined to identify gaps, and correspondents contacted again.

Not all proposed cases were accepted for the dataset, and rejections were made: (1) where there was no obvious link to agricultural sustainability; (2) where payments were used to encourage farmer participation, as there have long been doubts that ensuing improvements persist after such incentives end; (3) where there was heavy reliance on fossil fuel derived inputs, or only on their targeted use (this is not to negate these technologies, but to simply indicate that they were not the focus of this research); and (4) where the data provided were too weak or the findings unsubstantiated. We also acknowledge that just because projects/initiatives have been accepted for this dataset, this does not necessarily mean they will be sustained indefinitely. The problem of agricultural development activities not persisting beyond the end of projects has been widely analysed through post-project reviews (Bunch, 1983; Chambers, 1983; Cernea, 1991; Carter, 1995). However, we do have confidence about these projects/initiatives, as farmers are adopting novel technologies and practices on their own terms and because they pay – not because they are being offered distorting incentives to do what an external agency wishes.

The questionnaires were self-completed, so were subject to potential bias. We therefore established trustworthiness checks through checks with secondary data, by critical review by external reviewers and experts, and by engaging in regular personal dialogue with respondents. We verified projects by sending full details entered on the database to the named verifier on the questionnaire. We also sent batches of projects to key authorities to obtain a second or third view on the project. This research, therefore, comprises a purposive sample of existing 'best practice' projects/initiatives explicitly addressing agricultural sustainability. It was not a random sample of all agricultural projects, and thus the findings are not representative of all developing country farms. Our aim was to discover the impacts of existing initiatives, to understand the processes and policies that encouraged or restricted them, and to indicate the potential for addressing food poverty through a focus on agricultural sustainability.

Survey results

This was the largest known survey of sustainable agricultural practices and technologies in developing countries, with 45 projects in Latin America, 63 in Asia and 100 in Africa, in which 8.98 million farmers have adopted more sustainable practices and technologies on 28.92 million hectares. As there are 960 million hectares of land under cultivation (arable and permanent crops) in Africa, Asia and Latin America, more sustainable practices are now present on at least 3.0 per cent of this land (total arable land comprises some 1600 million hectares in 1995/97, of which 388 million hectares are in industrialized countries, 267 million hectares in transition countries, and 960 million hectares in developing countries (FAO, 2000).

The most common country representations in the dataset are India (23 projects/initiatives); Uganda (20); Kenya (17); Tanzania (10); China (8); the Philippines (7); Malawi (6); Honduras, Peru, Brazil, Mexico, Burkina Faso and Ethiopia (5); and Bangladesh (4). The projects range very widely in scale – from 10 households on 5 hectares in one project in Chile to 200,000 farmers on more than 10 million hectares in southern Brazil. Most of the farmers in the projects surveyed are small farmers. Of farms in the total dataset, 50 per cent are in projects with a mean area per farmer of less than one hectare, and 90 per cent with less than or equal to 2 hectares (Figures 27.1a and b). Thus of the total, there are some 8.64 million small farmers practising forms of more sustainable farming on 8.33 million hectares. Most of these initiatives increasing agricultural sustainability have emerged in the past decade. Using project records, we estimate that the area a decade ago in these 208 initiatives was no more than 500,000 hectares.

Changes in farm and household food productivity

We found improvements in food production were occurring through one or more of four different mechanisms:

1   the intensification of a single component of a farm system, with little change to the rest of the farm, such as home garden intensification with vegetables or tree crops, vegetables on rice bunds, and introduction of fish ponds or a dairy cow;
2   the addition of a new productive element to a farm system, such as fish or shrimps in paddy rice fields, or trees, which provide a boost to total farm food production and/or income, but which do not necessarily affect cereal productivity;
3   the better use of natural resources to increase total farm production, especially water (by water harvesting and irrigation scheduling), and land (by reclamation of degraded land), so leading to additional new dryland crops and/or increased supply of additional water for irrigated crops (both increasing cropping intensity);
4   improvements in per hectare yields of staple cereals through introduction of new regenerative elements into farm systems, such as legumes and integrated pest management, and new and locally appropriate crop varieties and animal breeds.

Thus a successful project increasing agricultural sustainability may be substantially improving domestic food consumption or increasing local food barters or sales through home gardens or fish in rice fields, or better water management, without necessarily affecting the per hectare yields of cereals. Figure 27.2 illustrates the frequency of occurrence of each of these mechanisms in the dataset. The most common mechanisms were yield improvements with regenerative technologies and new seeds/breeds, occurring in 60

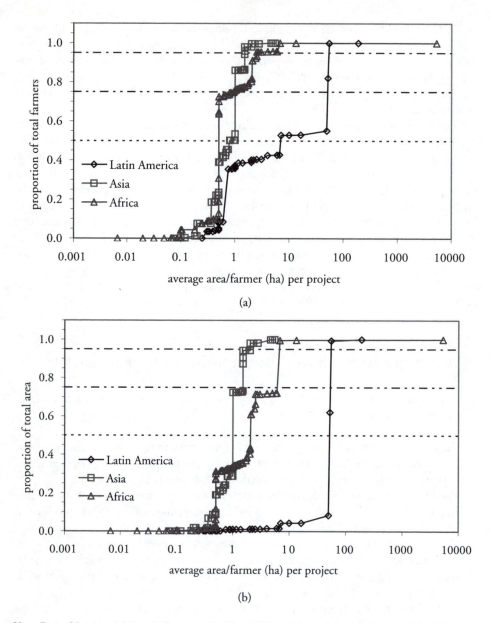

(a)

(b)

Note: Dotted horizontal lines represent 50, 75 and 95 per cent of total farmers in (a) and total area in (b).

Figure 27.1 *Cumulative proportion of (a) farmers and (b) total area by project size according to region*

per cent of the projects, by 56 per cent of the farmers and over 89 per cent of the area. Home garden intensification occurred in 20 per cent of projects, but given its small scale

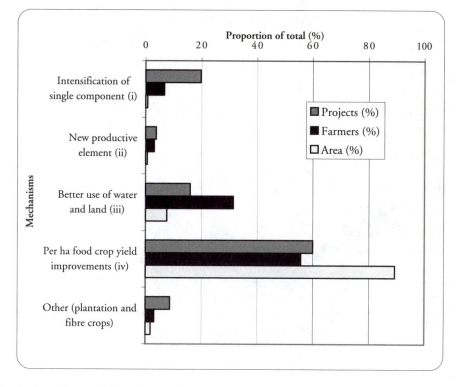

Source: Data from SAFE-World project

Figure 27.2 *Frequency of occurrence of each type of improvement mechanism by projects, farmers and areas*

only accounted for 0.7 per cent of area. Better use of land and water, giving rise to increased cropping intensity, occurred in 14 per cent of projects, with 31 per cent of farmers and 8 per cent of the area. The incorporation of new productive elements into farm systems, mainly fish/shrimps in paddy rice, occurred in 4 per cent of projects, and accounted for the smallest proportion of farmers and area.

As mechanism 4 was the most common, we analysed these projects in greater detail. The dataset contains 89 projects (139 entries of crop x projects combinations) with reliable data on per hectare yield changes with mechanism 4, and these are shown as relative yield changes over a baseline of 1.0 according to yields before or without the interventions (Figure 27.3). Agricultural sustainability has led to a 93 per cent increase in per hectare food production through mechanism 4 averaged across all projects. The weighted averages are a 37 per cent increase per farm household, and a 48 per cent increase per hectare in these projects. The relative yield increases are higher at lower yields, indicating greater benefits for poor farmers, most of whom have been missed by recent decades of agricultural development. We also analysed the yield data according to crop types (Figure 27.4). The largest relative increases in yield occur for vegetables and roots, and the smallest for rice and beans/soya/peas.

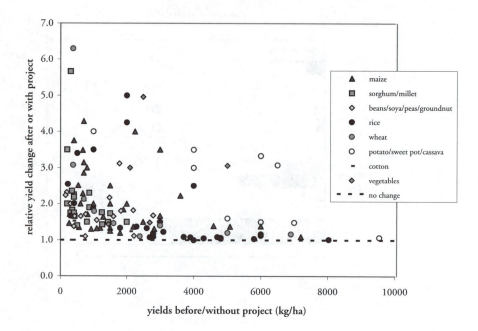

Source: Data from SAFE-World project

Figure 27.3 *Relative crop yield changes in agricultural sustainability projects/*
initiatives (89 projects)

We also calculated the marginal increase in food production per household for these 89 projects with reliable data on yields, area and numbers of farmers. Using the data for average farm size in each project, we calculated the average increase in annual food production per household after adoption of more sustainable practices and technologies (Figure 27. 5). In the 80 projects with small (< 5ha) farms where cereals were the main staples, the 4.42 million farms on 3.58 million hectares increased household food production by 1.71 tonnes per year (t yr^{-1}) (an increase of 73 per cent). In the 14 projects with roots as main staples (potato, sweet potato and cassava), the 146,000 farms on 542,000 hectares increased household food production by 17t yr^{-1} (increase of 150 per cent). In the four projects in southern Latin America with larger farm size (average size of 90ha per farm), household production increased by 150 t yr^{-1} (increase of 46 per cent).

These aggregate figures understate the benefits of increased diversity in the diet as well as increased quantity. Most of these agricultural sustainability initiatives have seen increases in farm diversity. In many cases, this translates into increased diversity of food consumed by the household, such as availability of fish protein from rice fields or fish ponds, milk and animal products from dairy cows, poultry and pigs kept in the home garden, and vegetables and fruit from home gardens and other farm micro-environments. Although these initiatives are reporting significant increases in food production,

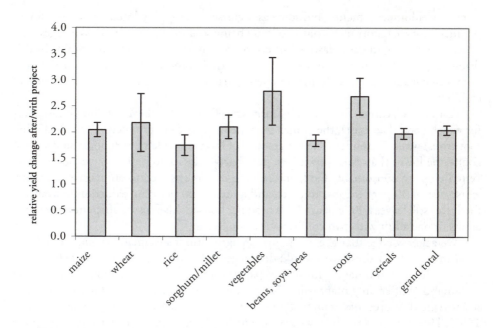

Source: Data from SAFE-World project

Figure 27.4 *Relative change in yield grouped by crop type (mean ± standard error of the mean)*

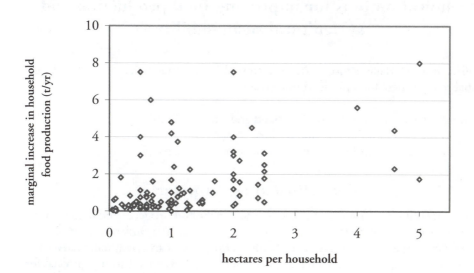

Source: Data from SAFE-World project

Figure 27.5 *Change in annual household food production with sustainable agricultural practices and technologies*

some as yield improvements, and some as increases in cropping intensity or diversity of produce, few are reporting surpluses of food being sold to local markets. We suggest this is because of a significant elasticity of consumption amongst rural households in this dataset experiencing any degree of food insecurity. As production increases, so also does domestic consumption, with direct benefit particularly for the health of women and children.

As indicated earlier, for an average farm size of 1.4ha (for the 4.4 million households for which good data exist), the annual increase in gross food production (not including root crops) was 1.71 tonnes. The net amount of food available to each household will, of course, be lower than this – owing to post-harvest losses to pests, conversion of harvested crops to consumable food, and feeding of some as feed to animals. Assuming a worst case of 30 per cent loss to pests, and a further 30 per cent reduction in available food, this still leaves 800kg of available food per household. This is sufficient to feed two adults or one adult with two children for a whole year.

We acknowledge that these findings on agricultural sustainability may sound too good to be true for those who would disbelieve these advances. Many still believe that food production and nature must be separated, that practices increasing agricultural sustainability offer only marginal opportunities to increase food production, and that industrialized approaches represent the best, and perhaps only, way forward (Avery, 1995). However, prevailing views have gradually changed in the last decade, and some sceptics are beginning to recognize the value of innovative capacity emerging from poorer communities in developing countries.

Technical options for improving food production and agricultural sustainability

We discern in the dataset three types of technical improvement that have played substantial roles in these food production increases:

1 more efficient water use in both dryland and irrigated farming;
2 improvements to soil health and fertility;
3 pest and weed control with minimum or zero-pesticide use.

More efficient use of water

Improvements in the efficiency of water use can benefit both irrigated and rainfed farmers by allowing new or formerly degraded lands to be brought under farming, and to increased cropping intensity on existing lands. In the projects analysed, water harvesting has been widely applied in dryland areas. The Indo-British Rainfed Farming project, for example, works with 230 local groups in 70 villages on water harvesting, tree planting, and grazing land improvements (Sodhi, 2001). Basic grain yields of rice, wheat, pigeon-peas and sorghum have increased from 400 to 800–1000kg ha^{-1}, and the increased fodder grass production from the terrace bunds is valued highly for the livestock. Improved water retention has resulted in water tables rising by one metre over 3–4 years, meaning that an extra crop is now possible for many farmers, thus turning an unproductive season into a productive one.

Women are major beneficiaries. Sodhi (2001) puts it this way: 'In these regions, women never had seen themselves at the front edge of doing things, taking decisions, and dealing with financial transactions. The learning by doing approach of the project has given them much needed confidence, skills, importance and awareness.' The wider benefits of a transformed agriculture are also evident, as 'the project has indirectly affected migration as people are gaining more income locally through the various enterprises carried out in the project. People are now thinking that they must diversify more into new strategies. There has also been a decline in drawing on resources from the forests.' Other projects in India have seen similar environmental and social changes (Devavaram et al, 1999; Lobo and Palghadmal, 1999).

In sub-Saharan Africa, water harvesting is also transforming barren lands. Again, the technologies are not complex and costly. In central Burkina Faso, 130,000ha of abandoned and degraded lands have been restored with the adoption of *tassas* and *zaï*. These are 20–30cm holes dug in soils that have been sealed by a surface layer hardened by wind and water erosion. The holes are filled with manure to promote termite activity and enhance infiltration. When it rains, water is channelled by simple stone bunds to the holes, which fill with water, and into which are planted seeds of millet or sorghum. Cereal yields in these regions rarely exceed 300kg ha^{-1}, yet these improved lands now produce 700–1000kg ha^{-1}. Reij (1996) calculated that the average family in Burkina Faso using these technologies had shifted from being in annual cereal deficit amounting to 650kg, equivalent to six and a half months of food shortage, to producing a surplus of 150kg yr^{-1}. Furthermore, *tassas* are best suited to landholdings where family labour is available, or where farm labour can be hired, so that this soil and water conservation method has led to a market for young day labourers who, rather than migrating, now earn money by building these structures.

Good organization also helps to improve irrigated agriculture. Despite great investment, many irrigation systems have become inefficient and subject to persistent conflict. Irrigation engineers assume that they know best how to distribute water, yet can never know enough about the specific conditions and needs of large numbers of farmers. Recent years, though, have seen the spread of programmes to organize farmers into water users' groups, and let them manage water distribution for themselves (Cernea, 1991; Uphoff, 2002). One of the best examples comes from the Gal Oya region in Sri Lanka. Before this approach, Gal Oya was the largest and most run-down scheme in the country. Now, farmers' groups manage water for 26,000ha of rice fields, and produce more rice crops per year and per unit of water. Moreover, when farmers took control, the number of complaints received by the Irrigation Department about water distribution fell to nearly zero (Uphoff, 1999). The benefits were dramatically shown during the 1998 drought. According to government, there was only enough water for irrigation of 18 per cent of the rice area. But farmers persuaded the Irrigation Department to let this water through on the grounds that they would carefully irrigate the whole area. Through cooperation and careful management, they achieved a better than average harvest, earning the country US$20 million in foreign exchange. Throughout Sri Lanka, 33,000 water users' associations have now been formed – a dramatic increase in local social organization that has increased farmers' own capacities for problem-solving and cooperation, and for using nature more efficiently and effectively to produce more food.

Improvements in soil health and fertility

Soil health is fundamental for agricultural sustainability, yet is under widespread threat from degradation processes (Cleaver and Schreiber, 1995; World Bank/FAO, 1996; Smaling et al, 1997; Hinchcliffe et al, 1999; Petersen et al, 2000; Koohafkan and Stewart, 2001). Agricultural sustainability starts with the soil by seeking both to reduce soil erosion and to make improvements to soil physical structure, organic matter content, water-holding capacity and nutrient balances. Soil health is improved through the use of legumes, green manures and cover crops, incorporation of plants with the capacity to release phosphate from the soil into rotations, use of composts and animal manures, adoption of zero-tillage, and use of inorganic fertilizers where needed (Reicosky et al, 1997; Sanchez and Jama, 2000). In projects in Central America, the incorporation of nitrogen-fixing legumes into agroecosystems has substantially affected productivity, particularly the velvetbean (*Mucuna pruriens*). This grows rapidly, fixes 150–200kg N ha^{-1} yr^{-1}, suppresses weeds, and can produce 35–50 tonnes of biomass ha^{-1} yr^{-1} (Bunch, 2000; Anderson et al, 2001). Addition of this biomass to soils substantially improves soil organic matter content, and has helped to increase cereal productivity for some 45,000 families in Guatemala, Honduras and Nicaragua.

In the past decade, Latin American farmers have found that eliminating tillage can be highly beneficial for soils. After harvest, crop residues are left on the surface to protect against erosion and, seed is directly planted into a groove cut into the soil. Weeds are controlled with herbicides or cover crops. The fastest uptake of minimum till systems has been in Brazil, where there are now 15 million hectares under *plantio direto* (also called zero-tillage even though there is some disturbance of the soil) mostly in three southern states of Santa Caterina, Rio Grande do Sul and Paraná, and in the central Cerrado. In neighbouring Argentina, there are more than 11 million hectares under zero-tillage, up from less than 100,000 hectares in 1990, and in Paraguay, another one million hectares of zero-tillage (Sorrenson et al, 1998; Petersen et al, 1999; de Freitas, 1999; Peiretti, 2000; Landers et al, 2001).

Elsewhere in Brazil, the transformations in the landscape and in farmers' attitudes are equally impressive. The Cerrado is a vast area of formerly unproductive lands colonized for farming in the past two decades. These lands needed lime and phosphorus before they could become productive, and now zero-tillage is being widely adopted (Landers et al, 2001). In the early days, there was a widespread belief that zero-tillage was only for large farmers. That has now changed, and small farmers are benefiting from technology breakthroughs developed for mechanical farming. A core element of zero-tillage adoption in South America has been adaptive research – working with farmers at microcatchment level to ensure technologies are fitted well to local circumstances. There are many types of farmers groups: from local (farmer microcatchment and credit groups), to municipal (soil commissions, Friends of Land clubs, commercial farmers and farm workers' unions), to multi-municipal (farmer foundations and cooperatives), to river basin (basin committees for all water users), and to state and national level (state zero-tillage associations and the national zero-tillage federation).

Farmers are now adapting technologies – organic matter levels have sufficiently improved that fertilizer use has been reduced and rainfall infiltration improved, such that some farmers are removing contour terraces. Other side-effects of zero-tillage include reduced siltation of reservoirs, less flooding, higher aquifer recharge, lowered costs of water treatment, cleaner rivers, and more winter feed for wild biodiversity (Landers et al, 2001). However, there is still controversy over zero-tillage, as some feel the use of herbicides to control weeds, or of genetically-modified crops, means we cannot call these systems sustainable. However, the environmental benefits are substantial, and new research is showing that farmers have some effective alternatives, particularly if they use cover crops for green manures to raise organic matter levels. Using 20 species of cover crops and green manures, Petersen and colleagues have shown how small farmers can adopt zero-tillage systems without herbicides (Petersen et al, 2000; von der Weid, 2000).

A public good is also being created when soil health is improved with increased organic matter. Soil organic matter contains carbon, and soils with above-ground biomass can act as 'carbon sinks' or sites for carbon sequestration (Reicosky et al, 1995; Smith et al, 1998; Sanchez et al, 1999; Watson et al, 2000; Pretty and Ball, 2001). Conservation tillage systems and those using legumes and/or cover crops contribute to organic matter and carbon accumulation in the soil.

In the Sahelian countries of Africa, the major constraints to food production are also related to soils, most of which are sandy and low in organic matter. In Senegal, where soil erosion and degradation threaten large areas of agricultural land, the Rodale Institute Regenerative Agriculture Resource Center works closely with farmers' associations and government researchers to improve the quality of soils. The primary cropping system of the region is a millet-groundnut rotation. Fields are cleared by burning, and then cultivated with shallow tillage using animals. But fallow periods have decreased dramatically, and inorganic fertilizers do not return high yields unless there are concurrent improvements in organic matter, which helps to retain moisture. The Center collaborates with 2000 farmers organized into 59 groups on improving soil quality by integrating stall-fed livestock into crop systems, adding legumes and green manures, increasing the use of manures, composts and rock phosphate, and developing water-harvesting systems. The result has been a 75 per cent-190 per cent improvement in millet and groundnut yields – from about 300 to 600–900kg ha^{-1}. Yields are also less variable year on year, with consequent improvements in household food security (Diop, 1999).

Thus if the soil is improved, the whole agricultural system's health improves. Even if this is done on a very small scale, people can benefit substantially. In Kenya, the Association for Better Land Husbandry found that farmers who constructed double-dug beds in their gardens could produce enough vegetables to see them through the hungry dry season. These raised beds are improved with composts, and green and animal manures. A considerable investment in labour is required, but the better water holding capacity and higher organic matter means that these beds are both more productive and better able to sustain vegetable growth through the dry season. Once this investment is made, little more has to be done for the next two to three years. Women in particular are cultivating many vegetable and fruit crops, including kale, onion, tomato, cabbage, passion fruit, pigeon pea, spinach, pepper, green bean and soya. According to one review of 26 communities, 75 per cent of participating households are now free from hunger during the year, and the proportion having to buy vegetables had fallen from 85 per cent to

11 per cent. For too long, agriculturalists have been sceptical about these organic and conservation methods. They say they need too much labour, are too traditional, and have no impact on the rest of the farm. Yet the spin-off benefits are substantial, as giving women the means to improve their food production means that food gets into the mouths of children. They suffer fewer months of hunger, and so are less likely to miss school (Hamilton, 1998).

Pest control with minimal or zero-pesticide use

Modern farmers have come to depend on a great variety of insecticides, herbicides and fungicides to control pests, weeds and diseases, and each year, some 5 billion kg of pesticide active ingredients are applied to farms (BAA, 2000). But farmers in these projects have found many effective and more sustainable alternatives. In some crops, it may mean the end of pesticides altogether, as cheaper and more environmentally benign practices are found to be effective.

Many projects in our survey reported large reductions in pesticide use in irrigated rice systems. Following the discovery that pest attack on rice was proportional to pesticide use (Kenmore et al, 1984), Farmer Field Schools were later developed to teach farmers the benefits of agrobiodiversity. In Indonesia, one million farmers have now attended 50,000 field schools, the largest number in any Asian country. In Vietnam, two million farmers have cut pesticide use from more than 3 sprays to 1 per season; in Sri Lanka, 55,000 farmers have reduced use from 3 to 0.5 per season; and in Indonesia, one million farmers have cut use from 3 sprays to 1 per season. In no case has reduced pesticide use led to lower rice yields (Eveleens et al, 1996; Heong et al, 1998; Mangan and Mangan, 1998; Desilles, 1999; Jones, 1999). Amongst these are reports that many farmers are now able to grow rice entirely without pesticides: 25 per cent of field school trained farmers in Indonesia, 20–33 per cent in the Mekong Delta of Vietnam, and 75 per cent in parts of the Philippines.

If pesticides are removed, then fish can be reintroduced. In Bangladesh, an aquaculture and integrated pest management programme implemented by CARE has completed 6000 farmer field schools, resulting in 150,000 farmers adopting more sustainable methods of rice production on about 50,000 hectares. The programme emphasizes fish cultivation in paddy fields, and vegetable cultivation on rice field dykes. Rice yields have improved by 5–7 per cent, and costs of production have fallen owing to reduced pesticide use. In addition, each hectare of paddy yields up to 750kg of fish, a significant increase in system productivity for poor farmers with few resources (Rashid, 2001).

In Kenya, intercropping of local legumes and grasses with maize has been found to reduce stem borer (*Chilo* spp) attack through interactions with parasitic wasps. Researchers call their redesigned and diverse maize fields *vutu sukuma* (push-pull systems). More than 2000 farmers in western Kenya have adopted maize, grass-strip and legume-intercropping systems, and have increased maize yields by 60–70 per cent. The official advice to maize growers in the tropics has been to create monocultures for modern varieties of maize, and then apply pesticide and fertilizers to make them productive. Yet this very simplification eliminated vital and free pest management services produced by the grasses and legumes. *Vutu sukumu* systems are complex and diverse, and are cheap as they do not rely on costly purchased inputs (Khan et al, 2000).

Another project in Yunnan, China has shown the value of mixtures of rice, both in reducing disease incidence and increasing yields (Zhu et al, 2000; Wolfe, 2000). Researchers working in ten townships on 5350 hectares encouraged farmers to switch from growing monocultures of sticky rice to alternating rows of sticky rice with hybrids. The sticky rice brings a higher price, but is susceptible to rice blast, which is generally controlled through applications of fungicides. But planting mixtures in the same field reduced blast incidence by 94 per cent and increased total yields by 89 per cent. By the end of two years, it was concluded that fungicides were no longer required.

Impacts on rural livelihoods and economies

Rural people's livelihoods rely for their success on the value of services flowing from the total stock of natural, social, human, physical and financial capital (Coleman, 1990; Putnam, 1993; Costanza et al, 1997; Carney, 1998; Scoones, 1998; Pretty and Ward, 2001). A number of examples can be extracted from the dataset to show that agricultural sustainability projects and initiatives have been able to contribute to the accumulation of locally valuable assets. A selection of the impacts reported in these sustainable agriculture projects and initiatives include:

- improvements to natural capital, including increased water retention in soils, improvements in water table (with more drinking water in the dry season), reduced soil erosion combined with improved organic matter in soils, leading to better carbon sequestration, and increased agrobiodiversity (Hinchcliffe et al, 1999; Watson et al, 2000; McNeely and Scherr, 2001; Pretty and Ball, 2001);
- improvements to social capital, including more and stronger social organizations at local level, new rules and norms for managing collective natural resources, and better connectedness to external policy institutions (see Uphoff, 1999; Pretty and Ward, 2001);
- improvements to human capital, including more local capacity to experiment and solve own problems; reduced incidence of malaria in rice–fish zones, increased self-esteem in formerly marginalized groups, increased status of women, better child health and nutrition, especially in dry seasons, and reversed migration and more local employment (Li Kangmin, 1998; Shah and Shah, 1999; Bunch, 2000; Regasamy et al, 2000).

The empirical evidence indicates that some improvements in agricultural sustainability have had positive effects on regional economies. In the Ansokia Valley, Ethiopia, one project increased annual food production from 5600 to 8370 tonnes in six years, at the same time as the population increased from 36,000 to 45,000. The project turned around an annual food regional deficit of – 2106 tonnes to a surplus of 372 tonnes year^{-1}. In Bushenyi, Uganda, formerly experiencing substantial food shortages during the months of October to December, one project so increased banana and cattle production that the region could sells 330 tonnes of bananas and 2.7 tonnes of meat each week. In En Nahud, Sudan, the 10,000 tonnes of additional food produced by 15,000 households were consumed by local people. None found its way into national statistics.

There is also evidence that productivity can increase over time as natural and human capital assets increase. If agricultural systems are low in capital assets (either intrinsically low, or have become low because of degradation), then a sudden switch to 'more sustainable' practices that have to rely on these assets will not be immediately successful. In Cuba, for example, urban organic gardens produced 4200 tonnes of food in 1994. By 1999, they had greatly increased in per area productivity – rising from 1.6kg m^{-2} to 19.6 kg m^{-2} (Murphy, 1999; Funes, 2001). Increasing productivity over time has also been noted in fish-ponds in Malawi. These are typically some 200–500m^2 in size. Researchers compared the performance of 35 fishponds over six years: in 1990 yields were 800kg ha^{-1}, but rose steadily to 1450kg ha^{-1} by 1996. This is because fishponds are integrated into a farm so that they recycle wastes from other agricultural and household enterprises, leading to steadily increasing productivity over time as farmers themselves gain understanding (Brummet, 2000).

Confounding factors and trade-offs

What we do not yet know is whether moving to more sustainable systems, delivering greater benefits at the scale occurring in these projects, will result in enough food to meet the current food needs in developing countries, let alone the future needs after continued population growth and adoption of more urban and meat-rich diets (Delgado et al, 1999). But what we are seeing should be cause for cautious optimism, particularly as evidence indicates that productivity can grow over time if natural, social and human assets are accumulated (McNeely and Scherr, 2001).

A more sustainable agriculture which improves the asset base can lead to rural livelihood improvements. People can be better off, have more food, be better organized, have access to external services and power structures, and have more choices in their lives. But like all major changes, such transitions can also provoke secondary problems. For example, building a road near a forest can help farmers reach food markets, but also aid illegal timber extraction. Projects may be making considerable progress on reducing soil erosion and increasing water conservation through adoption of zero-tillage, but still continue to rely on applications of herbicides. If land has to be closed off to grazing for rehabilitation, then people with no other source of feed may have to sell their livestock; and if cropping intensity increases or new lands are taken into cultivation, then the burden of increased workloads may fall particularly on women. Also additional incomes arising from sales of produce may go directly to men in households, who are less likely than women to invest in children and the household as a whole.

There are also a variety of emergent factors that could slow the spread of agricultural sustainability. First, practices that increase the asset base may simply increase the incentives for more powerful interests to take over, such as landlords taking back formerly degraded land from tenants who had adopted more sustainable agriculture. In these contexts, it is rational for farmers to farm badly – at least they get to keep the land. The idea of sustainable agriculture may also appear to be keeping people in rural areas away from centres of power and 'modern' urban society, yet some rural people's aspirations may precisely to be to gain sufficient resources to leave rural areas. Agricultural sustainability also implies a limited role for agrochemical companies, who would not be predicted to accept such losses of market lightly. It also suggests greater decentralization of power to local communities and groups, combined with more local decision making, both of which might be opposed by those who would benefit from corruption and non-transparency in private and public organizations. Research and extension agencies will have to change too, adopting more participatory approaches to work closely with farmers, and so must adopt different measures for evaluating job success and the means to promotion. Finally, social connectivity, relations of trust, and the emergence of significant movements may present a threat to existing power bases, who in turn may seek to undermine such locally based institutions.

Further tensions arise over the balance between whether food production is more sustainable if for local markets alone, or whether poorer farmers and communities should be encouraged to access international markets, but with the result that food causes greater transport externalities through long-distance travel. Moreover, farms with increased productivity export increasingly large amounts of nutrients to be eaten elsewhere, and it will be vital to ensure that replacement occurs at sustainable rates. There will be some who dispute this evidence of promising successes, believing that the poor and marginalized cannot possibly make these kinds of improvements. But we believe there is hope and leadership in this evidence of progress towards sustainability. What is quite clear is that these technologies and practices offer real opportunities for people to improve their food production whilst protecting and improving nature.

Scaling up through appropriate policies

Three things are now clear from this dataset about spreading agricultural sustainability:

1   some technologies and social processes for local scale adoption of more sustainable agricultural practices are well-tested and established;
2   the social and institutional conditions for spread are less well-known, but have been established in several contexts, leading to very rapid spread in the 1990s;
3   the political conditions for the emergence of supportive policies are least well established, with only a very few examples of real progress.

As has been indicated earlier, agricultural sustainability can contribute to increased food production, as well as make an impact on rural people's welfare and livelihoods. Clearly much can be done with existing resources. A transition towards a more sustainable agriculture will not, however, happen without some external help and money. There are always transition costs in learning, in developing new or adapting old technologies, in learning to work together, and in breaking free from existing patterns of thought and practice. It also costs time and money to rebuild depleted natural and social capital.

Most agricultural sustainability improvements seen in the 1990s arose despite existing national and institutional policies, rather than because of them (Pretty, 1999; Pretty et al, 2001). Nonetheless, the 1990s have seen considerable global progress towards the recognition of the need for policies to support sustainable agriculture. Although almost every country would now say it supports the idea of agricultural sustainability, the evidence points towards only patchy reforms. Only two countries, Cuba and Switzerland, have given explicit national support for a transition towards sustainable agriculture – putting it at the centre of agricultural development policy and integrating policies accordingly. Cuba has a national policy for alternative agriculture; and Switzerland has three tiers of support for practices contributing to agriculture and rural sustainability (Swiss Agency for Environment, Forests and Landscape, 1999, 2000; Funes, 2001). Several countries have given subregional support, such as the states of Santa Caterina, Paraná and Rio Grande do Sul in southern Brazil supporting zero-tillage and catchment management, and some states in India supporting participatory watershed and irrigation management. A larger number have reformed parts of agricultural policies, such as China's support for integrated ecological demonstration villages, Kenya's catchment approach to soil conservation, Indonesia's ban on pesticides and programme for Farmer Field Schools, Bolivia's regional integration of agricultural and rural policies, Sweden's support for organic agriculture, Burkina Faso's land policy, and Sri Lanka and the Philippines' stipulation that water users' groups manage irrigation systems (Pretty, 2002).

A good example of a carefully designed and integrated programme comes from China. In March 1994, the government published a White Paper to set out its plan for implementation of Agenda 21, and put forward ecological farming, known as *Shengtai Nongye* or agroecological engineering, as the approach to achieve sustainability in agriculture. Pilot projects have been established in 2000 townships and villages spread across 150 counties. Policy for these 'eco-counties' is organized through a cross-ministry partnership, which uses a variety of incentives to encourage adoption of diverse production systems to replace monocultures. These include subsidies and loans, technical assistance, tax exemptions and deductions, security of land tenure, marketing services and linkages to research organizations. These eco-counties contain some 12 million hectares of land, about half of which is cropland, and though only covering a relatively small part of China's total agricultural land, do illustrate what is possible when policy is appropriately coordinated (Li Wenhua, 2001).

An even larger set of countries has seen some progress on agricultural sustainability at project and programme level. However, progress on the ground still remains largely despite, rather than because of, explicit policy support. No agriculture minister is likely to say they are against sustainable agriculture, yet good words remain to be translated into comprehensive policy reforms. Agricultural systems can be economically, environmentally and socially sustainable, and contribute positively to local livelihoods. But without appropriate policy support, they are likely to remain at best localized in extent, and at worst simply wither away.

Conclusions

This empirical study shows that there have been promising advances in the adoption and spread of more sustainable agriculture. The 208 projects/initiatives show increases in food production over some 29 million hectares, with nearly 9 million households benefiting from increased food production and consumption. These increases are not yet making a significant mark on national statistics, as we believe there is a significant elasticity of food consumption in many poor rural households. They are eating the increased food produced, or marketing small surpluses to other local people. We cannot, therefore, yet say whether a transition to more sustainable agriculture, delivering increasing benefits at the scale occurring in these projects, will result in enough food to meet the current food needs of developing countries, the future basic needs after continued population growth, or the potential demand following adoption of more meat-rich diets. Even the substantial increases reported here may not be enough. There should be cautious optimism, as the evidence indicates that productivity can increase steadily over time if natural, social and human capital assets are accumulated.

Increased agricultural sustainability can also be complementary to improvements in rural people's livelihoods. It can deliver increases in food production at relatively low cost, plus contribute to other important functions. Were these approaches to be widely adopted, they would make a significant impact on rural people's livelihoods, as well as on local and regional food security. But there are clearly major constraints to overcome. There will be losers along with winners, and some of the losers are currently powerful players. And yet, social organization and mobilization in a number of contexts is already leading to new informal and formal alliances that are protecting existing progress and developing the conditions for greater spread. Improving agricultural sustainability clearly will not bring all the solutions, but promising progress has been made in recent years. With further explicit support, particularly through international, national and local policy reforms, these benefits to food security and attendant improvements to natural, social and human capital could spread to much larger numbers of farmers and rural people in the coming decades.

Acknowledgements

We are grateful to three organizations for their support for this research: the UK Department for International Development (DFID), Bread for the World (Germany), and Greenpeace (Germany). We are grateful to Thomas Dobbs, Per Pinstrup-Andersen, Hiltrud Nieberg, Roland Bunch and Vo-Tung Xuan for substantive comments on the project report, together with feedback from participants at the St James's Palace, London 2001 conference on *Reducing Poverty with Sustainable Agriculture*. Two anonymous referees gave additional useful comments. We are indebted to 358 people directly associated with sustainable agriculture projects who have given their valuable time to send material, to complete questionnaires, to verify findings, and to advise on the wider project. In the final project report, we thank them all by name (at URL www2.essex.ac.uk/ces).

References

ACC/SCN (2000) *4th report on The World Nutrition Situation*, UN Administrative Committee on Coordination, Sub-Committee on Nutrition, in collaboration with IFPRI. United Nations, New York

Altieri, M. (1995) *Agroecology: The Science of Sustainable Agriculture.* Westview Press, Boulder, Colorado

Anderson, S., Gündel, S. and Pound, B. (2001) *Cover Crops in Smallholder Agriculture: Lessons from Latin America,* IT Publications, London

Avery D. (1995) *Saving the Planet with Pesticides and Plastic,* The Hudson Institute, Indianapolis

BAA (2000) *Annual Review and Handbook,* British Agrochemicals Association, Peterborough

Balfour, E. B. (1943) *The Living Soil,* Faber and Faber, London

Brummet, R. (2000) 'Integrated aquaculture in Sub-Saharan Africa', *Environmental Development and Sustainability,* vol 1, pp315–321

Bunch, R. (1983) *Two Ears of Corn,* World Neighbors, Oklahoma City

Bunch, R. (2000) 'More productivity with fewer external inputs', *Environmental Development and Sustainability,* vol 1, pp219–233

Bunch R. and López G. (1996) 'Soil recuperation in Central America: Sustaining innovation after intervention', *Gatekeeper Series SA 55,* International Institute for Environment and Development, London

Carney, D. (1998) *Sustainable Rural Livelihoods,* Department for International Development, London

Carson, R. (1963) *Silent Spring,* Penguin Books, Harmondsworth

Carter, J. (1995) *Alley Cropping: Have Resource Poor Farmers Benefited?* ODI Natural Resource Perspectives No 3, London

Cernea, M. M. (1991) *Putting People First,* 2nd edition, Oxford University Press, Oxford

Chambers, R. (1983) *Rural Development: Putting the Last First,* Longman, London

Cleaver, K. M. and Schreiber, G A. (1995) *The population, agriculture and environment nexus in Sub-Saharan Africa,* World Bank, Washington, DC

Coleman, J. (1990) *Foundations of Social Theory,* Harvard University Press, Massachusetts

Conway, G. R. (1997) *The Doubly Green Revolution,* Penguin, London

Conway, G. R. and Pretty, J. N. (1991) *Unwelcome Harvest: Agriculture and Pollution,* Earthscan, London

Costanza, R., d'Arge, R., de Groot, R., Farber, S., Grasso, M., Hannon, B., Limburg, K., Naeem, S., O'Neil, R.V., Parvelo, J., Raskin, R.G., Sutton, P. and van den Belt, M. (1997) 'The value of the world's ecosystem services and natural capital', *Nature,* vol 387, pp253–260

de Freitas, H. (1999) 'Transforming microcatchments in Santa Caterina, Brazil', in Hinchcliffe, F., Thompson, J., Pretty, J., Guijt, I. and Shah, P. (eds) *Fertile Ground: The Impacts of Participatory Watershed Development,* IT Publications, London

Delgado, C., Rosegrant, M., Steinfield, H., Ehui, S. and Courbois, C. (1999) *Livestock to 2020: The Next Food Revolution,* IFPRI Brief 61, International Food Policy Research Institute, Washington, DC

Desilles, S. (1999) 'Sustaining and managing private natural resources: The way to step out of the cycle of high-input agriculture', Paper for Conference on *Sustainable Agriculture: New Paradigms and Old Practices?* Bellagio Conference Center, Italy, 26–30 April, 1999

Devavaram, J., Arunothayam, E., Prasad, R. and Pretty, J. (1999) 'Watershed and community development in Tamil Nadu, India', in Hinchcliffe, F., Thompson, J., Pretty, J., Guijt, I. and Shah, P. (eds) *Fertile Ground: The Impacts of Participatory Watershed Development,* IT Publications, London

Diop, A. (1999) 'Sustainable agriculture: New paradigms and old practices? Increasing production with management of organic inputs in Senegal', *Environmental Development and Sustainability,* vol 1, pp285–296

EEA (1998) *Europe's Environment: The Second Assessment. Report and Statistical Compendium,* European Environment Agency, Copenhagen

Eveleens, K. G., Chisholm, R., van der Fliert, E., Kato, M., Nhat, P. T. and Schmidt, P. (1996) 'Mid Term Review of Phase III Report. The FAO Intercountry Programme for the Development and Application of Integrated Pest Control in Rice in South and Southeast Asia', FAO, Manila and Rome

FAO (1999) *Cultivating Our Futures: Taking Stock of the Multifunctional Character of Agriculture and Land*, FAO, Rome

FAO (2000) *Agriculture: Towards 2015/30*, Global Perspective Studies Unit, FAO, Rome

Funes, F. (2001) 'Cuba and sustainable agriculture'. Paper presented to St James's Palace conference *Reducing Poverty with Sustainable Agriculture*, 15th January. University of Essex, Colchester

Hamilton P. (1998) *Goodbye to Hunger: A Study of Farmers' Perceptions of Conservation Farming*, ABLH Nairobi, Kenya

Heong, K. L., Escalada, M. M., Huan, N. H. and Mai, V. (1998) 'Use of communication media in changing rice farmers' pest management in the Mekong Delta, Vietnam', *Crop Management*, vol 17, pp413–425

Hinchcliffe, F., Thompson, J., Pretty, J., Guijt, I. and Shah, P. (eds) (1999) *Fertile Ground: The Impacts of Participatory Watershed Development*, IT Publications, London

Jones, K. (1999) 'Integrated pest and crop management in Sri Lanka', Paper for Conference on *Sustainable Agriculture: New Paradigms and Old Practices?* Bellagio Conference Center, Italy, 26–30 April, 1999

Kenmore, P. E, Carino, F. O., Perez, C. A., Dyck, V. A. and Gutierrez, A. P. (1984) 'Population regulation of the brown planthopper within rice fields in the Philippines', *Journal of Plant Protection in the Tropics*, vol 1, pp19–37

Khan, Z. R., Pickett, J. A., van den Berg, J. and Woodcock, C. M. (2000) 'Exploiting chemical ecology and species diversity: stem borer and Striga control for maize in Africa', *Pest Management Science*, vol 56, pp1–6

Koohafkan, P. and Stewart, B. A. (2001) *Water Conservation and Water Harvesting in Cereal-Producing Regions of the Drylands*, FAO, Rome

Landers, J. N., De C Barros, G. S.-A., Manfrinato, W. A., Rocha, M. T. and Weiss, J. S. (2001) 'Environmental benefits of zero-tillage in Brazil – a first approximation', in Garcia Torres, L., Benites, J. and Martinez Vilela, A. (eds) *Conservation Agriculture – A Worldwide Challenge Volume 1*, XUL, Cordoba, Spain

Li Kangmin (1998) 'Rice aquaculture systems in China', in Eng-Leong Foo and Tarcision Della Senta (eds) *Integrated Bio-Systems in Zero Emissions Application,* Proceedings of an internet conference on integrated biosystems, available at www.ias.unu.edu/proceedings/icibs

Li Wenhua (2001) *Agro-Ecological Farming Systems in China,* Man and the Biosphere Series Volume 26, UNESCO, Paris

Lobo, C. and Palghadmal, T. (1999) 'Kasare: A saga of peoples faith', in Hinchcliffe et al (eds) *Fertile Ground*, IT Publications, London

Mangan, J. and Mangan, M. S. (1998) 'A comparison of two IPM training strategies in China: The importance of concepts of the rice ecosystem for sustainable pest management', *Agriculture and Human Values*, vol 15, pp209–221

McNeely, J. A. and Scherr, S. J. (2001) *Common Ground, Common Future. How Ecoagriculture Can Help Feed the World and Save Wild Biodiversity*, IUCN and Future Harvest, Geneva

Murphy, B. (1999) *Cultivating Havana: Urban Agriculture and Food Security in Cuba*, Food First Development Report 12. Food First, California

Norse, D., Li Ji and Zhang Zheng (2000) *Environmental Costs of Rice Production in China, Lessons from Hunan and Hubei*, Aileen Press, Bethesda

NRC (2000) *Our Common Journey: Transition towards sustainability*, Board on Sustainable development, Policy Division, National Research Council, National Academy Press, Washington, DC

Peiretti, R. (2000) 'The evolution of the No Till cropping system in Argentina'. Paper presented to *Impact of Globalisation and Information on the Rural Environment*, 13–15 January, Harvard, Cambridge, MA

Petersen, P., Rardin, J. M. and Marochi, F. (1999) 'Participatory development of no-tillage systems without herbicides for family farming', *Environment, Development and Sustainability*, vol 1, pp235–252

Petersen, C., Drinkwater, L. E. and Wagoner, P. (2000) *The Rodale Institute's Farming Systems Trial. The First 15 Years*, Rodale Institute, Pennsylvania

Pimentel, D., Harvey, C., Resosudarmo, P., Sinclair, K., Kunz, D., McNair, M., Crist, S., Shpritz, L., Fitton, L., Saffouri, R. and Blair, R. (1995) 'Environmental and economic costs soil erosion and conservation benefits', *Science*, vol 267, pp1117–1123

Pingali, P. L. and Roger, P. A. (1995) *Impact of Pesticides on Farmers' Health and the Rice Environment*, Kluwer Academic Press, Dordrecht

Pinstrup-Anderson, P., Pandya-Lorch, R. and Rosegrant, M. (1999) *World Food Prospects: Critical Issues for the Early 21st Century*, IFPRI, Washington, DC

Popkin, B. (1998) 'The nutrition transition and its health implications in lower-income countries', *Public Health and Nutrition*, vol 1, pp5–21

Pretty, J. N. (1995) *Regenerating Agriculture: Policies and Practice for Sustainability and Self-Reliance*, Earthscan Publications, London; National Academy Press, Washington, DC; Action-Aid, Bangalore

Pretty, J. N. (1998) *The Living Land: Agriculture, Food Systems and Community Regeneration in Rural Europe*, Earthscan Publications Ltd, London

Pretty, J. N. (1999) 'Sustainable agriculture: A review of recent progress on policies and practice', United Nations Research Institute for Social Development (UNRISD), Geneva

Pretty, J. N. (2000) 'Can sustainable agriculture feed Africa?' *Environmental Development and Sustainability*, vol 1, pp253–274

Pretty, J. N. (2002) *Agri-Culture: Communities Shaping Land and Nature*, Earthscan, London

Pretty, J. N. and Ball, A. (2001) *Agricultural Influences on Carbon Emissions and Sequestration: A Review of Evidence and the Emerging Trading Options*, Centre for Environment and Society Occasional Paper 2001–3, University of Essex

Pretty, J. N. and Hine, R. (2000) 'The promising spread of sustainable agriculture in Asia', *Natural Resources Forum (UN)*, vol 24, pp107–121

Pretty, J. N. and Ward, H. (2001) 'Social capital and the environment', *World Development*, vol 29, pp209–227

Pretty, J. N., Brett, C., Gee, D., Hine, R., Mason, C. F., Morison, J. I. L., Raven, H., Rayment, M. and van der Bijl, G. (2000) An assessment of the total external costs of UK agriculture, *Agricultural Systems*, 65 (2), 113–136

Pretty, J. N., Brett, C., Gee, D., Hine, R., Mason, C., Morison, J., Rayment, M., van der Bijl, G. and Dobbs, T. (2001) 'Policy challenges and priorities for internalising the externalities of modern agriculture', *Journal of Environmental Planning and Management*, vol 44, pp263–283

Putnam, R. D., with Leonardi, R. and Nanetti, R. Y. (1993) *Making Democracy Work: Civic Traditions in Modern Italy*, Princeton University Press, Princeton, New Jersey

Rashid A. (2001) 'The CARE Interfish projects in Bangladesh', Paper presented to St James's Palace conference *Reducing Poverty with Sustainable Agriculture*, 15 January. University of Essex, Colchester

Reicosky, D. C., Kemper, W. D., Langdale, G. W., Douglas, C. L. and Rasmussen, P. E. (1995) 'Soil organic matter changes resulting from tillage and biomass production', *Journal of Soil and Water Conservation*, vol 50, pp253–261

Reicosky, D. C., Dugas, W. A. and Torbert, H. A. (1997) 'Tillage-induced soil carbon dioxide loss from different cropping systems', *Soil and Tillage Research*, vol 41, pp105–118

Reij, C. (1996) 'Evolution et impacts des techiques de conservation des eaux et des sols'. Centre for Development Cooperation Services, Vrije Univeriseit, Amsterdam

Regasamy, S., Devavaram, J., Prasad, R., Erskine, A., Balamurugan, P. and High, C. (2000) *The Land Without a Farmer Becomes Barren* (thaan vuzhu nilam thariso), *SPEECH*, Ezhil Nagar, Madurai, India

Sanchez, P. A., Buresh, R. J. and Leakey, R. R. B. (1999) 'Trees, soils and food security', *Philosophical Transactions of the Royal Society, London B*, vol 253, pp949–961

Sanchez, P. A. and Jama, B. A. (2000) 'Soil fertility replenishment takes off in east Southern Africa'. Paper for International Symposium on Balanced Nutrient Management Systems for the Moist Savanna and Humid Forest Zones of Africa, Cotonou, Benin, 9 October, 2000

Scoones, I. (1998) *Sustainable Rural Livelihoods: A Framework for Analysis*, IDS Discussion Paper, 72, University of Sussex, Brighton

Shah, P. and Shah, M. K. (1999) 'Institutional strengthening for watershed development: The case of the AKRSP in India', in Hinchcliffe F., Thompson J., Pretty, J., Guijt, I. and Shah, P. (eds) *Fertile Ground: The Impacts of Participatory Watershed Development*, IT Publications, London

Smaling, E. M. A., Nandwa, S. M. and Janssen, B. H. (1997) 'Soil fertility in Africa is at stake', in Buresh, R. J., Sanchez, P. A. and Calhoun, F. (eds) *Replenishing Soil fertility in Africa*, Soil Science Society of America Publication No 51, SSSA, Madison, Wisconsin

Smil, V. (2000) *Feeding the World: A Challenge for the 21st Century*, MIT Press, Cambridge, Massachusetts

Smith, P., Powlson, D. S., Glendenning, M. J. and Smith, J. U. (1998) 'Preliminary estimates of the potential for carbon mitigation in European soils through no-till farming', *Global Change Biology*, vol 4, pp679–685

Sodhi, P. S. (2001) 'Livelihood improvements in the KRIBCHO project'. Paper presented to St James's Palace conference *Reducing Poverty with Sustainable Agriculture*, 15 January, University of Essex, Colchester

Sorrenson, W. J., Duarte, C. and Portillo, J. L. (1998) *Economics of no-till compared to conventional systems on small farms in Paraguay*, Soil Conservation Project MAG-GTZ, Eschborn, Germany

Steiner, R., McLaughlin, L., Faeth, P. and Janke, R. (1995) 'Incorporating externality costs in productivity measures: A case study using US agriculture', in Barbett, V., Payne, R. and Steiner, R. (eds) *Agricultural Sustainability: Environmental and Statistical Considerations*, Wiley, New York, pp209–230

Swiss Agency for Environment, Forests and Landscape (1999) *The Environment in Switzerland: Agriculture, Forestry, Fisheries and Hunting*, Berne, Switzerland.

Swiss Agency for Environment, Forests and Landscape and Federal Office of Agriculture (2000) *Swiss Agriculture on Its Way to Sustainability*, SAEFL and FOA, Basel

Thrupp, L. A. (1996) *Partnerships for Sustainable Agriculture*, World Resources Institute, Washington, DC

UN (1999) *World Population Prospects – The 1998 Revision*, United Nations, New York

Uphoff, N. (1999) 'Understanding social capital: Learning from the analysis and experience of participation', in Dasgupta, P. and Serageldin, I. (eds). *Social Capital: A Multiperspective Approach*, World Bank, Washington, DC

Uphoff, N. (ed) (2002) *Agroecological Innovations*, Earthscan, London

von der Weid, J. M. (2000) *Scaling up and Scaling Further Up*, AS-PTA, Rio de Janeiro

Waibel, H. and Fleischer, G. (1998) *Kosten und Nutzen des chemischen Pflanzenschutzes in der Deutsen Landwirtschaft aus Gesamtwirtschaftlicher Sicht*, Vauk-Verlag, Kiel

Watson, R. T., Noble, I. R., Bolin, B., Ravindranath, N. H., Verardo, D. J. and Dokken, D. J. (eds) (2000) , IPCC Special Report on Land Use, Land-Use Change and Forestry', approved at Intergovernmental Panel on Climate Change (IPCC) Plenary XVI (Montreal, 1–8 May, 2000). IPCC Secretariat, c/o World Meteorological Organization, Geneva

Wood, S., Sebastien, K. and Scherr, S. J. (2000) *Pilot Analysis of Global Ecosystems*, IFPRI and WRI, Washington, DC

World Bank/FAO (1996) *Recapitalisation of Soil Productivity in Sub-Saharan Africa*, Washington, DC and Rome

Wolfe, M. (2000) 'Crop strength through diversity', *Nature*, vol 406, pp681–682

Zhu, Y., Chen, H., Fen, J., Wang, Y., Li ,Y., Chen, J., Fan, J., Yang, S., Hu, L., Leaung, H., Meng, T. W., Teng, A. S., Wang, Z. and Mundt, C. C. (2000) 'Genetic diversity and disease control in rice', *Nature*, vol 406, pp718–722

Index

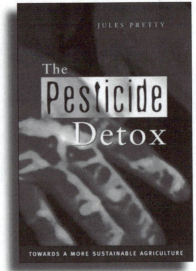

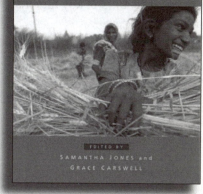

The Atlas Series from
EARTHSCAN

Each book in the Atlas series includes:

- 50 full-colour global and regional maps
- essential facts and figures
- extensive graphics
- historical backgrounds
- expert accounts of key regions, issues and political relations
- world table of statistical reference

'The State of the World Atlas is something else – an occasion of wit and an act of subversion. These are the bad dreams of the modern world, given colour and shape and submitted to a grid that can be grasped instantaneously'
NEW YORK TIMES on *The State of the World Atlas*

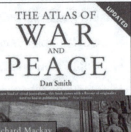

Order online at
www.earthscan.co.uk